The Rhetorical Nature of X

The Rhetorical Nature of XML is the first volume to combine rhetoric, XML, and knowledge management in a substantive manner. It serves as a primer on XML and XML-related technologies, illustrating how the naming of XML elements can be understood as a rhetorical act, and detailing the essentials of knowledge management practices that illustrate the need for intelligently conceived databases in organizations. Authors J.D. Applen and Rudy McDaniel explain how technical knowledge and rhetorical knowledge are symbiotic assets in the modern information economy, emphasizing that skilled professionals and apprentice learners must not only adapt to and become adept with new technological environments, but they must also remain aware of the dynamic social and technological contexts through which they communicate. Applen and McDaniel use this subject as a catalyst to encourage interdisciplinary connections and projects between experts in fields such as technical communication, digital media, library science, computer science, and information technology.

The authors demonstrate techniques for working with XML in interdisciplinary projects with attention to single sourcing and content management. Interviews with practitioners working with XML for research and in industry are also included, to illustrate how XML is currently being used in a variety of disciplines, such as technical communication and digital media. Combining applied theory and XML technology to solve real-world problems in technical communication and digital media, this work provides an entry point for students and practitioners who do not have an extensive background in markup languages, enabling them to begin developing user-centric projects using XML.

Visit the book's companion web site: http://rhetoricalxml.com/

J.D. Applen is an associate professor of English at the University of Central Florida. His scholarly interests include XML and archiving, knowledge management, hypertext theory, the history of texts and technology, and the rhetoric of science and technology. He received his doctorate from the University of Arizona.

Rudy McDaniel is an assistant professor of Digital Media at the University of Central Florida. His research interests include XML, narrative theory, video game technologies, and knowledge management frameworks. He received his doctorate from the University of Central Florida's Texts and Technology program.

The Rhetorical Nature of XML

Constructing Knowledge in Networked Environments

J.D. Applen
Rudy McDaniel
University of Central Florida

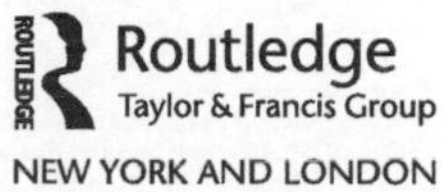

NEW YORK AND LONDON

First published 2009
by Routledge
711 Third Ave, New York, NY 10017

Simultaneously published in the UK
by Routledge
2 Park Square, Milton Park, Abingdon, Oxon OX14 4RN

Routledge is an imprint of the Taylor & Francis Group, an informa business

Typeset in Sabon by
Florence Production Ltd, Stoodleigh, Devon

Library of Congress Cataloging in Publication Data
Applen, J.D.
The rhetorical nature of XML/J.D. Applen, Rudy McDaniel.
p. cm.
1. XML (Document markup language). 2. Rhetoric.
3. Technical communication. I. McDaniel, Rudy. II. Title.
QA76.76.H94A69 2009
006.7′4—dc22 2008046966

ISBN13: 978–0–8058–6179–2 (hbk)
ISBN13: 978–0–8058–6180–8 (pbk)
ISBN13: 978–1–4106–1536–7 (ebk)

ISBN10: 0–8058–6179–3 (hbk)
ISBN10: 0–8058–6180–7 (pbk)
ISBN10: 1–4106–1536–7 (ebk)

Contents

Illustrations

Figures

Tables

Preface

In *The Rhetorical Nature of XML* we describe and connect eXtensible Markup Language (XML) authoring, the processes that show how we acquire and share knowledge, and rhetoric. We do this to show how these practices, both humanistic and technological, allow us to thoughtfully name, arrange, and distribute the information that we store in archives. This book is interdisciplinary and we imagine our audience will include students, professors, and professionals in the fields of technical communication, digital media, information technology (IT), library science, and management information systems. In this effort, we have employed selected ideas from theorists and practitioners working in these fields.

In the field of technical communication and rhetorical studies, we have many thoughtful colleagues who have challenged us to consider our audience(s) and have demonstrated how the social construction of language shapes our communication practices. Others have reminded us of the need for technical communicators to expand their skills and understand how to use, and not be used by, new communication technologies. Additionally, technical communicators are mindful of how the advent of single sourcing has changed the way we produce and reuse documentation. We have also seen why it is important to understand classification systems and our colleagues in the field of library science have much to offer us as we are all being called on to organize and archive ever larger collections of documents.

Under the aegis of the World Wide Web Consortium (W3C), XML technologies have been evolving and there are many books and online sources that illustrate how to produce syntactically correct code. Like HTML, much of XML is about using tags to mark up documents for display on the World Wide Web. Unlike HTML, though, XML places the responsibility for the naming of tags on the document author. This enables the technology to go far beyond what HTML is able to do as it is not merely a presentation language, but rather a *semantic* language. With HTML, Internet documents have formatting and display instructions. With XML, Internet documents have *embedded meanings.*

These semantic capabilities of XML are what make this language so important for technical communicators, media professionals, archivists, and business practitioners to understand. They are also what make the application of rhetoric to a computer language a useful pursuit. By understanding the nature of classification and rhetorical strategy, and by controlling the representation of meaning within documents, one can create powerful systems for packaging data and transmitting it from one computer to another in any number of domains.

There are numerous models and descriptions that show how XML can be manipulated, and we are seeing more documentation on how eXtensible Style Sheet Language Transformations (XSLT) and Cascading Style Sheets (CSS) can transform code to produce accessible and visually appealing documents. Document Type Definitions (DTDs) and XML schema provide us with additional powers of validating and standardizing data for exchange over computer networks. We study the phenomena of XML-encoding, XML-transformation, and XML-parsing from a rhetorical slant.

We draw significantly from prior work in knowledge management. Theorists have given us insights into how knowledge is acquired, generated, modified, applied, and passed on to others and they have provided us with a number of case studies that describe how this has worked. The value of tacit knowledge has been recognized, and, coupled with explicit knowledge, we can better understand it as an important element of all that we bring to our work. The ecology of workplace environments has been detailed, as has the manner in which communities of knowledge workers work to apply new technologies. We can learn much from these practices as we strive to create more meaningful and useful systems of classification and categorization using XML.

It is our hope that *The Rhetorical Nature of XML* allows people to better see the connections between these fields. We believe that there is a place in technological studies for rhetoric, and that there is a place in the humanities for XML.

J.D. Applen and Rudy McDaniel

Acknowledgments

The authors wish to thank several anonymous reviewers who provided input and feedback on early drafts. This feedback helped us enormously with the revision and editing process. We also want to thank our editor, Linda Bathgate, for her enthusiasm for this project and for her overall dedication to the field of technical communication. Jenny Draper provided additional proofreading for the final draft, which we appreciate.

J.D. Applen: For their support and encouragement, I would like to thank my family members who live too far away from me on the West coast—from San Diego, California, to Olympia, Washington, and points in between. I would also like to thank my colleagues in the Technical Communication Program at the University of Central Florida for their advice and counsel: Melody Bowdon, Paul Dombrowski, Madelyn Flammia, Mary Ellen Gomrad, Dan Jones, Karla Kitalong, and Blake Scott.

Rudy McDaniel: I would first like to thank my family and especially my wife, Carole, for her assistance with graphic design and even more so for putting up with me during the writing of this book. I also want to thank my two-year-old son, Brighton, for reminding me of both the importance of clear communication and the joy of technological discovery. Many folks at UCF have been influential to me during this project and I owe gratitude to Clint Bowers, Terry Frederick, Jon Friskics, Charlie Hughes, Bob Kenny, José Maunez-Cuadra, Mike Moshell, Joe Muley, Dan Novatnak, Phil Peters, Sae Schatz, Eileen Smith, Stella Sung, Natalie Underberg, David Vickers, and Jeff Wirth for good conversations and helpful ideas. Steve Fiore continually reminds me of the importance of interdisciplinarity and, even though he is a philosopher, I very much enjoy our conversations. Dawn Trouard has been a mentor to me from the beginning and I am grateful for her many years of tutelage. From my years working in the English Department, I am also thankful for the many technical communicators J.D. mentions above, especially Paul Dombrowski, my graduate advisor, for letting me write an English Department dissertation half filled with XML code.

Abbreviations

AA	Arthur Andersen
ACHRE	Advisory Committee on Human Radiation Experiments
AIDS	Acquired Immune Deficiency Syndrome
AJAX	Asynchronous JavaScript and XML
API	application programming interface
CMPM	Cast Member Performance Management
CMS	content management system
CPMS	Cast Performance Management System
CSS	Cascading Style Sheets
DITA	Darwin Information Typing Architecture
DOE	Department of Energy
DOM	Document Object Model
DTDs	document type definitions
E4X	ECMAScript for XML
EA	Electronic Arts
EAD	Encoded Archival Description
EPSS	electronic performance support system
FIEA	Florida Interactive Entertainment Academy
FTP	File Transfer Protocol
GDP	Gross Domestic Product
GRID	Gay-Related Immune Disorder
GUID	globally unique identifier
HHS	U.S. Department of Health and Human Services
ICD	International Classification of Diseases
IDE	integrated development environment
ISARs	information storage and retrieval systems
IT	information technology
KXDC	Knowledge Extraction from Document Collections
MORPGs	multi-player online role playing games
NLP	natural language processing
NIC	Nursing Interventions of Classification
OASIS	Organization for the Advancement of Structured Information Standards
OeB	Open eBook

PHP	Hypertext Preprocessor
PHP-FI	Personal Home Page Tools / Forms Interpreter
POS	point of sales
QMS	Quality Management Standards
RAX	Ad Hoc Rhetorical Analysis of XML
RSS	Really Simple Syndication (also Rich Site Summary)
RTF	Rich Text Format
SAX	Simple API for XML
S-FTP	Secure File Transfer Protocol
SGML	Standard Generalized Markup Language
SIDS	sudden infant death syndrome
SME	subject matter experts
SVG	scalable vector graphics
TEI	Text Encoding Initiative
URI	Uniform Resource Indicator
URL	Uniform Resource Locator
URN	Uniform Resource Name
VRML	Virtual Reality Modeling Language
W3C	World Wide Web Consortium
WHO	World Health Organization
XBRL	eXtensible Business Reporting Language
XDA	XML Document Design Architecture
XDM	XPath data model
XHTML	eXtensible Hypertext Markup Language
XML	eXtensible Markup Language
XPath	XML Path Language
XSD	XML Schema Definition
XSL	eXtensible Style Sheet Language
XSL-FO	eXtensible Style Sheet Language Formatting Objects
XSLT	eXtensible Style Sheet Language Transformations

Introduction
XML, Knowledge Management, and Rhetoric

Purpose

This book is about three subjects which may at first seem very different from one another: the eXtensible Markup Language (XML), knowledge management, and rhetorical theory. Our goal in writing this book is not to claim that these topics are all exactly alike, but rather to explain how technical knowledge and rhetorical knowledge can work together in knowledge management practices for the modern information economy. The modern workplace is a challenging environment. Not only must skilled professionals and apprentice learners constantly adapt to and become adept with new technological frameworks, but they must also consistently be aware of the dynamic social and political contexts through which they communicate. The semantic power of XML, coupled with an understanding of knowledge management and the rhetorical situation, is something that can be harnessed in order to create more humanistic and compelling frameworks for information and knowledge exchange.

Our premise in writing this book is quite simple: although there are many excellent texts on learning to develop using XML, these texts are generally syntax-centric and do not integrate the broader rhetorical context of user needs and the social context of information. Our goal is to use this subject matter as a catalyst to encourage interdisciplinary connections and projects between experts in a diverse array of fields. We mention several of these fields in the Preface. It is also our hope that this book will provide an entry point for students and practitioners without an extensive background in markup languages to begin developing projects using XML. A prior knowledge of HTML is helpful, but not absolutely necessary as we spend some time in Chapter 2 addressing the differences between these two markup languages.

The subject of XML is vast in scope, and we do not cover every component of the XML specification in exhaustive detail. We believe that we provide enough information about the technology to allow readers to complete creative and useful projects using XML, and we also include XML activities or questions at the end of each chapter which encourage discussion and synthesis of these ideas. These questions and activities can also be used as starting points for semester-long projects in both the theoretical (e.g., rhetorical analysis) and applied (e.g., technical production) domains.

While our aim has been to make this book as accessible as possible, we do cover some more technical material in several project examples that we discuss in Chapter 6. In this chapter, we discuss some basic programming concepts and explain why these skills are important for those wishing to learn about XML parser design. Since we are dealing with interactive technologies and XML, which was actually designed as a language for *computers* to communicate back and forth with, the humanistic component of computer-mediated communication is only one side of the interactive conversation that occurs between human beings and information technologies (IT). It is important that we understand, or at least be aware of, the computational side of things along with the rhetorical and the social context of information.

When we do discuss programming, however, we annotate and describe these programs in great detail and we offer the source files for download on our accompanying website. Though applied computer science is not a domain traditionally associated with information design (Albers "Introduction" 4–6), we feel that this basic competency with programming concepts is important both for the holistic reason stated above *and* because it allows one to design her own customized information and knowledge management systems rather than having to rely on existing products which may not be flexible enough to accommodate certain types of projects. Because we do focus rather heavily on the design elements of informational systems in the second half of the book, we extract and apply ideas from the domains of both information design and information architecture, which in and of themselves already share a good deal of overlap (Mazur 25).

The first five chapters of this book focus on the ways we can think about data and then organize this data in rhetorically meaningful ways using XML. By understanding the basic rules of syntax and validation, one gains an understanding of how data is typically packaged and processed in IT systems and how XML can be used as a tool for computer-to-computer communications. By understanding the complex social and psychological spaces through which meaning is negotiated in rhetoric, one becomes aware of the limitations of purely data-driven approaches and of the complexities involved in the relationships between data, information, knowledge, and cognition. The latter two chapters then focus on more specific examples of XML and XML parsers at work in real world types of problem solving and present some advice from practitioners working with XML in industry and research environments.

Chapter Overviews

Chapter 1 discusses knowledge management from various perspectives and explains how both explicit and tacit knowledge can be transferred using social networks, computational technologies, or combinations of these two methods of communication. In addition to a review of literature from knowledge management and technical communication, we also discuss perspectives on socially constructed knowledge and consider applied knowledge management at work in large corporations. In order to demonstrate how human expertise and technology can be paired to improve efficiency and communication in technical environments, we discuss successful applications of this methodology such as the Eureka project used by Xerox.

To help us build the types of knowledge transfer systems we write about in Chapter 1, we need to understand both top-down and bottom-up models of design. We describe these approaches conceptually and study the nuts and bolts of XML and information encapsulation in Chapters 2 and 3. Chapter 2 serves as a gentle introduction to XML and explains the differences between elements, attributes, namespaces, and parsed or nonparsed character data. We also explain the differences between well-formed and valid XML and provide several examples of XML documents in order to illustrate the semantic capabilities of this language. Additionally, we show how XML can impose a hierarchical structure on collections of data.

Chapter 3 then discusses more sophisticated implementations of XML and applies some of the organizing heuristics from library science, knowledge management, and technical communication to guide how we think about the artificial formation of elements and attributes based on our real world observations. Using historical examples, we describe in significant detail the practice of naming and arranging objects in classification schemes. How a group of professionals decides to name objects and then arrange them is a reflection of the rhetorical conventions that channel the thinking of these professionals and shapes their fields. We can also see how the granularization process works and how we can combine our knowledge of rhetoric and XML to solve real world problems using applied methodologies such as single sourcing.

Chapter 4 extends the rhetorical analysis of XML to the visual domain and considers specific formatting technologies such as Cascading Style Sheets (CSS) and eXtensible Stylesheet Language (XSL) as transformative tools for the document designer and information architect. We consider some fundamental ideas about visual rhetoric as it relates to XML and provide several tutorials that illustrate how to use style sheets and style sheet transformations with XML documents. We discuss why knowledge of visual style can be as important as the knowledge of the actual data.

Chapter 5 is focused on advanced technologies and the ways in which specialized constructs such as namespaces can be used to counter problems of recognition and collision. This chapter also includes a discussion of emerging XML technologies such as the specialized languages for linking and searching within XML documents. A majority of this chapter is dedicated to the discussion of schema definitions, which enable communication professionals to use a special syntax for verifying the integrity of their documents. Chapter 5 also includes a discussion of DocBook and DITA, the

Darwin Information Typing Architecture, two particular frameworks that are often used by technical communicators working with XML.

Chapter 6 serves as a synthesis of many of the rhetorical theories and XML technologies and applies them to specific informational problems. Even with the power that comes from knowing about the rhetorical situation and how elements of environment, user needs, and content interact to form complex relationships of interaction and communication, one cannot simply build an XML database with carefully crafted entities and wait for the problem to be solved. Instead, one must also carefully consider the manner in which these elements are to be manipulated and acted upon by human or technological agents. As Michael Albers notes:

> As an immature technology, research into all aspects of XML are needed, but we must not forget that the real usefulness lies not in pulling data out of a database, but in using it afterwards in a manner which fits the user's real-world goals and information needs.
>
> ("Complex Problem Solving": 266)

So, while it is important to recognize and understand the intricacies of knowledge representation through XML, the distributional and transactional aspects of XML are also important to consider. By distributional, we mean the function performed by software that allows XML data to be distributed from one computer to another, or from a content author to her audience(s) in a computer-mediated fashion. We use the term transactional to refer to the process by which XML content is packaged, processed, and transmitted using collections of logical rules and conditional processing. Both of these roles are fulfilled by a special software system known as a parser. This chapter applies three different rhetorical models: an ad hoc analysis, a three-stage information design analysis, and a bottom-up approach, in order to guide the construction of customized parsers for the purposes of newsfeed distribution, content management, and single sourcing.

The last chapter, Chapter 7, examines some uses of XML in the real world by various types of technical professionals. In this chapter, we include interviews from several types of professionals who have used XML in their own careers for a variety of knowledge management purposes. We also discuss how XML might play a role in several different types of careers and provide summaries of each chapter in condensed format.

Website

To meet the technological needs of our readers, a repository of XML tools and tutorials, links of interest, and sample documents is housed on our accompanying website at www.rhetoricalxml.com. Please visit this website to obtain the code examples and XML tools mentioned in the chapters.

References

Albers, Michael J. "Introduction." *Content & Complexity: Information Design in Technical Communication*. Eds. Michael J. Albers and Beth Mazur. Mahwah, NJ: Lawrence Erlbaum Associates, 2003. 1–13.

—— "Complex Problem Solving and Content Analysis." *Content & Complexity: Information Design in Technical Communication*. Eds. Michael J. Albers and Beth Mazur. Mahwah, NJ: Lawrence Erlbaum Associates, 2003. 263–83.

Mazur, Beth. "Information Design in Motion." *Content & Complexity: Information Design in Technical Communication*. Eds. Michael J. Albers and Beth Mazur. Mahwah, NJ: Lawrence Erlbaum Associates, 2003. 15–38.

1 Knowledge Management and Society

Evaluating the Convergence of Knowledge and Technology

Chapter Overview

The advent of technology has altered the working environment of the technical communicators, information technologists, and library scientists. In this chapter, we will show how these communication professionals can expand their skillsets and better integrate themselves in their organizations. Additionally, communication professionals can be more critical about the way they can use technology and be more active consumers of it. One of the insights we will provide is how knowledge is socially constructed in a community of professionals, and this will allow these stakeholders to claim a greater role in the workplace.

We will also describe the basics of knowledge management and show, using case studies, how it has been successfully implemented. To do this, we will discuss how the tacit and explicit knowledge(s) that people in a community possess can lead to the generation of new knowledge. Additionally, the metaphor of "information ecology" will be examined to deepen our understanding of how knowledge can be better managed and produced.

Understanding Technology and Information

In an attempt to empower technical communicators, Johnson-Eilola, Selber, and Selfe have encouraged this community to see the rapid advances in computer technology as an opportunity to effect change, and that to accomplish this change, they need to think in critically informed ways about how they use this technology. This need is driven by the fact that the pace of deploying these technologies in the workplace is accelerating and this phenomenon challenges technical communicators to hurriedly adopt new applications, thus cutting into their ability to examine their practices critically. Moreover, it is becoming clear that the key factors associated with computers depend on the histories, contexts, and relationships that people who design and use these products have with technology as much as on the hardware and software themselves.

For example, hypertext is a medium that allows for a new emphasis on the roles of the reader and the writer and it can function as a contractive or expansive communication technology (Johnson-Eilola and Selber 125). A contractive technology assumes that in a communication between sender and receiver, information is packaged into "discrete, ideally unambiguous chunks," and that the reader is essentially a passive receiver of information; the reader has no choice but to read the "chunk" and understand it as a self-contained unit that has reference outside it. In contrast, the expansive mode of communication is in effect when the transfer of information is a process in which readers construct and deconstruct pieces of information, putting them into a context that is in part a function of the social and political environment in which they are working. Expansive communication is a recursive process that takes the "user, designer, technology and context" into consideration. Not to see communication technology in this recursive perspective suggests a "technological determinism" where technology can only be used to send and receive one kind of message regardless of the needs of the people involved in the process (Johnson-Eilola and Selber 121).

Based on the premise that the design of a system has already been set by the engineers who built it, technical communicators have been traditionally relegated to learning just enough about a part of the system so they can explicate it clearly. Technical communicators are usually not understood as "authors"; at best, they are translators of the information that has been generated by others before them. This perception results from our culture's practice of attributing single ownership to those who are believed to have invented, designed, or written something that is not to be altered. However, when technical writers engage in their work, it is understood that they are always "adding, deleting, changing, and selecting meaning" (Slack, Miller, and Doak 31) within the context of the communication medium they are employing to meet the needs of their audience. Their work is not neutral; they are authors, even when they write in a manner that makes their articulations seem invisible. It is as if the writers of technical documentation never really had something to do with constructing the meaning of the technology and how to use it, that this has already been decided by those who initially designed and constructed the technology. However, authorship comes with some responsibility. When they are articulating meaning, they need to think critically about the ethical responsibility of their work. Who they work for and what they communicate also matters (Slack, Miller, and Doak 32).

If technical communicators do not understand this and continue to document practices as mere translators, the ability of the user to employ the technology in an expansionist sense will be undermined (Johnson-Eilola 247). For example, hypertext products are generally framed in industry in "strictly automating terms"; users are asked to follow a series of steps to get a predetermined result that the technology can offer without teaching and encouraging its use for potential tasks not imagined by its designers (Johnson-Eilola and Selber 124). Traditional forms of documentation manual production also assume that the same practice will be followed by the passive readership of the documentation.

Echoing Robert Reich, Johnson-Eilola asks that technical writers work to promote themselves as "symbolic-analysts" where their skills include the manipulation of information, which requires a greater understanding of it in the abstract (245–6). This challenge is built on the premise that we have moved from an industrial to an information economy where technical writers are to a greater degree producing knowledge products. Instead of taking the traditional approach by breaking down problems into small and discrete parts, such as a list of tasks that many software documentation professionals are asked to describe, a symbolic-analyst would work to make meaning out of information with an awareness of the larger system and its ability to serve them and the people who use their products. We feel that one way of describing the work of symbolic-analysts is: 1) identifying what constitutes relevant and meaningful information, 2) breaking this information down into specific elements, 3) providing names for these elements, and 4) contextualizing these elements of information to best meet the rhetorical needs of their audiences. While XML technology by itself does not perform this work, it does serve as a robust tool that would allow for the storage and transmission of this kind of work performed by symbolic-analysts, and we will be demonstrating this in the chapters that follow.

Michael Hughes describes how technical writers can have a greater role in an organization by engaging in "critical reverse engineering." For example, if a technical communicator is asked to document a software application, she would interact with the screen elements and think about how the consumer's similar interaction would add value for the user and make the tasks the user engages in easier and more efficient (281). By "critical reverse engineering" a product, the technical communicator could raise some questions and even challenge some of the software designer's assumptions. These questions might also better allow the designer to more clearly explain the design of the software so it can be better documented for the end user.

If technical communicators are engaged in this kind of process, they become part of the overall design team that works towards a final product. Hughes has observed that what is initially thought to be an end product evolves into something different because of the many different people and groups of people with different skills who work on a product from its initial conception to its final state. Because these groups do not work on the product at the same time, but work on it in different combinations of individuals at different stages, understandings of what the end product is often vary. Because of this, emergent understandings of the product come about, but not by everyone at the same time. This allows some people who receive a partially conceived and constructed product to ask questions of those who have already done some work on the project. When a technical writer asks "Why does this do that?" to a designer in an effort to just get the documentation right, the technical writer is working in the "information domain" (275). When she does it in an effort to create a dialogue with the other members of the design team, she is contributing to the thinking of the design team as a whole and working with its members to create new knowledge, a greater challenge and opportunity than just transferring information (Hughes 282–3). This socially constructed knowledge can ultimately lead to a better product for the consumer, the end user.

Communication professionals need to ask themselves how the social and political context affects the use of the technology they are using, whether it be hypertext or XML, and how this context supports their ability to use this technology in an expansive manner. Wick (2000) challenges technical communicators to claim their role in the knowledge management game by emphasizing their considerable theoretical understanding of rhetoric and ability to communicate within, between, and across different sectors of an organization. To expand on Wick's premise, we believe that because technical writers are professionals who can go out, acquire information from people skilled in disciplines different from their own, then synthesize, organize, and explicate this information for different audiences in such a manner that people can understand and use it to meet their unique needs, they are at the center of an organization's knowledge and can be knowledge managers.

To become more proficient at knowledge management, communication professionals have to look beyond their roles as architects of documents and developers of technological applications that are employed to produce and add value to texts. They also need to recognize their ability to help professionals throughout organizations interact with each other and to utilize knowledge from others inside and outside their organization. This knowledge can be shared in a manner that enhances or leverages not just the physical and financial resources of an organization, but its knowledge capital as well: "information assets and knowledge capital seem to be governed by a different law of economic returns: investment in every additional unit of information or knowledge created and utilized results in a higher return" (Malhotra "Knowledge" 13). The more people who contribute to and take from a knowledge network, the greater the value of the network.

Social Construction and Paradigms

Johnson-Eilola, Selber, and Selfe (1999) have asked technical writers to adopt a more critical understanding of communication technologies so they can be wiser consumers of these products and assume the roles of more effective symbolic-analytic workers. To better engage in this pursuit, they can benefit from examining how the use of tools and their relationship to the materials, assumptions, and methods of the scientific community contribute to the culture of research activity.

In the scientific community, Kuhn tells us that there is something invisible at work beneath the thinking of any group of scientists who agree on a methodology that can be used to explore the perplexing structure of nature:

> Though many scientists talk easily and well about the particular individual hypotheses that underlie a concrete piece of current research, they are little better than laymen at characterizing the established bases of their field, its legitimate problems and methods. If they have learned such abstractions at all, they show it mainly through their ability to do successful research. (47)

What Kuhn calls "normal science"—the science that uses the existing and agreed upon tools, materials, procedures, and assumptions to answer the kinds of questions and produce the appropriate results within the context of the reigning socially constructed paradigm—can take place without scientists realizing what are the bases of their belief systems. Things just seem "right" in the context of a socially constructed paradigm. In modern science, the paradigms that scientists utilize are often based on what the tools and materials they are employing can measure.

For Kuhn (185–7), each scientific paradigm shared by a scientific community is based on four socially constructed elements:

- Every community member believes in the same "symbolic generalizations." The mathematical entity $F = ma$ (force equals mass times acceleration) establishes that there are three algebraic variables that can be manipulated for the puzzle-solving operations of modern physics that would reveal the way nature operates.
- Every community also has a set of "shared commitments," or a belief in models that furnish the group with, among other things, the type of metaphors and analogies that can be accepted. Aristotelian mechanics found that movement of material bodies could be adequately explained by just assuming that it was in their "nature" to do so, as if material bodies had a *natural* attraction to what was understood as "the center of the universe." What really made these bodies "fall" was never entertained as a legitimate scientific question in Aristotle's day.
- The "shared values" of a community describe the social beliefs that scientists in a community agree on. These could be something as straightforward as the argument(s) that "science should (or need not) be socially useful."
- "Shared exemplars" in a community of scientists are the agreed upon examples and "problem-solutions." These include the laboratory exercises and homework or exam questions used in the education of students.

Kuhn sees great value in the establishment of "normal science" paradigms that are built on these four elements. They allow a scientist to look at nature and test with great precision, and, more importantly, they provide a scientist with a theoretical basis from which to generate hypotheses. When considering a certain body of data, scientists engage in activity that "tells a story" that makes sense within the context of prevailing theories that allows them to produce the claims or results that seem sensible. If they can produce such claims, they can assert that their work has scientific value and thus get it published or receive further support from corporate management. Without these hypotheses, scientists would not know what to test for or how to interpret their results; they cannot even begin to practice science.

However, Kuhn draws our attention to these elements above as they often go unchallenged, thus producing an invisible set of practices that channel the way scientists think. If we still adhered to Aristotle's idea that the movement of material bodies could be adequately explained by referring to their "nature," scientists would not be able to develop the hypotheses and perform the experiments that allowed them to accurately gauge the effects of gravity and relativity. Another example of a socially constructed paradigm can be seen in the practice of biology and the way it is affected by Darwin's Theory of Evolution. Biologists who are interested in understanding the distribution of organisms in the field often design their experiments around the underlying assumption that species live where they live because they have out-competed rival species. It is often difficult for well-trained biologists to think of other reasons that might describe why certain species exist where they are found because of the powerful influence of evolutionary theory, another "shared commitment."

Every once in a while, a perceptive scientist, engineer, or other professional who is well trained in a certain way of thinking and performs an experiment using the ideas that she has internalized via the "normal science" paradigm comes up with an answer that does not fit neatly within it. She might perform the experiment several more times and still come up with the same anomaly. If this scientist points out this anomaly and suggests that the reigning paradigm in the field must be modified or entirely thrown out in favor of a new paradigm that explains some element of nature better, she is often met with resistance. A few scientists might also join in and support this scientist's arguments, while others in the field cling to the existing paradigm. This is when we see major arguments and changes within a field; this is where we see a scientific revolution. This is not to say that normal science is necessarily detrimental; in fact, it is through the use of the methods of normal science that we find anomalies. Normal science gets in the way when we cling to it and fail to realize that from time to time we are going to have to reexamine our assumptions and modify our theoretical approaches.

Similarly, the writers cited in the first section of this chapter ask technical writers to examine the unchallenged assumptions they adhere to as they use communication tools and methods of documentation that explain communication technologies and products to others. For example, Selfe and Selfe have suggested that IBM's DOS system leads us to believe that hierarchical ways of organizing knowledge are better than more intuitive methods (491). We could extend this insight to understand how this "logical" system or "shared commitment" would lead communication professionals to believe that the only information that they should be documenting is information that can be captured in a table, ordered list, or set of definitions, not the kind of tacit knowledge that people possess which connects or explains procedures and ideas that are difficult to reduce to a numbered list. Tacit knowledge is what we learn from making mistakes, developing our own workarounds that better describe how something can be done, and learning how to survive and even flourish by experiencing the complexity of our work cultures, things that cannot be taught to us by a book or a trainer. If technical writers assume the role of symbolic-analytic workers, they can better understand how what they are told to believe is true is in fact a social construction. This construction allows technical writers to convey knowledge in a certain way, but perhaps keeps them from a critical understanding of their present methods and from employing alternative techniques that might better allow them to share other kinds of knowledge.

As certain tools or materials become adopted by community, whether they be Microsoft® Office 2007 applications or a certain organism, their use instills a series of socially constructed or unquestioned assumptions that allows those who use them to be able to better articulate the claims they make; the "*logic* of justification or discovery" seems more apparent or "right" (Griesemer 52). For example, the common fruit fly, Drosophila, is still widely used in university laboratories as the primary tool for teaching students of biology just how genetic traits are passed from generation to generation; Drosophila has become, in Kuhn's terminology, a "shared exemplar." Microsoft® Office tools have now become shared exemplars in today's universities: students learn how to produce spreadsheets, present tables, plot graphs, and crunch large bodies of numerical data using Microsoft Excel®.

Clark and Fujimura, like Kuhn, ask scientists to expand their vision of the scientific process by paying attention to all of the elements of research; "tools, jobs, and rightness" are a function of the socially constructed situation in which science is done (6). In fact, they are co-constructions of the interplay between the elements of the work situation, and sometimes the distinction between these elements becomes blurred. For example, Clark and Fujimura ask "What is a technology versus a material versus a theory?" and "When is a scientist a technician, when is a technician a scientist, and when are both technologies?" We could extend this to the field of technical communication by asking "What is a technology versus information versus an organizational strategy?" and "When is a technical communicator a technician, and when is her work shaped by the technologies she employs?" (6).

Analogously, communication professionals should know how the business practices of Microsoft® executives have allowed their Windows® operating system and applications to gain a near monopoly as the tools used in government, education, and industry, and why they are now the primary tools used by technical communicators. They should also understand how tacit knowledge can be either captured or excluded by these tools.

Tacit Knowledge

Michael Polanyi believes that much of the valuable knowledge we possess is tacit knowledge, which is a kind of knowledge that we cannot really convey to others in totality and perhaps it is a kind of knowledge we take for granted. For example, we know how to recognize the subtleties of someone's mood by the way their face appears, but it is difficult to explain how we are able to do this ("Tacit" 5), or as Polanyi famously declares, "we can know more than we can tell" ("Tacit" 4). After working in an organization for a while, we just know how things are done. This is apparent when someone new comes into our workplace and we can tell by the questions she asks of us that we have learned something while employed there, but what we have learned is not from a manual we read or a training session we attended. It is also something we would not have thought to write down.

Polanyi held that tacit knowledge has greater value than explicit knowledge, knowledge that has been written down and codified, as all knowledge stems from it. Tacit knowledge comes about from the process of indwelling, where we engage ourselves in a problem and cull from large body of information and sensory stimuli what we feel we need to know about it. It is important to understand that while tacit knowledge might come to us through sources we do not always readily identify, acquiring tacit knowledge is an active process.

We might have been taught to believe that the scientific method is based solely on objective knowledge and techniques. When followed step-by-step and in a completely lucid fashion, this method has lead scientists to the great discoveries that have changed our civilization. While Polanyi makes the point that people actively search for answers to difficult questions, he characterizes the method of scientific problem solving as one that is infused with intuition: "The scientist's intuition can integrate widely dispersed data, camouflaged by sundry irrelevant connexions, and indeed seek out such data by experiments guided by a dim foreknowledge of the possibilities that lie ahead" (Polanyi "Science" 31). Science, according to Polanyi, who was an accomplished physical chemist before he became interested in philosophy, is more subjective than many of its practitioners will admit, and, in some ways, Polanyi's view of science is similar to Kuhn's in this regard.

In his study of the manner in which managers make their decisions, Baumard asks "Why do some managers consider far fewer alternatives than others when arriving at their decisions? Is it a question of flair (intuition) or expertise?" (Baumard 65). He supports his discussion by explicating how the Greeks used the term "metis" to describe conjectural or tacit knowledge (Baumard 65). Metis is that type of knowledge that is constantly shifting and ambiguous, a knowledge that comes from a source that is not readily identifiable. Greek philosophers "usually passed over" any mention of how metis might have inspired their own ideas "in silence or hostility" because to admit to it would undermine the seemingly rigorous or rational arguments that shaped the existing communicative genres they were required to use in order to be recognized in their profession. The Greek philosophers were famous for using the Socratic method—where they went through a series of careful "logical" steps—to come to their conclusions. However, they did acknowledge that metis constituted a possible way of knowing (Detienne and Vernant 3).

As an instance of metis in action, Baumard cites the *Intelligence Newsletter*, which has published "difficult-to-obtain insights" about the world of espionage (Baumard 139). Without the resources of the CIA, this small French publication bases its research on a careful examination of major newspapers and press releases from government officials throughout the world. These sources can be "full of contradictions, internal opposition and cliques," and that "what is not said allows one to follow what is meant" (Baumard 142). One accomplished contributor to the newsletter, Maurice Botbol, navigates these resources with "a 'floating eye'—not necessarily attached to the pursuit of anything precise, but navigating left and right, collecting elements of sense; always uncertain, unstable, even disordered" (Baumard 147). Utilizing his tacit assumptions, Botbol was able to make a connection between an ambassador with dubious credentials and the assassination of an opposition leader in the Seychelles. This demonstrates how information that initially seems ambiguous can be shaped using our conjectural thinking into meaningful ideas.

Knowledge Management

Because of the regimentation and rules of many organizations, they may not be the most suitable settings for augmenting the tacit knowledge of individuals. However, they still value these nonexpressed knowledges of individuals, and this is most evident when a person leaves an organization. When she leaves, she takes with her a know-how that cannot be captured in a job description or manual. Thus organizations do value tacit knowledge and work to understand the questions below (Baumard 77):

- How can tacit knowledge be used or exploited?
- How can tacit knowledge be systematized?
- How can organizations "protect, enrich, and use tacit knowledge" when they are faced with an ambiguous or perplexing situation?

Organizations also maintain large bodies of tacit knowledge through their "ceremonies, rites, traditions, and communities of practice" (Baumard 77). The maintenance of this tacit knowledge base allows a community to gain an identity and to be aware of how this identity changes with the addition of new members and the exit of existing members. Choo refers to this aspect of tacit knowledge as "cultural knowledge," where an organization's identity is framed for its members with the following questions: "'What kind of organization are we?' 'What knowledge would be valuable to the organization?' and 'What knowledge would be worth pursuing?'" (396). These questions allow an organization's members to examine not only what they know, but how their organization can look forward and grow.

All too often, enabling knowledge creation has been relegated to deploying better IT and management tools. To think of knowledge creation in management terms implies the attitude that managers can control the process, thus undermining the ability of people to bring the best of their own insights and creativity to a situation (Von Krogh, Ichijo, and Nonaka 4).

People who want to take action need to feel that they can take advantage of their own tacit assumptions and create knowledge for their own specific requirements, and, for this to occur, an enabling context needs to be in place (Von Krogh, Ichijo, and Nonaka 178). Von Krogh, Ichijo, and Nonaka employ the term "ba" to describe this enabling context, a "shared space" that can be thought of as a "network of interactions" (178). The concept of ba unifies mental, virtual, and physical spaces and differs from "ordinary" interactions in that it allows for the potential generating "individual and or collective knowledge creation." To make the environment stimulating, there should be a degree of variety, redundancy, and "creative chaos" (178); people need to hear an array of ideas, revisit ideas in different contexts, and understand that ambiguity is not readily resolved with ideas that fit easily into a preconceived pattern.

One of the key features of an enabling context is that a non-competitive atmosphere is created. In a hypercompetitive environment, individual members are more likely to behave in a critical manner which would undermine the ability for a supportive exchange of ideas between people (46). Instead, an environment that allows for people to "live" with an ongoing exchange of ideas that are related to the general subject area or mission of a corporation needs to be fostered. Working in this environment, also referred to as "indwelling," allows people to "dwell" on more than one perspective at the same time, thus enabling them to commit themselves to "an experience, an idea, to a concept, or to a fellow human being" (Von Krogh 58).

Explicit information provided by organizations in manuals, memos, and training sessions eventually becomes internalized by their members and the personal modification of this explicit information turns it into tacit information. Eventually, this knowledge becomes part of the conversation carried within an indwelling community and opens up new perspectives and exchanges. Thus, managers need to understand that not every piece of corporate information can or should be codified in an explicit fashion (Von Krogh, Ichijo, and Nonaka 182). This will better enable the ongoing accumulation of new knowledge (183) and application of it to completely new fields (185).

These features of an enabling environment are summarized below. An enabling environment:

- Allows for a dynamic network of interaction;
- Creates a non-competitive atmosphere;
- Values both explicit and implicit information.

While these criteria are designed to frame the interactions of people in business settings, they can be readily transferred to a Web environment to enable patrons who desire to seek, create, exchange, and perform actions tailored to their own agendas.

Knowledge Creation

In the *The Knowledge Creating Company*, Nonaka and Takeuchi describe in some detail through case studies how both tacit knowledge of employees and explicit knowledge—knowledge such as statistics and company policies that are formally documented and in place—all come together to increase the knowledge base of a company's ongoing "knowledge spiral." They discuss four different kinds of knowledge transfer: tacit knowledge to tacit knowledge through socialization, tacit knowledge to explicit knowledge through externalization, explicit knowledge to explicit knowledge through combination, and explicit knowledge to tacit knowledge through internalization. Below is a summary of these four phenomena, and in a few places, we have added some of our own insights and examples to enhance what Nonaka and Takeuchi write.

Tacit Knowledge to Tacit Knowledge by Socialization

Experience is how we acquire tacit knowledge, and socialization is the process by which we pass on ideas or shared mental constructs and technical know-how. Often we acquire tacit knowledge without being told in a formal setting how to do something or hearing someone else explain how things are best done. You might have taken on a new job, gone through some training, and then begun your work in earnest only to find out that many things you needed to know were not passed on to you during training. Instead, by talking to and working with your co-workers, you learned exactly how to do your job best.

Nonaka and Takeuchi have reported on several practices that exemplify socialization. For example, it is the practice at Honda to conduct "brainstorming camps" where members of a corporate unit who are engaged in developing a new product meet outside the workplace, usually at a resort inn, to discuss the project (63). The participants socialize at this site; they eat, drink, and engage in other group activities. This allows for a more informal atmosphere that enhances group discussions about how people see the problems they are facing and how they might be solved. Anyone can offer criticism, but only if the criticism comes with constructive advice about how something might be done better or thought of in a different way. These events bring to light new ways of thinking about a problem as they allow for the harvesting of unique and tacit insights from people who might be reluctant to voice their ideas in a more traditional setting.

Learning how customers feel about a company's products has better enabled NEC to produce computers that are more likely to fit the needs of its consumers, another example of socialization. At BIT-INN, NEC's display service center in the part of Tokyo where many electronic retailers set up shop, the concerns, needs, and ideas of the customers who frequented this store and were interested in improved NEC computer products were recorded and passed on to management.

Tacit knowledge can also be acquired by means other than dialogue between people. Discovering and internalizing tacit knowledge can be done by observing and engaging in the practices of others. In order to design a better bread-making machine, top executives and engineers for the Matsushita Electric Industrial Company asked a master bread baker if they could apprentice with him. By observing and imitating the master baker, they discovered that the best way to prepare the dough for bread was to add a little twist to it when stretching it, and this practice was replicated in the machine that they eventually designed. The master baker did not explain what had taken him years to learn; like many physical skills that seasoned professionals possess, the subtle details were such that the Matsushita professionals could only learn them by engaging in a process and observing his technique (Nonaka and Takeuchi 63–4).

Tacit Knowledge to Explicit Knowledge by Externalization

Externalization is where we take the tacit knowledge that we possess and convert it into explicit forms that are expressed in forms such as metaphors, concepts, and models. Perhaps the most common method of externalization is writing.

We all know that writing enables us to communicate to the outside world the thoughts that we have. The most brilliant ideas die with the genius that came upon them if she is incapable of conveying these ideas to others. However, writing is not only the means of explaining forcefully and cogently the ideas we have, but also a way of finding out what we know. We see writing as a way of ordering our thinking, of visually presenting in front of us a systematic pattern of the tacit and explicit information we hold, information that is a part of us, concerning a certain subject or idea. Furthermore, writing is thinking, and this thinking can lead us beyond what you already know. Writing thoughtfully demands that we take the abstract thoughts and intuitive tacit knowledge that have been bouncing around behind our eyes and then convert them into language. Because the thinking required in good writing is so absorbing, new ideas are often generated. This is thinking of the highest order. Not only is writing a means of communication, it is an extremely powerful catalyst for our thoughts.

Externalization is the primary method of knowledge creation; from tacit knowledge comes explicit knowledge—knowledge that is easier to identify, know, and use (Nonaka and Takeuchi 66). As we have shown, the process of writing is a solitary means of externalizing our ideas, and we believe that Nonaka and Takeuchi extend this basic process by illustrating how ideas can be generated by group dialogue or reflection. Both deductive and inductive reasoning methods can inform this process of externalization. For example, Mazda executives decided that the general guideline for all Mazda products was that all of their products should "create new values and present joyful driving pleasures" (qtd. in Nonaka and Takeuchi 64), and at the same time, they also wanted to produce a new vehicle for the U.S. market that would make Americans feel that Mazda was an innovative company. Using the deductive technique where a general idea contributes to a specific concept or practice, this general Mazda maxim was refined by the research and development group to produce a more specific guideline that they should build an "authentic sportscar" that was "exciting and comfortable to drive," which eventually began the process that led to the Mazda RX-7. Additionally, induction—the process of taking specific pieces of information and contributing to a general idea—also contributed to this product. A dialogue with both customers and car experts who offered their specific concerns allowed Mazda engineers to understand what they generally desired in a car that, up until this point, was not available to them. The specifics were the conversations and ideas of the experts and consumers, and this supported the general idea of producing a comfortable sports car.

Nonaka and Takeuchi like to point out how metaphors in general and a kind of metaphor, an analogy, can help us tease out associations between things and thereby let us better understand them. People use analogies to point out the concrete similarities between two ideas to better make a point or understand a concept. For example, physician William Hervey, the first to devise a theory of the way blood circulates in humans, compared this process with the way sap circulates in a tree to keep it alive. At the time, people knew about the way trees sustained themselves, and therefore, it was a reasonable and understandable extension of their thinking to imagine a similar process for humans (Crowley and Hawhee 176). An analogy allows us to bring reason to problems we are trying to solve by allowing us to see the similarities between things.

Nonaka and Takeuchi point out most other kinds of metaphors make more intuitive connections between things and work on a more symbolic level (66). We draw attention to perhaps one of the most common metaphors, "Love is like a rose." While we cannot make as neat a correspondence between two things with a standard metaphor like we can with an analogy between the circulation of sap and blood, metaphors do allow us to find abstract similarities between two things. For example, love is extraordinarily beautiful like a rose, but its thorns, like love, can hurt us at times. Applying this metaphor perhaps better allows us to understand how these two almost contradictory elements can have something to do with this complex emotion.

When Canon decided to produce a personal copier, the design team members knew that they would have to produce them at a much lower cost than commercial copiers. To do this, they would have to make the copying drum of the new machine replaceable as this was the source of ninety percent of maintenance problems. For design and economic purposes, the drum needed to be made of aluminum and there were drawn out discussions on how this could be produced efficiently. Finally, the head of the design team handed everyone a beer can one day and asked the engineers to compare the process currently used to produce this container, which was rather inexpensive, to one that they might use for the copying drum. This analogy gave them a starting point; they could quickly see the similarities between the two processes, but they could also begin identifying the differences and what would need to be changed to produce this item. This use of analogy engaged the tacit assumptions held by the engineers as they imagined applying the process of building an inexpensive beverage container to a reasonably priced part of a complex industrial product.

The concept of metaphor was used to imagine how cars should be constructed at Honda. The engineers decided to think of an automobile not as a construct of human engineering, but as an "organism" that would "evolve" over time to best suit the needs of humans. At the time, Detroit manufacturers designed cars to look a certain way—to be long, low, and stylish. This would best draw customers to them, and this way of thinking worked well for some time. However, the ability to ask "What will the automobile eventually evolve into?" (Nonaka and Takeuchi 65) allowed the Honda professionals to come up with the idea that cars are built for human comfort, that a "minimum space be given to mechanics and the maximum space for passengers" (Nonaka and Takeuchi 65).

Much of tacit knowledge can be turned into explicit knowledge through the use of figurative language such as analogies and metaphors, and Nonaka and Takeuchi see the progression from metaphor to analogy to a model as the most effective way to achieve this (66). Starting with the metaphor of the automobile as an organism that evolves to better meet the needs of humans, the concepts of "man-maximum, machine-minimum" and the geometrical shape that best allows for the most volume with the least amount of surface area, the sphere, were put into play as an analogy. The process of exploring the commonalities of these two concepts and reconciling some of the differences between them literally reshaped the way cars should be built and produced a new model. Instead of long and low, cars should be shorter and taller or more spherical, and the physics of this shape allowed Honda to build cars that were lighter, stronger, more comfortable, and less expensive to manufacture.

Explicit Knowledge to Explicit Knowledge by Combination

Taking explicit knowledge—knowledge that is already known and written down or recorded in some other way—and reconfiguring it for different purposes into other forms of explicit knowledge so it can be more readily used, is what Nonaka and Takeuchi refer to as combination.

The explicit knowledge that one starts with in this process can be knowledge from sources such as databases, documents, and intranets. By identifying existing explicit knowledge sources and reorganizing them for different uses from which these sources were originally intended, we can see how combination works. For example, knowledge learned at universities can be thought of as combined knowledge as it is usually found in textbooks that have been produced by professors who have read many scholarly articles and other books in a field and then shaped and distilled and thus recombined the knowledge in them for students.

Combination in the business world is often seen when middle-level managers reconfigure data in corporate databases or reshape a corporate vision generated by upper level executives so that new practices are put into play. Middle managers at Kraft Foods have demonstrated this practice of combination by reusing data that was originally for one purpose so they could accomplish another task. Their POS or point of sales database, which was designed to determine what products are selling well in various stores or regions, is also used to devise new sales techniques. By reformulating the data from categories that describe pricing, space needs, consumers, and merchandising criteria, this has allowed for sales promotions in certain markets and the designation of certain mixes of Kraft products in a practice known as "micro-merchandising" that reveals the demands of customers for each store or region. (68)

Middle-level professionals can also take corporate visions and combine them with their immediate tasks to hone the corporate vision from which the tasks are built on. Nonaka and Takeuchi detail how top-level executives at Asahi breweries decided that the phrase "live Asahi for real people" and "Asahi will provide natural and authentic products and services for those who seek active minds and active lives" would be their overarching vision (68). Middle-level engineers then developed Asahi Super Dry beer based on their concept of "richness and sharpness" which was their interpretation of the grand corporate vision. This in turn enhanced and made more specific the larger vision and also enabled the company to adopt new production techniques based on combining techniques; the information that was extracted from the corporate vision was adopted for the specific needs of technical professionals who used it to develop a new product, and the information based on the "richness and sharpness" concept further shaped the company's original vision.

Explicit Knowledge to Tacit Knowledge by Internalization

When we take in explicit knowledge by "embodying" it or "learn by doing," we are making it our own so we can use it; we are internalizing it (69). Tacit knowledge can be gained through the processes mentioned above: socialization, externalization, and combination. To enable an organization to have the most robust knowledge management culture, all tacit knowledge gained and that which resides in people needs to be passed on to others through internalization.

Documented knowledge can lead people to understand, to internalize, what it is to have been involved at a personal level in some process or field of information. For example, GE documented help requests and questions from customers who called in to its Answer Center in Louisville, Kentucky. By reviewing these complaints, a GE product development team could later acquire tacit knowledge as they would learn like those originally taking the calls in the Answer Center would. This knowledge enabled the team to build a database of over 1.5 million potential problems and their solutions. Additionally, new product development specialists would also have conversations with veteran members of their team and visit and speak with representative telephone operators who originally fielded the complaints, thus re-experiencing the knowledge of these two groups so they could internalize this information.

During hard economic times in the early 1990s, Osamu Tanaka voiced the fears of the executives of Matsushita Electrical Industrial Corporation, who had become concerned that their employees were working too many overtime hours that did not yield significant productivity or creativity: "How can anyone be creative if he works until twelve midnight everyday? People's sense of value is rapidly changing. You cannot make original products just by looking at plans at the office every night" (qtd. in Nonaka and Takeuchi 118).

To internalize the idea for the employees that they could be more efficient by eliminating overtime hours and increasing personal time, they had their hours reduced to 150 hours a month and, ultimately, to 1,800 hours a year. Some of the major tacit realizations of the Matsushita employees were that they could not continually make design changes in products being developed and that face to face meetings had to be cut back in favor of computerized communications. More than policy changes, these new "sense of values" or new embodied practices became part of the way work was done (119).

Oftentimes we hear stories or anecdotes about a practice or insight from others who are clearly articulating their knowledge, and then we take these stories and construct our own work habits. Nonaka and Takeuchi point to the many "mental models" that are captured by articles and books written by Japanese executives and business journalists about their lives and the culture of their corporations. In the United States today, there are whole sections of bookstores where one can find the memoirs, how-to books, and biographies written by or about contemporary business executives and culture that reflect the same kinds of tacit knowledge that, when studied, can become embodied by those seeking to adopt new strategies and practices.

Technical Communication and Knowledge

Michael Hughes expands on the theories of tacit and explicit knowledge by using the work of W.C. Howell which describes these types of knowledge on the learning process. Additionally, Hughes connects these ideas to technical communication. There are four major states in the learning process according to Howell (Hughes 278):

- "Unconscious incompetence" is when we are not even aware that we lack a certain kind of competence or skill, or that one even exists.
- "Conscious incompetence" describes the position we are in when we know we lack a certain skill.
- "Conscious competence" illustrates how we identify knowledge we have discovered and then use it or document it.
- "Unconscious competence" is when we know how to do something and apply it without even thinking about it.

Hughes asks us to replace the following words in Howell's depiction of the learning process: "competence" should be replaced with "knowledge," "incompetence" should be inserted in place of "ignorance," "conscious" should be substituted with "explicit," and "tacit" should take the place of "unconscious."

Thus we have the following comparisons:

- Unconscious incompetence becomes tacit ignorance.
- Conscious incompetence becomes explicit ignorance.
- Conscious competence becomes explicit knowledge.
- Unconscious competence becomes tacit knowledge.

In doing this, Hughes gives us another way of looking at the range of learning states that describe the technical communicator's need to be aware of what are more in line with what Nonaka and Takeuchi theorize. At one end of the spectrum, tacit ignorance, we have some end users or customers who buy products who do not even know where to start learning about something or what really needs to be known, and technical writers are often challenged to derive information from experts who are at the other end of the spectrum, tacit knowledge, who really know how to do something, but really do not understand that they know this. These subject matter experts (SME) just take it for granted that what they know is simple and that everyone else knows it, and it is the job of technical communicators to harvest this information and explain it for beginners (Hughes 278).

Hughes applies the theories of Nonaka and Takeuchi by showing how technical writers can contribute to an organization's knowledge spiral.

Nonaka and Takeuchi describe the durability of knowledge; this is how fixed the knowledge is within the organization. For example, if someone possesses a lot of tacit knowledge and then leaves the organization, the knowledge leaves the company and is thus not very durable. When knowledge is socially constructed by the interactions of group members within an organization, if one person leaves the company, the others remaining will still possess much of the knowledge this person held and they can pass on information to others (Hughes 279).

When knowledge is passed on from individuals to groups and from groups to entire organizations, this "escalation" makes the knowledge more durable, allows the knowledge to be scrutinized by more people and thus perhaps challenged and validated, and it also allows for greater knowledge transfer within an organization. In this last example, Hughes suggests that if technical communicators produce a style template for documenting knowledge, this template can by adopted by other members in the organization who are not technical communicators. This process of tacit knowledge that moves from individuals and becomes the explicit knowledge that resides in groups ultimately increases the organization's knowledge assets and produces more durable knowledge.

Deploying Knowledge Management: Intranets and Extranets

An intranet can be likened to an Internet that is deployed within an organization. The advantage of an intranet is that it can be protected by a security firewall that excludes people from other organizations who might be trying to steal ideas from their competitors. In contrast to the Internet, firewalls can also be better maintained by IT professionals and they can be set up to exclude "malicious or intrusive" software. Also, intranets allow for the storage of company documents in a well-organized manner and reduce the need for paper-based distribution of material which can allow for better access to material and save an organization money (Callaghan 3–4).

An extranet is an intranet that has its protective firewall set up so some people who are not organization members have access to some or all of the content on the corporation's intranet. This might be for customers who need extra information about an organization's products (Callaghan 4).

It is easy to see how these systems can support knowledge management activity as they can store, arrange, and allow easy access to information and knowledge in an organization as they are dedicated to an organization's specific needs.

Corporate-Wide KM Systems

To stay competitive with other corporations by being more efficient in disseminating information, CoreTech introduced an intranet in the mid-nineties that eventually crossed all layers of the organization (Callaghan 67). The company is in the IT solutions business and is hierarchically structured;

the information flowed from senior management, to middle-level management, and then was disseminated by the managers of each individual work unit, oftentimes verbally. Additionally, there was information such as project reports that were exchanged horizontally between individual team members, their immediate manager, and their group members. One of the reasons that the intranet grew was that most employees were technically savvy as the company's main line of work is to help other companies find creative solutions to IT problems.

Before the intranet was introduced, information was presented in paper documents, which can hamper productivity as paper documents are not really organized in a database system; they just pile up on the desks of most people, are often thrown out when the information in the document might be useful at a later date, and are not automatically indexed where they can be retrieved later if needed.

The implementation of the intranet was devised by the CoaT office, which was dedicated only to this project. This was beneficial in that the structure of the early intranet at CoreTech did not need to have to go through committees in all branches of the corporation for approval; it was just put in place by the CoaT people.

The CoreTech intranet allows people to publish their own websites so personnel information could be catalogued, and also gave access to Quality Management Standards (QMS) of specific work area standards, organizational charts, summaries of company business strategy, and company news (74). Ongoing projects, outcomes of previous projects, and information on directors and senior managers were also included. A "Desktop Directory" was implemented on the intranet so specific employees, their job titles, and their location and work areas could also be found out. The "whereabouts" or what each employee would be doing for the following four weeks could also be discovered (75).

Seventy-eight percent of the CoreTech's members agreed that the intranet allowed them to work more efficiently. It can be argued that the most important element of the intranet was that all employees with access to it had "equal access" to whatever they felt they needed; the previous hierarchical system that was in place was bypassed as people did not have to wait for their managers to disseminate important information after it came down to them from above. The middle-level managers were not in control of what information their employees would be able to know. Also important was that employees could contact senior managers directly, bypassing middle-level managers.

During a period when there was a possible merger of CoreTech and another company, information on the ongoing negotiations was distributed to the employees so they could feel that they were in the loop, which was very effective according to a senior level manager:

> [I]t was allowing people to share in where the company was going and to share in where the company was going fairly swiftly after the decisions

> had been made, decisions were being made in the morning and the information was coming out either that afternoon or the next morning on the intranet. So people were getting to know very, very quickly and I think they felt part of the company and that was information management at its best, it was the impact of IT at its best, it was the opening up where the company was. (qtd. in Callaghan 76–7)

That all of the workers could know more than just hearsay made them feel more connected to the company. The Director of Global Sales liked the intranet as he used his "Ask Bill" element of his intranet website to get feedback from anyone in his division, regardless of position. The intranet also allowed people to be more proactive in finding others in the company who might be able to assist them, and it broke down barriers between levels in the overall organization. One manager commented that it put people in contact with information and knowledge, but it also put "people in contact with people" (qtd. in Callaghan 102).

There was a downside to this more open flow of information. For some middle-level managers, they felt less control over their employees as they lost the power one receives when she is given control over choosing what information to convey from the top of the company. However, one middle-level manager commented that it gave him more time to concentrate on getting their teams to accomplish their goals.

The initial intranet had many websites constructed by employees who produced information that they thought would be of value to other members of CoreTech. In April of 1999, it was decided that all of the company's Web pages would be migrated to a new server, and only those websites that met a specific official criteria would be allowed on the intranet. This produced a negative response by some members of the organization who were concerned that they were never consulted about what information standards should be adhered to. They made the point that official organization policy needed to be standardized, but that did not mean that other contributions had to be kept off of the intranet. One engineer defended the right for all contributions by employees be accessible by saying: "[W]hat will it take to make 'the powers that be' listen to the USERS, not providers of the intranet, and actually give them what they want?" (qtd. in Callaghan 92).

One of the counterarguments made by management of this policy was that all users of the intranet needed to be sure that the information is up to date and accurate. Another comment was that while the Internet is anarchic, intranets need to be more controlled and that there needs to be a balance between the creative ideas of people and some degree of control (94). While this section on intranets and extranets shows how knowledge management practices can be actually put into place, the next section demonstrates a framework for thinking about how information can be better balanced for the needs of people.

Information Ecology

In biological studies, ecology is the study of the interrelationships of the many living species, both plant and animal, found in an ecosystem. In their book *Information Ecology*, Nardi and O'Day expand this concept into complex systems of information in a way to frame the relationships between people with different skills, technology, values, and the individual human relationships that people have with the IT they use every day. What is key here is that all of these elements have an effect on the uses of technology and that they cannot be studied independently to understand how an ecosystem of information functions.

They use this metaphor as all too often we look at one element of technology and are often unaware of how it really is used by people in certain contexts. To say that an organization will now be functioning at a higher capacity because a new software system is put into play or that a new generation of computers and the way they are networked will improve efficiencies is not necessarily true. We all have our own relationship with our computers; we set our computers up to work in the way we feel they meet our needs. The interaction between people regarding their technologies is also a key factor. People will use technology in different ways, and people also learn from each other when they engage in work using their computers. The use of technology is employed in specific ways in specific settings: educators, students, engineers, and executives all have their own way of using technology.

It is important to view the IT we use as "tools." The use of the word tool for a technology is important as it challenges designers of the tools to imagine how the end users might use them when they design them (Nardi, 30). All too often, designers forget this and design overly complicated tools that really do not replace traditional technologies efficiently. Some people come to meetings with laptops and take notes with them, while others are just as comfortable applying pen or pencil to paper. Some people do their taxes using the many available software packages available, while others are comfortable just using a calculator. Some educators have experimented with building a curriculum around the use of students doing all of their work on laptops, but this does not take into account all of the problems that laptops might bring with them.

In a New York public school system, long term studies revealed that the basic learning achievement scores of students who were provided with laptops were no better than those who engaged in traditional learning technologies: blackboards, pens, books, and paper. With laptops, the networks broke down, students tended to wander off on their own, teachers had to work too hard teaching the basic software packages and this got in the way of teaching their course material, students could more readily exchange answers on tests, and laptops broke down all too often. The cost of classes built around students with individual laptops was also much higher than the cost of a traditionally outfitted school (Hu). Designing a tool that has great potential does not necessarily mean that in certain technical ecosystems it will really prove to be useful.

Nardi and O'Day tend to be optimistic about the use of technology in certain ecosystems, but they also make a point that they also review the spectrum of theorists who are concerned about the implementation of technology into our lives. Employing the ideas of Bruno Latour, Nardi and O'Day point out that when we use machines to communicate, "Prescriptions are written into technologies when they are designed." As Selfe mentioned earlier, Nardi and O'Day tell us that technologies have an "authority and presence" by the way they are designed (32). They ask certain things of us. Sometimes, there is a large gap between what the intentions of the designers of the technology imagine how a technology can and should be used and how it is actually employed.

Nardi and O'Day point out some of the concerns of Jacques Ellul, who is troubled by the institution of technology. More specifically, technology integrates what he refers to "technique" in our culture, and technique in Ellul's mind creates a mindset in our culture that we need to always be working towards greater efficiency, but are doing so blindly, and, because of this, we do not consider the potential negative side effects:

> Technique has become autonomous; it has fashioned an omnivorous world which obeys its laws and which has renounced all tradition. Technique no longer rests on tradition, but rather on previous technical procedures, and its evolution is too rapid, too upsetting, to integrate into older settings. (qtd. in Nardi 34)

In the example above regarding the New York school system, many traditional elements might have been lost had there not been some studies done to support the idea that notebook pedagogy was working. Students might learn to type better, but they also might have lost the ability to look at a teacher, one human working hard to help students understand what they need know to be successful, and concentrate and listen to what the teacher is trying to say. Birkerts has posited the idea that individual subjectivity, being aware of ourselves and our thinking relative to others, is diminished in this kind of network environment:

> One day we will conduct our private and public lives within networks so dense, among so many channels of instantaneous information, that it will make almost no sense to speak of the differentiation of subjective individualism. (267)

Some might argue that the technology could be improved—that the computers could be sturdier, and the software easier to manipulate by the students and teachers—but we would just be adding on to "previous technical procedures" and the valuable lessons that we have learned from listening to people and taking directions in "real time" would still be lost.

As mentioned before, when Nardi and O'Day liken systems of IT to an ecosystem, they mean to point out that the relationships of many different entities depend on one another for that system to form and remain in place. As a technological metaphor, this does not mean that there are certain entities at the top of a technological food chain who survive by eating smaller entities. What they mean is that people with a diversity of skills and positions in organizations play complementary roles. Without designers of software or IT professionals, IT would not exist or be put in place. If it were not for people who were served as the consumers of technology, there would never be a market and demand for such services, nor there any feedback loops that would allow the designers and technicians to improve and implement the technologies. In this last example, we see the concept of coevolution in operation.

When a new technology is put in place in an organization such as a new intranet that allows everyone to be connected and to send messages and share ideas with each other, the members who are asked to use the system learn some new skill and evolve to a more technically proficient level. When the system is flawed to some extent, the designers become more aware of how humans use machines and strive harder to make their products more convenient and useful.

The metaphor of information ecology asks that we see that everyone in a system is of value, but there are certain keystone species that are needed to make the system work. One example of a keystone species, according to Nardi and O'Day, would be the teachers who train employees how to use a newly implemented technology. Without them, the employees would face too steep a learning curve to even begin using some technologies.

In some biological ecosystems, one species might in fact become extinct, but this does not necessarily mean that the ecosystem will fall apart. In some business organizations, one unit in the organization might be phased out; its members might be given their notices or retrained to do another kind of work. However, in both examples, a keystone species is essential for the ongoing evolution of the system, be it a biological or information ecosystem.

Buchanan further refines the notion of keystone species by noting that they are usually connected to many of the other entities in an ecosystem (153). In a biological system, a keystone species might be one that many other species within the ecosystem depend on for food. If the environment was significantly altered by some change in the climate and this species died out, so would the other species who were connected to it. In an organization, those who teach the technology well might be thought of as connected to many others in an organization. In addition to training the end users of a technology, they might also serve as the liaisons between executives in organizations, the middle-level managers, and the designers of the technology as they are in the best position to see what are the strengths and weaknesses of the information systems that are deployed in an organization. Regarding information systems, they are a keystone species. In Chapter 3, we will discuss the role of communication professionals who assume the position of knowledge managers by implementing XML technologies across an organization's branches. We will show how these knowledge managers can be likened to keystone species.

Technology also acquires its value in different "localities" by where it is placed and who uses it and how it is set up to fit within an organization's environment. How a technology is utilized is not so much determined by who sets up the technology, but by how the end users use it. They assume the "responsibility" and take advantage of the "opportunity" to use technologies in the environment they work in. They construct the "identity" of the technology via these practices (Nardi and O'Day 55). For example, a desktop computer that works at the same speed and has the same office-use styled software is referred to differently in varying environments. The same kind of computer can be used for payroll in a small business, and a computer in a library can function as a replacement for a card catalogue and provide library access to Internet resources that are related to their research. These computers in these two environments would be referred to differently, even given different names, as how they are named "identifies what it means to people who use it" (54).

When contrasted to the often used term "community," Nardi and O'Day assert that the ecological metaphor allows them to better illustrate how a technology is employed in an organization (56). "Community" is often used to describe a group of people who have more similarities than differences, whereas an ecological system built around communication technologies implies that there are many different kinds of people who might be working to the same end, but who play different roles and have different skill sets. Ecological systems are dynamic in that while they might be complex and in equilibrium, they are always changing or evolving. In a complex biological system with many species, the relative numbers of each species might change, but as long as the keystone species remain in place, they will be in equilibrium. In the complex ecology of a large organization, people might be asked to move to a different division and/or be retrained to take on a different task to meet the organization's needs in a dynamic economy.

Nardi and O'Day also acknowledge that the presence of a technology can impose itself on people who use it as Ellul would suggest, but feel that if the members of each local element of the larger ecological system are aware of this and take the initiative to use the tool in the way that meets their needs, that they can claim a technology as their own and to make it fit it in with their own needs. A new technology should not be met with "resistance"; end users should be invited to come to a new technology with the spirit of "engagement" and "participation" (57).

While it seems obvious to say that librarians are a keystone species in a library, Nardi and O'Day go to some length to describe how they occupy this position and how their work extends deep into the information ecology of several industrial settings, including the Hewlett Packard Library and the

Apple Research Library. One of the points they make is that when these librarians work with non-library professionals in these organizations, their work practices are often "idiosyncratic"; what they do cannot be boiled down to a set of hard and fast rules that can be described in a simple operations manual as every patron these librarians work with have very specific needs and personal work habits (83). This also challenges the librarians to better tailor the available technologies to the specific needs of their clients.

In both libraries, librarians would receive requests from the professionals they worked with, sometimes by phone, fax, or e-mail, and sometimes in person. They would interview their clients and ask for some specifics that helped them better refine their research. The librarians would make an initial search to cull the information they felt the client could use, and, if needed, they would do a second round of searching. Oftentimes clients needed guidance in choosing more specific search terms as the fields these professionals engage in are cutting-edge and the new search terms come into play. More specific search terms lead to more focused results. In this way, librarians helped professionals "understand their own needs" better (88); the nature of searching for information often helped the clients understand "more about what they want[ed]" (89). Oftentimes the needs of the professionals might be a bit vague at first and they needed to engage in conversations with the librarians about their needs, and the recursive nature of a search for specific information. The going back and forth from client to search technologies, and then back to the client, exemplifies how the information ecology of these organizations works.

The information ecology of these organizations is built around professionals such as engineers and business people working with librarians who are continually learning how to use and teach the use of new information resources. The librarians need the stimulation of their patron's requests to compel them to stay relevant in an organization, and the professionals who work with the librarians do not need to spend all of their time learning about what new databases have come into play in their fields.

Oftentimes, librarians will specialize in certain fields of inquiry such as chemistry or business (93) and will acquire working relationships with their patrons where they begin to acquire knowledge of what each specific client is interested in (96–102). When not performing a specific search for a client who needs information right away, librarians can be on the lookout for new databases and setting up "canned searches" with "agents" for new information they feel might be of value to clients and then alerting them to it. This personal engagement meets the needs of many organizations that are on the cutting-edge of research and development:

> Because access to information is a fundamental need in today's world, it must be supported to the fullest, which means a living, breathing community of helpful people at the ready. The human touch will become more, not less, important as online information resources grow and information resources grow and information access tools proliferate. (92)

The point here is that no one can do it alone; there is a symbiotic relationship between all of these professionals as they all need each other to advance to organization's overall needs.

Chun Wei Choo describes in greater detail the process through which information professionals can work with subject matter experts to assist in adding to the knowledge of an organization. For example, "harvesting" the tacit knowledge that an organization's employees possess can start with something as simple as collecting and updating the resumes of an organization's employees so they can seek out in house help if they need to. They can also alert employees to upcoming conferences, training courses, projects, and assignments. Information professionals could take it even farther by creating an "electronic yellow pages" that consisted of the skillsets of each employee and put them in an organization's intranet. They could ask for some indication of the "breadth and depth" of knowledge that each person possesses regarding each skill that is listed and the contact information for each employee (367).

In addition to formally classifying the skills of an organization's employees, Choo sees information specialists as those who can also perform the following tasks:

- Writing and editing "raw knowledge" so it is accessible to others.
- Indexing, producing subject headings, using links for cross referencing, and designing metadata.
- Packaging and publishing and then distributing through corporate sources such as CD-ROMs, intranets, and subject-oriented search software.
- Constructing and managing the information architecture for the content while paying special attention to the style and classification schemes so that they are appropriate for the organization. (398)

All of the skills listed become part of the overall information ecology of an organization; while performing these tasks, information specialists are working with other professionals to grow and shape the organization's knowledge base. The metadata—or data about data—that information professionals would be working with could include such things as "project names, project stage, product names or categories, authors, departments, [and] dates" (399), and we see these as elements that could be coded and warehoused using XML.

While information specialists can perform all of the tasks above and need to understand the basic kinds of information that each division within an organization needs, Owen makes it clear that these specialists need to understand that people should not depend on them for all their knowledge needs; instead, they are there to enhance the information ecology of an organization by getting professionals to help develop and contribute to their own systems of knowledge so they can exchange information with each other.

The traditional role of a librarian-styled knowledge broker needs to be downplayed because it suggests that an organization's professionals must always go to this person to find "available knowledge on a single item basis . . . to provide a knowledge item whenever required" (Owen 13). This is based on what Owen sees as a "fundamental principle of knowledge management," which is to remove the "intermediary" between the knowledge and the user of the knowledge. For example, engineers should be exchanging tacit knowledge generated by other engineers, and managers should do the same with managers: "knowledge management is more focused on the flow and interchange of knowledge, and therefore on knowledge channels and networks, than on the management of 'knowledge objects' as distinct entities" (Owen 14). If possible, there should be a robust ecology where people can look to ideas outside their work units without always having to go through an information professional.

The most explicit kinds of information—facts, figures, and documented corporate rules—are not really what a knowledge worker needs when she is not able to move forward and solve a problem. Knowledge management workers need access to more tacit forms of knowledge such as background information on a project or client, examples of analyses, best practices, arguments about relevant corporate projects and issues, opinions, methods, and procedures (Owen 12). This is where the tacit knowledge that can be used to solve thorny problems is needed when professionals are at a loss and need help, and perhaps this kind of knowledge is best understood and conveyed by the professionals who specialize in specific work practices.

In addition to the information specialist tasks that Choo provides, Owen suggests some more specific activities:

- Generate a typology or systematic classification scheme of employee profiles, current projects, lists of clients and contacts, external knowledge databases.
- Provide indexing standards for subject domain models or subject descriptors such as keywords and a thesaurus.

(Owen 15)

Information specialists can also create knowledge profiles or "cluster" knowledge objects for the following:

- Combine all knowledge that pertains to a specific employee or general job designation that describes the work of more than one employee.
- Cluster knowledge that is relevant for a project team.
- Identify knowledge that would aid an organization's members in their encounters with external contacts.
- Find all general knowledge that would be relevant to an organization's employees.

(Owen 15)

While Owen does not use the phrase "information ecology" as Nardi and O'Day do, he is clearly describing how an information specialist can help implement one. Again, his emphasis is more on inviting organization members to make it a habit to input or "publish" their tacit knowledge when they have something to offer and extracting tacit knowledge from the resources identified above when they need it, to create "a 'knowledge chain' which links authors with users, through which knowledge flows from one to another" (14). While we can see information specialists working to help combine knowledge specific to organization members, their activities are to facilitate knowledge building and dissemination between organization members.

Information and its Integration into Social Systems

Echoing Nardi and O'Day, Brown and Duguid are interested not so much in the growth of technology, but rather ask that we pay more attention to how technology is deployed in its social context before we design it, thus making it more useful:

> The ends of information, after all, are human ends. The logic of information must carry the logic of humanity. For all of information's independence and extent, it is people, in their communities, organizations, and institutions, who ultimately decide what it all means and matters. (18)

They point to the fact that some technologies, while perhaps looked upon by their designers as elegant and valuable, really create more problems than they solve. Some organizations implement software that is so complex that it takes more time to learn how to use it than the value its designers believe it will deliver to the end user. Additionally, one of the problems with knowledge management is that all too often organizations are able to warehouse large quantities of information, but, in doing so, they fail to make distinctions between quantity versus quality information. Brown and Duguid point to one study that shows that "information production" grows at fifty percent a year, but our personal consumption of information is growing at 1.7 percent a year, and this latter figure has "natural limits" (XIII).

The overall point that Brown and Duguid emphasize in their book, *The Social Life of Information*, is that we have to look beyond the technology and see how individuals and groups of individuals encounter it, and how the social nature of information exchange between people supports the use of technology.

In the early nineties, business reprocessing engineering was thought to be the method that saved companies that were not as efficient as they could be. This reeingineering trend looked at companies linearly; one part of a process is done by one part of the company, then another takes on its part of the project, then another does the same until the organization finally has the product out the door and to the consumer. Each sector of a corporation was examined in terms of its inputs and outputs, but little attention was paid to the actual workings within each sector and how they related to other sectors. This worked well for some workers who engaged in procurement, shipping, and receiving where the work tasks are relatively straightforward and easily measured. However, problems arose in the divisions within corporations where the tasks that needed to be done were more complicated, ones that were less "clearly defined" (93).

In the health insurance claims sector, the claims adjusters who worked with the patients who applied for benefits were commanded to do their jobs in a certain manner from executives in sectors of the organization that were above them. These directions did not take into account the many nuances of these tasks. For example, the claims adjusters saw that sometimes similar claims provided different reimbursements and there were difficulties in figuring just who qualified for claims. These difficulties were "traced to clashes over meaning and sense making" (96).

While the process model might look efficient on a flow chart to company executives, it is difficult for those who have to deal directly with the public. Getting the job done for practicing claims adjusters can be just as much a "craft" as the work of the physicians who perform their services on the patients (97). In Nardi and O'Day's view, the local culture of the processors had been overlooked and not given any credibility, thus creating a flawed information ecology. The claims adjusters had to look to their local colleagues for guidance that might, at times, run against the directives aimed at them from on high (98). Practice at the local level needs to inform an organization's overall process, but this is often overlooked by managers who want a top-down process system that does not recognize the "lateral ties" between divisions and workers within divisions as these are considered "non-value adding" practices by management.

It is often the case that documentation produced to train people how to do their work is written by people who do not explain why workers are asked to perform specific tasks so they can learn to troubleshoot or deal with the unanticipated problems that always come up. In a study done on copy machine technicians who represented Xerox, as in the claims adjusters study, the technicians had to learn from each other. The adjusters in the health insurance agencies most likely work in the same workspace and can more readily find a peer to help them, but the Xerox technicians had to go out on their own into the offices of other organizations to service the machines, and thus their work was "highly individual" in that they were out in the field alone. As there was no scheduled time in each technician's day to meet each other, they began to meet for breakfast, lunch, or even dinner to exchange information about the way they figured out how to go beyond what the manuals said and fix the copy machines. In the process model of organization management, this social "chatter" was never built in; it was assumed that the technicians were proficient because they were trained and provided exemplary documentation. In reality, the ability of the technicians to do their work properly was based on the social life they established with each other (Brown and Duguid 102–3).

In both these examples above, the focus has been not so much on information, but in know-how or knowledge. Brown and Duguid are quick to point out that oftentimes knowledge management is only associated with giving people access to information. Knowledge is different than information in three ways in Brown and Duguid's estimation:

- First, knowledge implies that there is a knower. Someone has processed some information and really knows how to do something or how to get something done, not just learned a fact or figure.
- Second, knowledge is hard to pass on to someone else. In the Xerox examples above, the technicians had to meet often and discuss the problems they were encountering and then explain their reasoning to their colleagues. What they learned was hard, and it took the give and take of human interaction to allow them to transfer it.
- Third, knowledge is hard to assimilate. The Xerox technicians had to struggle with their tasks, to venture questions, to hear and also learn by explaining their own ideas to others to acquire the real knowledge that the manuals they were provided with could not (119–20).

Brown and Duguid see the shift from an information economy to a knowledge economy as one that is a shift towards people or knowers, something from which the process model of business, in seeking greater efficiency, was moving away from. Sometimes, people speak of knowledge management in terms of information, not intellectual knowledge capital, and this allows them to feel that they can store mere information in an online database or intranet and thus they have produced a robust knowledge management system. In the Xerox technician example, it was hard won knowledge that was gained through experience in the field and then transferred by human interaction that allowed for the increased competence of these technicians, not just information that consisted of what one would find in a manual written by non-practitioners.

Several Xerox engineers worked to implement a database system, the Eureka project, which sought to expand the best practices developed by the technicians in their local communities. Starting with technicians who serviced Xerox machines in France, they began to build a body of knowledge based on their insights that could be viewed by using the French Minitel system, the national phone company, which could be accessed with a small keyboard and viewed on a local display monitor. This was met with some resistance from the Xerox management. Even though there was documented success in the social meetings with the technicians, Xerox executives still maintained the process model that reinforced the value of sticking to the documentation provided to the workers and their initial training. Xerox managers were concerned that "a single flawed" tip that made its way into the system would prove too costly (57).

To counter this concern, inputs from any technician were screened by a validator, a product specialist for Xerox France, to make sure that it was of value. If the tip was deemed of value, the validator worked with the technician to make sure that all the relevant information was captured and explained properly (Bobrow and Whalen 51). In the following chapters, we will reveal how communication professionals can construct document type definitions (DTDs) and schema, both essential XML technologies, so that data entered into a content management system (CMS) better meets specific and valid standards. Below is a generic sample of a tip submitted by a technician that was stored in the database:

- *Diagnosing unusual, costly failures*—Bimetallic corrosions builds up on A and causes intermittent failures that seem to be in B. Replacing B makes the problem seem to go away because A is moved in installation. First clean A, and later replace by new gold-plated AA, available as Part #1234.
- *Workarounds*—Paper cut in a dry environment causes excessive jams on baffle Q. Putting Mylar tape from tool kit on edge will ease problems.
- *Easing the job*—To make it easier to adjust M, paint white-out on back wall near M.

(Bobrow and Whalen 52)

Part of the success of the Eureka project was due to pieces of information such as the one above. Depending on the product serviced, the efficiency of the technical support people improved by five to twenty percent when contrasted to other European divisions of Xerox. The technicians would often use the Eureka program before they even arrived on site because, based on initial information from the client, they were often able to make sure that they picked up a part that might be needed for the machine before they left. The implementation also decreased the amount of calls the technicians had to make to their supervisors and sped up the learning curve for new employees.

Because of this success, a similar system was deployed in a Xerox division based in Canada. To input and access data, the technicians used laptop computers. The Canadian technicians were provided with a monetary incentive to input a valid suggestion, whereas in France this was not the case. A single search engine was designed, dedicated to the Xerox technicians, called SearchLite, which was fast and easy to use. Following the success of this program, the Eureka system was deployed in the United States in 1997. By 1999, there were 2,000 validated tips put in place with 9,000 "solves" recorded based on information American and Canadian technicians were able to find using the system.

Eureka served as an able diagnostic tool for problems that arose in the field, and even if Eureka did not provide a "precise solution" for a particular problem, it did allow technicians to "rule out certain sources of trouble" when they collaborated with each other (Bobrow and Whalen 56). As one technician reported:

> Eureka isn't so much as an end, but a beginning. Someone will call over the radio with a fault code like "I'm having 12–142s," and I can look it up in Eureka and scroll through common causes. It's faster to find it in Eureka than it is to go in and fire up the documentation CD for the repair procedures here.
>
> (qtd. in Bobrow and Whalen 55)

The bottom-up approach is in play here; for this technician at least, the data that was warehoused by other like technicians was of greater value than the documentation the corporation had compiled. Not only did Eureka serve to solve problems in the field, it also served as a learning tool. One technician pointed out "Whenever I download new Eureka data, I like to see what guys are doing. I look through the tips and bulletins. It teaches me a lot" (qtd. in Bobrow and Whalen 56). In this way, Eureka served to further embed itself in the culture of Xerox technicians and helped produce an enabling environment that Von Krogh describes above.

Following this success, Eureka II was implemented, which was a worldwide deployment of this system via the Internet. Part of understanding the concept of the social life of information is that sometimes there is resistance by managers who oversee operations and who favor the top-down approach. Bobrow and Whalen make the point that they were able to finally implement a full fledged search system into Xerox's world operations because they started small and in foreign markets, working under the radar of the company's Worldwide Customer Service branch. When they implemented Eureka II, some managers wanted to use a standard Web browser such as Internet Explorer® in place of SearchLite, but this would have been a problem because there are so many versions of this. Additionally, when a world wide version of Eureka was implemented, Eureka II, the managers saw it as a large investment and wanted to see certain proven efficiencies met to meet their schedule. Some of these schedules could not be met in the initial stages. It was also difficult to get funding from some divisions to train the technicians to use Eureka, despite its promise of greater efficiency in the long run. The corporation also could have benefited more had there been an effort to take some of the better technician generated solutions back into versions of succeeding corporate documentation that was initially used to train and guide technicians. The Eureka II system and corporate documentation existed as separate bodies of information (Bobrow and Whalen 57–8).

However, Eureka II has proven to be successful in many ways. As of this writing, there are now 50,000 tips in the collection (PARC), and Eureka can store information in eight languages, validators are bilingual, and all tips that have been validated are translated into English (Barth 7). Eureka has saved the service definition as much as 100 million dollars in service costs (Brown and Duguid 112). The sales division of Xerox has also shown interest in developing its own Eureka-like system for its sales professionals.

Research on Knowledge Extraction from Document Collections (KXDC) is presently in play at Xerox. The focus at PARC is on natural language processing (NLP), language that technicians might put into database that is not a formal computer language, like the language we use when we use Google for a search. KXDC researchers are also working to clean up the database of tips by eliminating redundancy, obsolescence, and contradictory tips, and their long term goal is to create a "knowledge fusion" capability that would allow for the creation of composite documents from many different documents that list a series of possible tips that would relate to the needs of the technicians (PARC 2). We will show in Chapter 3 how documents that are tagged in XML code can serve as single source entries that organizations could employ to accomplish these tasks.

Discussion Questions

1. Using Kuhn's terminology, what are some of the "symbolic generalizations," "shared exemplars," "shared commitments," and "shared values" that the community of people whom you are working with or are studying with hold? Is there a "normal science" in your field that, while helpful, also goes unchallenged?
2. What tacit knowledge(s) do you possess that you have taken for granted and that could be of value to others? Whether you are a student or a professional, think of the things you already know that you have shared with others that you learned by doing something that was not explained to you explicitly before you attempted the task. What kinds of tacit knowledge have been passed on to you from other students or peers that have been of value to you?
3. How does the metaphor of "information ecology" better allow you to understand the nature of the way knowledge is produced and enhanced? How might it allow us to better challenge more hierarchical ways of distributing knowledge?
4. Describe what Brown and Duguid mean by the "social life of information." How does it better describe how knowledge is consumed and distributed by workers in large organizations? Think about the social dimension of knowledge in your life. What do you know that has been gained via social interactions with your peers?

References

Barth, Steve. "Eureka! Xerox Has Found It." *Field Force Automation Magazine*. April, 2000. http://choo.fis.utoronto.ca/mgt/KM.xeroxCase.html.

Baumard, P. *Tacit Knowledge in Organizations*. London, England: Sage Publications, 1999.

Birkerts, Sven. "Into the Electronic Millenium." *Cyberreader.* 2nd ed. Victor Vitanza, ed. Needham Heights, Massachussetts: Allyn and Bacon, 1999. 257–69.

Bobrow, Daniel G., and Jack Whalen. "Community Knowledge Sharing in Practice: The Eureka Story." *Reflections.* 4.2 (2002): 47–59.

Brown, John Seeley, and Paul Duguid. *The Social Life of Information.* Boston, Massachussetts: Harvard Business School University Press, 2000.

Buchanan, Mark. *Nexus.* New York: W.W. Norton, 2002.

Callaghan, James. *Inside Intranets and Extranets: Knowledge Management and the Struggle for Power.* New York: Palgrave Macmillan, 2002.

Chun, Wei Choo. "Working With Knowledge: How Information Professionals Help Informational Professionals Help Organizations Manage What They Know." *Library Management.* 21.8 (2000): 395–403.

Clark, Adele, and Joan Fujimura. "What Tools? What Jobs? Why Right?" *The Right Tools for the Job: At Work in Twentieth-Century Life Sciences.* Adele Clark and Joan Fujimura, eds. Princeton, NJ: Princeton University Press, 1994. 3–44.

Crowley, Sharon, and Debra Hawhee. *Ancient Rhetorics for the Modern Student.* 2nd ed. Boston: Allyn and Bacon, 1999.

Detienne, M., and J-P Vernant. *Cunning Intelligence in Greek Culture and Society.* Sussex, England: Harvester Press, 1978.

Griesemer, James. "The Role of Instruments in the Generative Analysis of Science." *The Right Tools for the Job: At Work in Twentieth-Century Life Sciences.* Adele Clark and Joan Fujimura, eds. Princeton, NJ: Princeton University Press, 1994. 47–76.

Howell, W.C. *Information Processing and Decision Making (Human Performance and Productivity).* Mahwah, NJ: Lawrence Erlbaum Associates, 1982.

Hu, Winnie. "Seeing No Progress, Some Schools Drop Laptops." *New York Times.* May 4, 2007.

Hughes, Michael. "Moving from Information Transfer to Knowledge Creation: A New Value Proposition for Technical Communicators." *Technical Communication.* 49.3 (2002): 275–85.

Johnson-Eilola, Johndan. "Relocating the Value of Work: Technical Communication in a Post-Industrial Age." *Technical Communication Quarterly* 5. 3.2 (1996): 245–70.

Johnson-Eilola, Johndan, and Stuart Selber. "After Automation: Hypertext and Corporate Structures." *Electronic Literacies in the Workplace: Technologies of Writing.* P. Sullivan and J. Dautermann, eds. Urbana, IL and Houghton, MI: National Council of Teachers of English and Computers and Composition Press, 1996. 115–41.

Johnson-Eilola, Johndan, Stuart Selber, and Cynthia Selfe. "Interfacing: Multiple Visions of Computer Use in Technical Communication." *Three Keys to the Past: The History of Technical Communication.* T. C. Kynell and T. Moran, eds. Stamford, CT: Ablex, 1999. 197–226.

Kuhn, Thomas. *The Structure of Scientific Revolutions.* 2nd ed. Chicago: University of Chicago Press, 1970.

Malhotra, Yogesh. "Knowledge Management for e-Business Performance: Advancing Information Strategy to 'Internet Time.'" *Knowledge Management and Business Model Innovation.* Yogesh Malhotra, ed. Hershey, PA: Idea Group Publishing, 2001. 2–15.

Nardi, Bonnie A., and Vicki L. O'Day. *Information Ecologies: Using Technology with Heart.* Cambridge, Massachusetts: MIT Press, 1999.

Nonaka, Ikujiro, and Hirotaka Takeuchi. *The Knowledge Creating Company: How Japanese Companies Create the Dynamics of Innovation*. New York: Oxford University Press, 1995.

Owen, John M. "Knowledge Management and the Information Professional." *Information Services and Use*. 10.1 (1991) 7–16.

PARC Research. "Knowledge Extraction from Document Collections." December 29, 2007. www.parc.xerox.com/research/projects/knowledge_extraction/default.html

Polanyi, Michael. *Science, Faith, Society*. Riddel Memorial Lectures. London: Geoffrey Cumberlege, Oxford University Press, 1946.

Polanyi, Michael. *Tacit Dimension*. Garden City, NY: Anchor Books, 1946.

Selfe, Cynthia, and Richard Selfe. "The Politics of Interface: Power and its Exercise in Electronic Contact Zones." *College Composition and Communication*. 45.4 (1994): 481–504.

Slack, Jennifer, David J. Miller, and Jeffrey Doak. "The Technical Communicator as Author: Meaning, Power, Authority." *Journal of Business and Technical Communication*. 7.1 (1993): 12–36.

Von Krogh, G, Ichijo, K., and Nonaka, I. *Enabling Knowledge Creation: How to Unlock the Mystery of Tacit Knowledge and Release the Power of Innovation*. Oxford, New York: Oxford University Press, 2000.

Wick, Corey. "Knowledge Management and Leadership Opportunities for Technical Communicators." *Technical Communication*. 47.4 (2000): 515–29.

2 Introduction to XML

A Primer on the eXtensible Markup Language

Chapter Overview

Thus far, we have discussed the basics of knowledge management and presented some case studies to show how it has been employed in organizational settings. In this chapter, we will discuss the basic concepts of XML technology so we can begin to see how we can store information using an object-oriented language. We will illustrate how basic XML elements need to be written and then demonstrate how they can be arranged in a DTD and why DTDs are of value. We will also show how we can use XML entities to breakdown a significant text document into modular units and then reconstitute parts of it as needed.

Because we will learn in this chapter how elements can be identified, named, and then organized or encapsulated relative to other elements in DTDs, we will be better prepared to discuss the rhetorical nature of XML that we cover in Chapter 3 and the chapters that follow. This will also enable us to understand how we can employ single sourcing methods that we will explain in Chapter 3.

The Value of XML

In this section, we will describe the differences between HTML and XML and show what tasks can be accomplished with the use of each respective language. HTML became very popular very quickly as it was easy to learn. Essentially, when one is using HTML, one is acting like a typesetter and layout artist of traditional printing mediums like newspapers, books, posters, and pamphlets. For instance, the HTML tags that you put around words make them look a certain size, character, color, and font. If we want to make the word "and" in a file appear in italics as *and*, we would place an opening and closing set of italic tags around it: <i>and</i>. Additionally, HTML can position these words on a website relative to other words and images; one could make the words and sentences span all the way across the page, put them in columns, or wrap them around an image.

The other great thing about HTML is the way it allows us to make links from one document to another document with such ease. This is perhaps the most defining feature of HTML. In traditional print materials, we could see that a writer was referencing a certain source with the use of formal MLA, APA, and CBE citation styles, and we could also be referred to an additional insight provided by a footnote at the bottom of a page, but we never really "go" to another page or source. HTML changed this and provided us with options to link to other sources that were available to us.

However, HTML has its limitations. We could produce these hypertext documents for the Internet so all with access to them could see them, but these documents were rather static in regard to their information or data. However, this is not the case with XML. While HTML is about typesetting text and images, XML is about identifying, separating, and recombining specific data that we might need for our own purposes. With XML, we can employ software that would allow us to search and retrieve specific data we are looking for if this data has been marked or coded in XML.

XML allows us to locate and extract or point to specific pieces of data that might exist in a large database or a website. In HTML, we cannot be very precise about the data we are trying to locate. In XML, we can create our own tags for the kinds of specific information we think is important and then reuse these tags for different purposes. We can also exchange this information with others with much more efficiency. If we just used HTML for our information, the best we can do is say something like "Go to the link on this website, look around for awhile, and find what you need."

When parts of a larger document have been coded using XML and they exist in an external source such as an organization's database, we can locate the specific parts or sections of the document and reuse them for more specific purposes. This does not work well with HTML. XML is sometimes referred to as an object-oriented language because the documents that are stored in external databases are often referred to as objects. When you produce an XML document, you are singling out or referring to these objects and using them as you need to for whatever project you are working on.

Unlike HTML, XML allows us to design our own markup tags. In the HTML example above, italics are marked with the <i> tags. This is an HTML tag that all browsers recognize and HTML professionals have used since the earliest days of the World Wide Web. However, we cannot define specific HTML tags for our own needs. In XML, if we want to mark up the "summary" sections of documents, or designate the "average temperature recorded in the city of Orlando in March," we could designate tags such as <summary> or <average_temperature_March_Orlando> tags. Note that we are not quite contrasting the same kinds of tags above because the italics tags used in HTML are for formatting text while the "summary" and "average temperature in a city in March" tags are for describing specific information or data. XML is about designating content, and HTML is about how the content is to be arranged and how it looks.

XML Declarations

You should start an XML document with an XML declaration. The following is the simplest form of a declaration:

```
<?xml version="1.0"?>
```

The reason you should use a declaration is to indicate that the document you are producing is an XML document and that the software that will be processing your files, such as a browser, will properly read it as such. This declaration is also saying that the XML version you are using is 1.0.

You can also add other instructions such as a standalone attribute:

```
<?xml version="1.0" standalone="yes"?>
```

You would use this to signify that the document you have written can stand by itself, that it is complete, and that it is not necessary to import other files outside this document. There is other information that you can add between the <? and the ?> tags that will be read as processing instructions. For example, style sheets are processing instructions that "tell" the software how your XML encoded information is to look on a screen. Style sheets will be covered in Chapter 4.

Elements

Elements are the basic coding units for XML documents. An element consists of tags that describe the actual information itself. For example,

```
<birthday>July 27, 1980</birthday>
```

is a complete element. The tags are <birthday> and </birthday>, and they label and characterize the information: July 27, 1980. Also, the closing tag is different from the opening tag in that it must include a forward slash or "/" before the name you have chosen to describe the information.

To choose a proper name, you have to start with a letter or an underscore character ("_"). Within the name you can use letters, numbers, underscore characters, hyphens ("-"), or dots ("."). Spaces between any of these letters, numbers, underscores, hyphens, or dots in the names you choose to represent information are not allowed; this violates the syntax rules of XML. You can put underscores between words so they make semantic sense as we do in the example, <average_temperature_March_Orlando>. We discuss other semantic naming strategies in Chapter 3 and naming strategies for coding in Chapter 6.

When you design XML element tags, you can use upper case letters, lower case letters, or you can mix them up. For example, you could use <DATE>, <date>, or <Date>. However, once you have decided on a tag, it must be written out consistently in the document. For example, using <birthday> as the opening tag and closing with </Birthday> will not work. You can get away with mixing up the case of letters in HTML, but not in XML. We will use various mixtures of lowercase and uppercase element names in the XML examples used in this book, but the opening tags and closing tags must always use the same patterns for case.

Organizing Elements

When organizing elements, they need to be properly nested. This means that if you code an opening element, and then code another element that follows, the second element must be closed off before closing off the first element. This is illustrated in the following example:

```
<personal_information>
<birthday>July 27, 1970</birthday>
</personal_information>
```

The second element, <birthday>, is closed off before the first element, <personal_information>, is closed.

Root Elements and Hierarchies of Elements

All XML documents have one root element, which is the element that contains all of the other elements. In the simple example below, the root element is "person."

XML is written in a tree form where there is a hierarchy of elements with some elements nested within other elements. The nested elements are called "child" elements, and they are nested or encapsulated within "parent" elements. There is not a limit on the size of an XML document; it can have as many parent and child elements as the situation requires.

```
<person>
<name>John Raymond</name>
<birthday>July 27, 1970</birthday>
<city_of_birth>Des Moines</city_of_birth>
<state_of_birth>Iowa</state_of_birth>
<immediate_family>
      <spouse>Donna Raymond</spouse>
      <daughter>Jennifer Raymond</daughter>
      <son>Jeffrey Raymond</son>
      <son>Casey Raymond</son>
</immediate_family>
</person>
```

The hierarchical tree for the XML code that represents the personal information for John Raymond is represented in Figure 2.1.

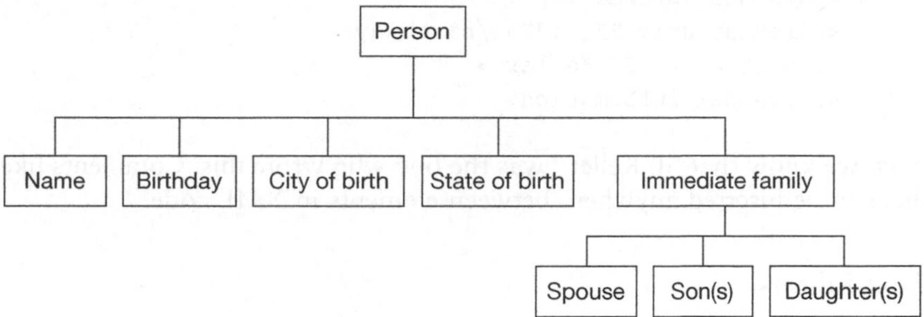

Figure 2.1 Personal Information Hierarchy

Note how this tree visually illustrates the organization rules we have discussed so far. All of this information is nested under one root element, "Person," and there is one parent element, "Immediate Family," that represents certain family members that are its child elements: "Spouse," "Son(s)," and "Daughter(s)." These three child elements are encapsulated within the parent element.

To be able to show up on a browser or other kind of interface, XML must usually be "well formed" which means that it must conform to the syntactic rules governing elements and other XML content. If we used the closing tag </immediate family> above without an underscore character between "immediate" and "family," our XML code would not be well formed and would not work.

When XML code is "well formed," this means something different than when XML code is "valid." Being valid means that all of the elements in an XML coded document have to conform to a "DTD." We will discuss DTDs later in this chapter.

Writing Comments

Sometimes it helps to be able to make some comment in the middle of the XML code that you are writing, but this is not information that you want to be represented on the screen as it is not part of the essential information your document was designed to provide. Any comment written between a "<!—" and an "—>" will only show up if you look at the actual XML code itself, not on a display screen such as a browser.

For example, if someone who produced the code wanted to leave a name in case someone like a fellow employee could contact him or her later if there were any questions, this person could write the following comment:

```
<!—written by J. Keller->
```

Here is how it looks in some code:

```
<personal_information>
<birthday>July 27, 1970</birthday>
<!—written by J. Keller—>
</personal_information>
```

Now we know that "J. Keller" was the one who wrote this. Comments like this can be inserted anywhere between elements in XML code.

Text Editors

As with HTML, when you write XML code, you will need a text editor. A text editor will allow you to save a file in plain text. This means it will not add any extraneous code to the code you write, and, therefore, the software that will eventually read your work will not be confused, so to speak.

There are many free text editors available. Perhaps the most popular is Microsoft® Notepad. We also recommend more robust text editors with syntax highlighting and other useful features, such as EditPlus, on the resources section of our website.

XML Parsers

A tree-based type of XML editor is even better suited to composing XML. Because the most minor error in XML will keep it from being read by whatever software you are using, it is a good idea to use an XML editor because it will more readily alert you to a problem you might have in your code. Additionally, most text editors use varying colors for different parts of your code and indent the code you produce to show how some elements are nested within other elements. Many XML editors, such as Microsoft® XML Notepad, are also free.

Other Software for Displaying XML

The most common software programs for displaying XML coded documents in raw form are most likely browsers such as Microsoft Internet Explorer® or Mozilla Firefox®. When you see the XML code displayed in raw form in these browsers, it will look very much like the XML you coded using a text editor without any distinguishing design features; it will still have the XML tags in place around the information in your XML coded document and you will not see any aesthetic features such as attractive fonts or thoughtfully employed white space used to separate different bodies of information. To format your XML code in a fashion that makes it look like you really want it to, you have to pair it with eXtensible Style Sheet Language Transformations (XSLT) code or a CSS style sheet, and this is something we will cover later on in Chapter 4.

Web browsers are not the only kind of software that will allow code to display properly. For example, some organizations have their own XSLT coding that allows for the information in your code to show up in a table or nicely rendered print document. For our purposes in this chapter, we will be displaying XML without any XSLT coding.

Writing and Viewing an XML Document

Now it is time to make a basic XML document and then view it using a Web browser. There are many kinds of software that will open up an XML document as described directly above, and we are just using a Web browser as virtually every computer has at least one. Below is a simple XML document. Open up Notepad or EditPlus and type in the following information.

```
<?xml version="1.0" standalone="yes"?>
<hello>
Hello world!
</hello>
```

After you have done this, you need to save this file. To do this properly, you need to save it as an XML file, or with an XML extension (.xml). Please follow these directions:

1. Pull up Notepad.
2. Go to “File” in the upper left hand corner of the Notepad screen, then hit “Save”.
3. Make sure that you get rid of any information in the “File name” box that is already there such as “*.txt”. This box should be empty.
4. In the file name box, type in “hello.xml”. Make sure that in the “Save as type” box that it says “Text Documents”.
5. Following this, hit the “Save” button in the lower right hand side of the Notepad screen. Make sure you save it in a place where you can find it. If you are using Microsoft Windows®, you can save it to the desktop.
6. Pull up a browser such as Microsoft Internet Explorer®.
7. Go to “File” in the upper left hand corner, and depending on whatever browser you have chosen, choose “Open” or “Open file”. Find your file and click on it.

You should see the XML displayed as shown in Figure 2.2.

As you can see, the browser is showing the XML code you wrote. If you want to see how this code looks with some simple style information, you need to add a little more XML code. You will also need to add a cascading style sheet (CSS) by producing the following file in an editor and saving it in the same file as your “hello.xml” document:

```
hello {display: block; font-size: 30pt; font-weight:
bold;}
```

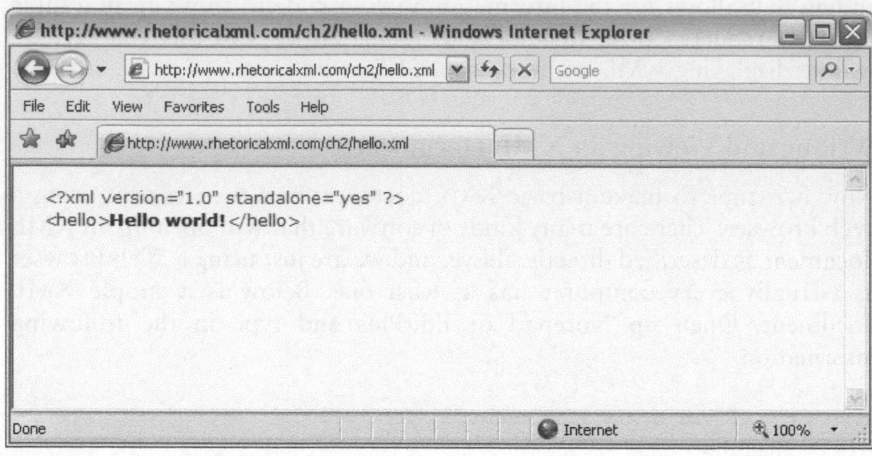

Figure 2.2 Hello World

Save this file separately as “hello.css”, and then add the following line to your hello.xml document: <?xml-stylesheet type=“text/css” href=“hello.css”?>. Thus, we have the following code:

```
<?xml version="1.0" standalone="no"?>
<?xml-stylesheet type="text/css" href="hello.css"?>
<hello>
Hello world!
</hello>
```

Save this as “hellostylesheet.xml” to differentiate it from the “hello.xml” document.

After saving this file, pull it up again and you will see how it will allow you to produce text in a browser without the surrounding XML tags (see Figure 2.3).

Now you see “Hello World!” without the XML tags because we added simple style sheet instructions. CSS will be covered in more depth in Chapter 4, and for the rest of this chapter we will not be employing them in our code.

Using Attributes in Elements

As mentioned previously, one of the strengths of XML is that we can carefully define our information by the way we choose to name our elements. We can even give our elements greater specificity by adding attributes within an element.

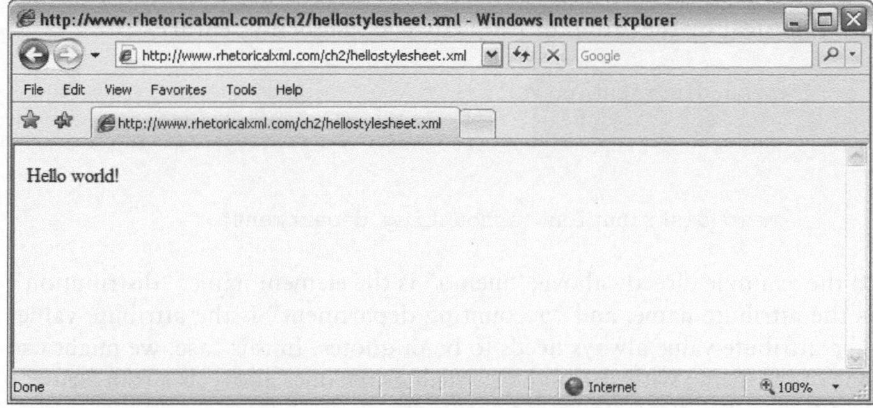

Figure 2.3 Hello World with Style Sheet

For example, here is one way of capturing the following information with XML elements:

```
<person>
  <gender>male</gender>
     <name>John Raymond</name>
</person>
```

We could also capture the same information by adding the attribute of gender to the person element:

```
<person gender="male">
     <name>John Raymond</name>
</person>
```

One advantage of doing this is that we can reduce the complexity of our element tree. We still can note that John Raymond is a male, but with fewer tags. Because we are using an attribute in the opening tag, "gender='male'", we do not have to put it in the closing tag, "</person>".

If we wanted to be able to distinguish different currency types when using XML code to designate the cost of something, we could write:

```
<cost_currency="dollars">250</cost>
```

or

```
<cost_currency="pounds">250</cost>
```

A document that we have stored in an XML database could be further distinguished by the group of people it was originally sent to:

```
<memo_distribution="all">
```

or

```
<memo_distribution="accounting department">
```

In the example directly above, "memo" is the element name, "distribution" is the attribute name, and "accounting department" is the attribute value. The attribute value always needs to be in quotes. In this case, we might use one of the memo with attribute tags such as the ones above as a root element and have other elements nested within it.

When naming attributes, the same rules apply as for naming elements properly. For example, an attribute name can only start with an underscore or a letter.

An element can contain more than one attribute, but you cannot have more than one attribute with the same name. If you use child elements, this is not the case. In the "Root Elements and Hierarchies of Elements" section above, you can see that the <son> element was used twice because in this example, there are two sons in the Raymond family.

As a general rule, attributes are best for self-referencing metadata, or metadata about particular elements. One can go into the code and see precisely what was meant by the element when there is an attribute. In the memo example above, we can tell by looking at the tags that it was sent to either "all" or to the "accounting department." We can also see that the "250" between the <cost_currency> tags are in dollars, not pounds or yen.

Empty Elements

Empty elements are used when you want to include information that does not have any text between the element tags but still needs to be included in the XML code to make it valid. Often it is used for images such as jpegs or gifs. For example, if you wanted to mark a place in your XML coding for an image of the Declaration of Independence, you would need to use the following empty element:

```
<image filename="Declaration_of_Independence.jpeg"/>
```

In this example, there is no need for closing tags. All we need is a name that describes what we want to include, "image filename", an equals sign, then the name of the graphic in quotes. In an empty element, you need to add a forward slash "/" before you close it off. To actually embed the image of the actual Declaration of Independence, we would have to use some other information that will be covered later when we discuss "unparsed entities."

Special Symbols

There are five symbols that you cannot write out in XML as you normally would. Instead, you have to write them out in the following manner:

- A left angle bracket or less than sign (<) needs to be written out as <
- A right angle bracket or greater than sign (>) needs to be written out as >
- An ampersand (&) needs to be written out as &
- A single quote or apostrophe (') needs to be written out as '
- A double quote (") needs to be written out as "

The following would not work:

```
<restaurant>Jake's Place</restaurant>
```

Instead, here is what you would write:

```
<restaurant>Jake&apos;s Place</restaurant>
```

There is only one ampersand in the text of the Declaration of Independence, and if we were to code it properly, the following would not work:

```
cruelty & perfidy
```

Instead, we would write the following:

```
cruelty &amp; perfidy
```

DTDs

In the previous sections we discussed how elements can be designed and organized in an effort to represent the information we want in an XML document. In these exercises, we did not base a code on an established DTD. However, it is generally better to have a DTD associated with your XML coding as it will better allow you to:

- view and more easily understand the basic structure of a body of data without looking at the actual data itself,
- construct standardized applications using XML documents,
- practice single sourcing by reusing and recombining parts of many documents,
- challenge people in an organization to write their XML code to conform to one well-designed structure,
- and enable professionals to better exchange XML documents with each other.

When the organization of your XML elements conforms to the structure of the DTD you or someone else defined, it is considered "valid." Note that this is different than being "well formed," which we mentioned earlier in the chapter. To be well formed, an XML coded document needs to conform to the basic syntactic rules of element design that we have already described. For example, a document that does not use "&" for an ampersand or does not properly close off elements would not be well formed. A document can be well formed, but not be considered valid, because the elements in it do not conform to the structure of a DTD.

This section will illustrate how DTDs specify the elements that will be allowed in documents and how these elements need to be related to one another. It is important to understand that DTDs contain no empirical data or information that might be needed; they contain only metadata, or data describing other data. For example, DTDs never house information such as the birthday or birthplace of a specific person. Instead, DTDs show us how to model knowledge and what is in fact the allowable information structure for an XML document. A DTD will not tell us about specific birthdates and birthplaces, but it *can* tell us that there is a place to include birthdates and birthplaces in a specified XML coded document with properly arranged elements. Valid XML code contains all the designated elements in the order required by the DTD. It forbids tagged elements that are not declared in the DTD, and a DTD describes the hierarchical structure that XML elements must adhere to so it can be read by an XML parser.

XML parsers in most Internet browsers and other software not necessarily related to the Internet will be able to search and represent specific data coded in XML. This is not the case for information that is coded in HTML code. In HTML, you might be directed to a website that has a large document in it, whereas XML will allow you to find one particular part of a document and extract it and use it for your own purposes. You will learn how to use some of these extraction techniques using associated XML technologies like XML Path Language (XPath), XLink, and XPointer in Chapter 5.

In HTML, you might be able to find an entire table of data, whereas if something is coded in XML, you can find and extract information in a specific cell in a table, say the data for rainfall in a specific city in a specific month. Browsers will be able to focus on the hierarchical structure designated by the DTD which lays out the strict syntactical ground rules for a specific application. The DTD prescribes and efficiently organizes the data in a document so they can be properly read or parsed by browsers and other forms of software.

For a simple memo, we might want to include in our DTD the elements such as author, addressee, subject, date, and text as these are the standard organizing features of this traditional business document:

```
<!ELEMENT memo (author, addressee, subject, date, line+)>
<!ELEMENT author (given_name, family_name)>
<!ELEMENT given_name (#PCDATA)>
```

```
<!ELEMENT family_name (#PCDATA)>
<!ELEMENT addressee (given_name, family_name)>
<!ELEMENT subject (#PCDATA)>
<!ELEMENT date (#PCDATA)>
<!ELEMENT line (#PCDATA)>
```

The DTD demands a strict hierarchy, and this is similar to how the elements described in the previous section describe information and how it should be nested in other information. For example, in this DTD, "memo" is the root element and "author", "addressee", "subject", "date", and "line+" reside under the memo element. Also, the "given name" and "family name" of the "authors" and "addressees" is embedded under the author and addressee elements.

In a DTD, the required elements must be named within "<!ELEMENT" and ">" code, the opening and closing tags.

The #PCDATA code that follows each named element stands for "parsed character data." PCDATA is data content and needs to be represented in alphanumeric characters which can include spaces along with entities such as "&". It is content data that the XML document holds and does not include any XML markup. The plus sign "+" that follows the designated element "line" in the first line of code above means that for a memo coded in XML that conforms to this DTD, there must be a least one line, and there could be more.

DTDs *declare* which elements must be in a document and the order in which these tags need to be embedded inside one another. In essence, they allow us to "say" things like "First you must designate that this is a memo. Then you must provide the author's name, but you do this in the following order: given name, then family name. Following this, you do the same for the addressee's name in the same order: given name, then family name. Continuing on, you must provide the subject, date, and the actual line or lines of the memo, and these three bodies of information must follow in this order and are separate from one another."

XML code that contains the actual information in a database such as "John Raymond" needs to follow the arrangement set by the DTD that defines the arrangement of a database's information. Below is some XML code using properly nested elements that would be used to store actual information in a fashion that would be considered *valid* by the hierarchy indicated by the DTD above:

```
<memo>
<author>
      <given_name>Condoleezza</given_name>
      <family_name>Rice</family_name>
</author>
<addressee>
      <given_name>Colin</given_name>
```

```
        <family_name>Powell</family_name>
    </addressee>
    <subject>Speech</subject>
    <date>September 22, 2001</date>
    <line>Please read the speech the President will give
    tonight.</line>
    <line>It might surprise you.</line>
    </memo>
```

Between the XML tags above we have actual textual content that is stored as data and can be extracted as information for a specific purpose. We know that this memo has been addressed to Colin Powell, and we can see that it was authored by Condoleezza Rice. We can see that the subject is Speech, and that the date is September 22, 2001.

Note the plus sign "+" that follows the designated element "line" in the first line in the DTD code above: <!ELEMENT memo (author, addressee, subject, date, line+)>. This indicates that for a memo coded in XML that conforms to this DTD, there must be a least one line. In this very short memo there are two lines that are separated by: "Please read the speech the President will give tonight" and "It might surprise you." The "+" allows us to add as many lines as we want.

Rules for Designing DTDs

DTDs are templates that list all of the elements that need to be in an XML document and how they are arranged or embedded relative to one another. Again, a DTD does not contain any real information that one might be looking for in a search; it is a structure that describes how the elements of information are to be arranged.

As you can see in the DTD for the memo example above, the first line is,

```
<!ELEMENT  memo (author, addressee, subject, date,
line+)>
```

What this first line is saying is that the root element of a document in this specific database must be <memo>, and that between the <memo> and </memo> tags, there also has to be opening and closing author tags, addressee tags, subject tags, date tags, and line tags. The comma "," that separates each of these items indicate that the XML coded elements in this line must follow this order.

We can also see from this DTD that there must be both <given name> and <family name> tags nested under both the author tags and the addressee tags:

```
<!ELEMENT  author (given_name, family_name)>
<!ELEMENT  given_name (#PCDATA)>
```

```
<!ELEMENT  family_name (#PCDATA)>
<!ELEMENT  addressee (given_name, family_name)>
```

In the fourth line that designates "addressee", we do not have to follow with the "given name" and "family name" declarations like we did with "author" because we have already described them under "author" and noted that they contain (#PCDATA). The XML parsers in the software that you use will recognize these two declarations.

We also need a few remaining elements for our XML coded documents to still be valid given this DTD, but they do not have any other elements embedded within them. The required elements are subject, date, and line:

```
<!ELEMENT  subject (#PCDATA)>
<!ELEMENT  date (#PCDATA)>
<!ELEMENT  line (#PCDATA)>
```

The lines that do have (#PCDATA) in them are ones that describe elements that will have data embedded in them. For example, "Speech" is information that can be typed between <subject> and </subject>, and "September 22, 2001" can be typed in between <date> and </date> elements because of the (#PCDATA) code. In the lines of the DTD above that do not have (#PCDATA) included, textual information cannot be included; these lines only describe how elements should be organized. In the element above, <!ELEMENT author (given_name, family_name) >, it tells us that the parent element is "author" and that there are two child elements, "given_name" and "family_name" that should be embedded within the "author" element. However, there is no (#PCDATA) in this line. This can also be said for the root element, memo, which contains no (#PCDATA).

In a previous example before the memo example directly above, we used the following information to describe a body of information that captured what we wanted to know about a "person":

```
<person>
<name>John Raymond</name>
<birthday>July 27, 1970</birthday>
<city_of_birth>Des Moines</city_of_birth>
<state_of_birth>Iowa</state_of_birth>
<immediate_family>
     <spouse>Donna Raymond</spouse>
     <daughter>Jennifer Raymond</daughter>
     <son>Jeffrey Raymond</son>
     <son>Casey Raymond</son>
</immediate_family>
</person>
```

The DTD for this example would be:

```
<!ELEMENT  person (name, birthday, city_of_birth,
state_of_birth, immediate_family)>
<!ELEMENT name (#PCDATA)>
<!ELEMENT birthday (#PCDATA)>
<!ELEMENT city_of_birth (#PCDATA)>
<!ELEMENT state_of_birth (#PCDATA)>
<!ELEMENT immediate_family (spouse*, daughter*, son*)>
<!ELEMENT spouse (#PCDATA)>
<!ELEMENT daughter (#PCDATA)>
<!ELEMENT son (#PCDATA)>
```

Note that in the line that describes the need for the "person" root element and the line that describes the "immediate family" element line, there is no (#PCDATA) coding. This line of coding only tells us what elements must be nested under it.

The asterisk "*" symbol that follows "spouse", "daughter", and "son" in the DTD above symbol specifies zero or more occurrences of an element. In this case, it means that if in fact the person being described does not have a spouse, son, or daughter, that this is not something that will be coded in the XML elements and is in fact still valid. To generalize, the asterisk sign means that an element can appear once or not at all. It can also indicate, as in the case of sons and daughters, that there can be more than one of these particular elements in valid XML elements.

We could rewrite the DTD above using the pipe or "|" sign if we wanted to designate that only one of the two elements listed in the DTD can show up. For example, if we wanted to designate that either the wife or husband be designated as opposed to spouse, this would be a more accurate way of capturing the specifics of a person's personal data. Instead of,

```
<!ELEMENT  immediate_family (spouse, daughter, son)>,
```

we could write,

```
<!ELEMENT  immediate_family (wife | husband, daughter,
son)>.
```

When the code shows up on a screen for viewing, it would designate that the person named was either the wife or husband of the person named, not just the spouse.

If we wanted to indicate in our DTD that there is an element that does not necessarily have to be included in XML coded information, we would put a question mark "?" behind it. For example, if we had an ELEMENT in the code above for "ethnicity", a piece of data that is oftentimes not filled

Table 2.1 Syntax Options for Child Elements

*	Can have zero or more children
+	Can have one or more children, but, if it does, it must only occur once
?	Does not have to occur at all
|	Either one or the other, but not both

out on forms by people for personal or political reasons, we could code our first DTD line like this:

```
<!ELEMENT  person (name, ethnicity?, birthday, city_of_birth,
state_of_birth, immediate_family)>
```

Table 2.1 shows the syntax options for designating the number of children in the <!ELEMENT > tags of a DTD.

Combining Document Type Declarations, Document Type Definitions, and XML Coded Information

Now we will code a basic XML file with a document type declaration that designates a specific DTD that our XML code will be validated against. To be clear, a document type declaration is *not* the same thing as a document type definition or DTD, which we have covered above. This might sound confusing, but this is something that needs to be understood as this is part of the XML universe.

In this example, we are going to declare an internal DTD with our document type declaration. This means that we are going to indicate what specific DTD we want our XML code to be validated against. It is a way of our asking if the actual code we produced measures up to the rigor of the DTD we originally designed.

Now we are going to add an XML declaration that we covered in a previous section. For this document it will be "<?xml version="1.0" standalone="yes"?>".

In addition, we have to add a document type declaration that will specifically designate the DTD that our XML code will have to be validated against. Here are the basics of a document type declaration:

```
<!DOCTYPE memo [
]>
```

The "!DOCTYPE" is telling the parser that this is a document type declaration, and that the root element of the DTD that it is referring to is "memo." Between the brackets "[ ]", we put the DTD that we have designed that tells us in our memo example just what elements belong and how they should be ordered. Taking the group of DTD <!ELEMENT> tags from above, we would have:

```
<?xml version="1.0" standalone="yes"?>
<!DOCTYPE memo [
<!ELEMENT memo (author, addressee, subject, date, line+)>
      <!ELEMENT author (given_name, family_name)>
      <!ELEMENT given_name (#PCDATA)>
      <!ELEMENT family_name (#PCDATA)>
      <!ELEMENT addressee (given_name, family_name)>
      <!ELEMENT subject (#PCDATA)>
      <!ELEMENT date (#PCDATA)>
      <!ELEMENT line (#PCDATA)>
]>
```

Next we want to add the XML coded data to complete our XML file. Again, taking from above and adding our data to the DTD, our complete XML document would look like this:

```
<?xml version="1.0" standalone="yes"?>
<!DOCTYPE memo [
<!ELEMENT memo (author, addressee, subject, date, line+)>
      <!ELEMENT author (given_name, family_name)>
      <!ELEMENT given_name (#PCDATA)>
      <!ELEMENT family_name (#PCDATA)>
      <!ELEMENT addressee (given_name, family_name)>
      <!ELEMENT subject (#PCDATA)>
      <!ELEMENT date (#PCDATA)>
      <!ELEMENT line (#PCDATA)>
]>
<memo>
      <author>
            <given_name>Condoleezza</given_name>
            <family_name>Rice</family_name>
      </author>
      <addressee>
            <given_name>Colin</given_name>
            <family_name>Powell</family_name>
      </addressee>
      <subject>Speech</subject>
      <date>September 22, 2001</date>
      <line>Please read the speech the President will give
      tonight.</line>
      <line>It might surprise you.</line>
</memo>
```

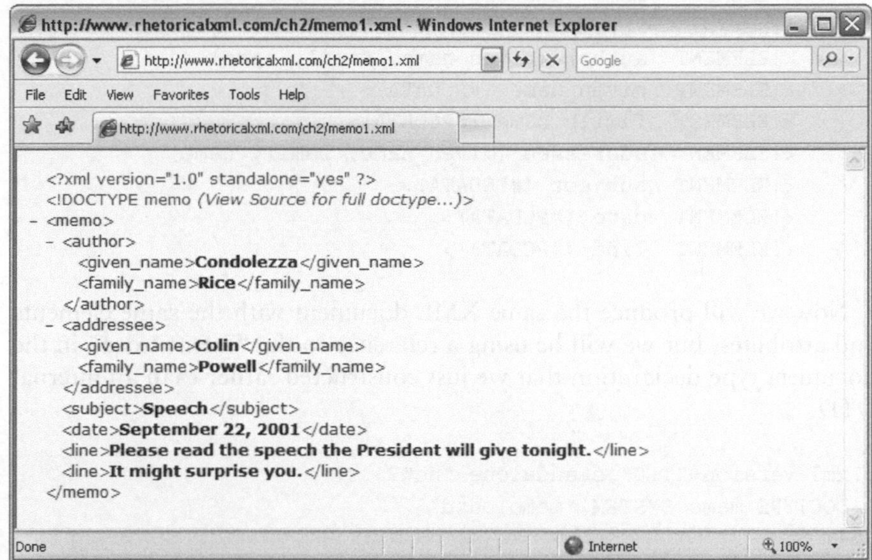

http://www.rhetoricalxml.com/ch2/memo1.xml - Windows Internet Explorer
http://www.rhetoricalxml.com/ch2/memo1.xml
Google
File Edit View Favorites Tools Help
http://www.rhetoricalxml.com/ch2/memo1.xml
<?xml version="1.0" standalone="yes" ?>
<!DOCTYPE memo (View Source for full doctype...)>
- <memo>
- <author>
<given_name>Condolezza</given_name>
<family_name>Rice</family_name>
</author>
- <addressee>
<given_name>Colin</given_name>
<family_name>Powell</family_name>
</addressee>
<subject>Speech</subject>
<date>September 22, 2001</date>
<line>Please read the speech the President will give tonight.</line>
<line>It might surprise you.</line>
</memo>
Done
Internet
100%

Figure 2.4 Standalone Memo

Save this file as a text file, and make sure you add an .xml extension to it. Name it "memo1.xml". You should also note that we added one extra line to our XML file. Next, pull this up in a browser or an XML Editor.

If you go to "View", then to "Source" in Microsoft Internet Explorer®, you will see not only what you see in the Figure 2.4, but also the DTD that we wrote for this memo.

External DTDs

In the XML document above, everything we needed was provided in the same file: the document type declarations, DTDs, and XML coded information. This is why we wrote in the XML declaration that it was a standalone document. We did this when we wrote "yes" after "standalone". However, it is possible to designate a DTD that is stored as a separate file on a hard drive, and thus it would not be combined in the same document declarations and XML coded data we wrote above in the "standalone" document. This separate DTD could also be stored on an Internet server. To do this we have to make an external DTD file.

First, we have to save the following DTD that was presented before in an editor such as Notepad, but instead of a .txt extension that we would use for text file, or an .xml extension we would use for XML data, we need to save our file with a .dtd extension so that the software we will be using will know that this is in fact a DTD. Type in the following DTD in an editor and save it as "memo1.dtd".

```
<!ELEMENT  memo (author, addressee, subject, date,
line+)>
<!ELEMENT  author (given_name, family_name)
<!ELEMENT  given_name (#PCDATA)>
<!ELEMENT  family_name (#PCDATA)>
<!ELEMENT  addressee (given_name, family_name)
<!ELEMENT  subject (#PCDATA)>
<!ELEMENT  date (#PCDATA)>
<!ELEMENT  line (#PCDATA)>
```

Now we will produce the same XML document with the same elements and attributes, but we will be using a reference to the "memo1.dtd" in the document type declaration that we just constructed rather than an internal DTD.

```
<?xml version="1.0" standalone="no"?>
<!DOCTYPE memo SYSTEM "memo1.dtd">
<memo>
    <author>
        <given_name>Condoleezza</given_name>
        <family_name>Rice</family_name>
    </author>
    <addressee>
        <given_name>Colin</given_name>
        <family_name>Powell</family_name>
    </addressee>
    <subject>Speech</subject>
    <date>September 22, 2001</date>
    <line>Please read the speech the President will
    give tonight.</line>
    <line>It might surprise you.</line>
</memo>
```

Save this file as written above as "memo1standalone.xml". In this file, we changed just two things. We changed the XML declaration from

```
<?xml version="1.0" standalone="yes"?>
```

to

```
<?xml version="1.0" standalone="no"?>
```

We also changed the document type declaration (not to be confused with a DTD) to

```
<!DOCTYPE memo SYSTEM "memo1.dtd">
```

We got rid of all of the <!ELEMENT> tagged data, and added,

```
SYSTEM "memo.dtd"
```

The inclusion of the word SYSTEM asks the software we will be using to refer to the "memo1.dtd" we have stored in the same file. If you open the "memo1standalone.xml" file in Microsoft Internet Explorer®, it should have the same XML content displayed as the "memo1.xml" document.

External DTDs on a Website Server

We can also place a DTD on a server on the Internet and refer to the URL where it has been placed. For example, we placed the "memo1.dtd" document on this book's accompanying website as a file, and in the coding below, you can see that we referred to the URL after SYSTEM. The title for the document below is "memo1.xml".

```
<?xml version="1.0" standalone="no"?>
<!DOCTYPE memo SYSTEM "www.rhetoricalxml.com/ch2/memo1.dtd">
<memo>
      <author>
            <given_name>Condoleezza</given_name>
            <family_name>Rice</family_name>
      </author>
      <addressee>
            <given_name>Colin</given_name>
            <family_name>Powell</family_name>
      </addressee>
      <subject>Speech</subject>
      <date>September 22, 2001</date>
      <line>Please read the speech the President will
      give tonight.</line>
      <line>It might surprise you.</line>
</memo>
```

The advantage of this external file approach is that you can have easy access to the DTD, and anyone who might also be working with you can access this DTD by just referring to a website where it was stored. We have already pointed out that making a DTD available to a group of people allows for an organization to have a model for a well-designed document. Again, if you pull this up in Microsoft Internet Explorer®, it will look the same as the "memo1standalone.xml" and "memo1.xml".

Including Simple Attributes and Attributes with Unique Values

Earlier we described some simple attributes, which we likened to metadata for elements. We can also use attributes in DTDs to better represent the information we are coding. Below is a DTD followed by some XML coded data that does not contain any attributes.

```
<?xml version="1.0" standalone="yes"?>
<!DOCTYPE person [

<!ELEMENT  person (year_recorded, number_of_immediate_family_
members, name, birthday, city_of_birth, state_of_birth,
immediate_family)>
<!ELEMENT number_of_immediate_family_members (#PCDATA)>
<!ELEMENT  name (#PCDATA)>
<!ELEMENT  birthday (#PCDATA)>
<!ELEMENT  city_of_birth (#PCDATA)>
<!ELEMENT  state_of_birth (#PCDATA)>
<!ELEMENT immediate_family (number_of_immediate_family_members,
spouse, daughter+, son+)>
<!ELEMENT number_of_immediate_family_members (#PCDATA)>
<!ELEMENT spouse (husband | wife)>
<!ELEMENT  husband (#PCDATA)>
<!ELEMENT  wife (#PCDATA)>
<!ELEMENT  daughter (#PCDATA)>
<!ELEMENT  son (#PCDATA)>
]>
<person>
      <name>John Raymond</name>
      <birthday>July 27, 1970</birthday>
      <city_of_birth>Des Moines</city_of_birth>
      <state_of_birth>Iowa</state_of_birth>
      <immediate_family>
            <number_of_immediate_family_members>
            5
            </number_of_immediate_family_members>
            <spouse>
                  <wife>Donna Raymond</wife>
            </spouse>
            <daughter>Jennifer Raymond</daughter>
            <daughter>Ashley Raymond</daughter>
            <son>Jeffrey Raymond</son>
            <son>Casey Raymond</son>
      </immediate_family>
</person>
```

Save this, the DTD and the XML code above, as "immediatefamily.xml".

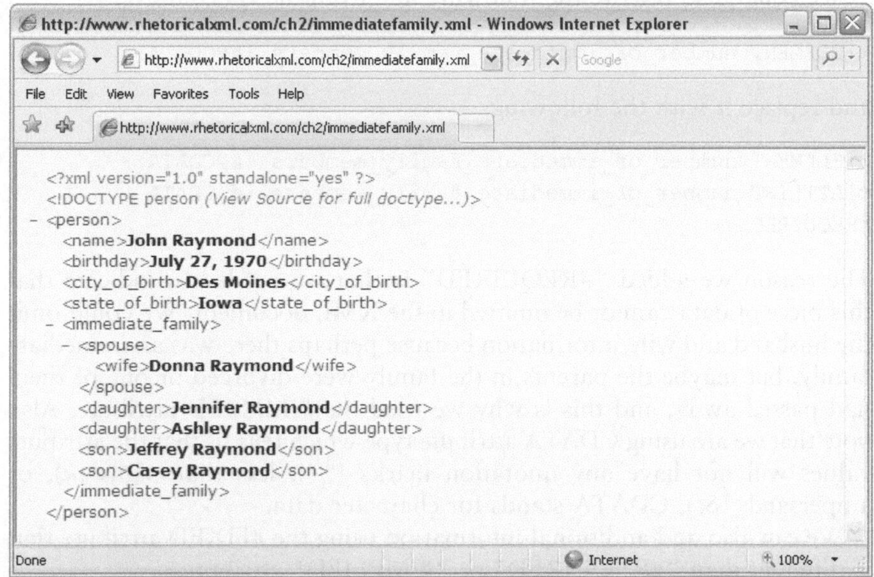

Figure 2.5 "Immediatefamily.xml" in Microsoft Internet Explorer®

In Figure 2.5, Donna Raymond shows up as the "wife" of John Raymond. In the DTD above, note the elements that we used to describe the gender distinctions for spouses:

```
<!ELEMENT spouse  (husband | wife)>
<!ELEMENT husband (#PCDATA)>
<!ELEMENT wife (#PCDATA)>
```

The pipe element "|" in this DTD dictates that if there is a spouse, there must be just one, and that it will be either a husband or wife. We can replace the three line part of the DTD above with a reconfigured <!ELEMENT> line that we see in the first line below, followed by an attribute, which is the second line below:

```
<!ELEMENT spouse  (#PCDATA)>
<!ATTLIST spouse id ( husband | wife ) #IMPLIED >
```

To write an attribute for a DTD, we start with "<!ATTLIST" tag, then we follow it with the name of the tag that will identify the additional information. In this case it is "spouse id". We need to follow this with the value or values which will be "husband" or "wife," and we will use the pipe "|" symbol again to distinguish them. You can think of the pipe symbol as the "or" in "husband or wife." Then we will follow with "#IMPLIED" to say that there does not necessarily have to be a husband or wife for the "person" we are describing, and close it off with a ">".

We could also rewrite the following line from the DTD above,

```
<!ELEMENT number_of_immediate_family_members (#PCDATA)>
```

and replace it with the following:

```
<!ELEMENT number_of_immediate_family_members (#PCDATA)>
<!ATTLIST number_of_immediate_family_members id CDATA
#REQUIRED>
```

The reason we added "#REQUIRED" is that we needed to indicate that this piece of data cannot be omitted in the XML document. We could omit the husband and wife information because perhaps there was an immediate family, but maybe the parents in the family were divorced or one of them had passed away, and this is why we used the #IMPLIED attribute. Also note that we are using CDATA attribute type which tells us that the attribute values will not have any quotation marks (""), less than signs (<), or ampersands (&). CDATA stands for character data.

We can also add additional information using the #FIXED attribute that is different than "#REQUIRED" or "#IMPLIED" attributes:

```
<!ELEMENT  year_recorded (#PCDATA)>
<!ATTLIST year_recorded id CDATA #FIXED "2006">
```

The code that shows up as a #FIXED attribute needs to be present in a DTD even though all it is telling us is that all of the data for this file was acquired in 2006 and no other year. When one reviews this DTD that has this particular piece of information, it does state to the people who originally designed the DTD or people who are adding to it that all of the data for this "person" DTD was acquired and coded for the year 2006. This fulfills the original goal of adding some attributes that provide to professionals some more information about the data's elements. Who knows, the Raymond family might have had a new family member added after this year, so knowing that this data might be dated is of value. Additional uses of fixed attributes might include XML code written for a company intranet that would state the name of the company and/or the division within the company.

What follows is the revamped version of the XML document above with attributes:

```
<?xml version="1.0" standalone="yes"?>
<!DOCTYPE person [
<!ELEMENT  person (year_recorded, name, birthday,
city_of_birth, state_of_birth, immediate_family)>
<!ELEMENT  year_recorded (#PCDATA)>
<!ATTLIST year_recorded id CDATA #FIXED "2006">
<!ELEMENT  name (#PCDATA)>
<!ELEMENT  birthday (#PCDATA)>
```

```
<!ELEMENT  city_of_birth  (#PCDATA)>
<!ELEMENT state_of_birth (#PCDATA)>
<!ELEMENT immediate_family (number_of_immediate_family_
members, spouse, daughter+, son+)>
<!ELEMENT number_of_immediate_family_members (#PCDATA)>
<!ATTLIST number_of_immediate_family_members id CDATA
#REQUIRED >
<!ELEMENT  year_recorded (#PCDATA)>
<!ATTLIST year_recorded id CDATA #FIXED "2006">
<!ELEMENT spouse (#PCDATA)>
<!ATTLIST spouse id ( husband | wife ) #IMPLIED >
<!ELEMENT  daughter (#PCDATA)>
<!ELEMENT  son (#PCDATA)>
]>
<person>
      <year_recorded>2006</year_recorded>
      <name>John Raymond</name>
      <birthday>July 27, 1970</birthday>
      <city_of_birth>Des Moines</city_of_birth>
      <state_of_birth>Iowa</state_of_birth>
      <immediate_family>
      <number_of_immediate_family_members>5</number_of_
      immediate_family_members>
      <spouse>
      <wife>Donna Raymond</wife>
      </spouse>
      <daughter>Jennifer Raymond</daughter>
      <daughter>Ashley Raymond</daughter>
      <son>Jeffrey Raymond</son>
      <son>Casey Raymond</son>
      </immediate_family>
</person>
```

Save this file as "immediatefamilyattributes.xml" and view it using Internet Explorer®. Then select the "View Source" option in the browser and you can see the altered DTD with the attributes that reveal more about the specific nature of the XML coded data.

Defining Attributes with Unique Values using ID Attributes

Sometimes we might have a large XML database with some values that are identical and we will need to differentiate between them. In the examples above, we might have more than one John Raymond in our database, and, therefore, we will need to designate between all of the "John Raymonds." We could also use designations that suggest unique values for information such as customer codes; we could designate customers who have the same first and last name as separate pieces of data.

Relational database systems handle the potential problems of data duplication by using what they define as a "primary key" for each individual row or record in a database. In XML, we can use a similar approach. To do this, we would add "ID" to reference a particular value that we do not want to repeat in the database, one that is unique. Using the ID attribute does not work with a #FIXED designation. Usually, it is associated with #REQUIRED attribute, but it can also work with the #IMPLIED attribute.

In the "person" DTD example we have been using, we are going to use the ID attribute for only the following elements: name, spouse (husband or wife), daughter, and son. Here is how they would now be coded:

```
<!ELEMENT  name (#PCDATA)>
<!ATTLIST  name id ID #REQUIRED>
<!ELEMENT spouse (#PCDATA)>
<!ATTLIST spouse id ID ( husband | wife ) #IMPLIED >
<!ELEMENT  daughter (#PCDATA)>
<!ATTLIST  daughter id ID #REQUIRED>
<!ELEMENT  son (#PCDATA)>
<!ATTLIST  son id ID #REQUIRED>
```

Now we are going to add some specific designations to the XML coded elements in the "person" file. The attribute value must follow the conventions of XML; it can be anything we would like as long as we use only alphanumeric symbols (letters or numbers) without any whitespace between them. Use an underscore "_" if you do want to use an attribute value that would normally have whitespace between letters, words, or numbers.

```
<name ID="ra01">John Raymond</name>
<wife ID="ra02">Donna Raymond</wife>
<daughter ID="ra03">Jennifer Raymond</daughter>
<daughter ID="ra04">Ashley Raymond</daughter>
<son ID="ra05">Jeffrey Raymond</son>
<son ID="ra06">Casey Raymond</son>
```

Putting it all together in a standalone XML document with these changes in the DTD and XML code, we would have the following file:

```
<?xml version="1.0" standalone="yes"?>
<!DOCTYPE person [
<!ELEMENT  person (year_recorded, name, birthday,
city_of_birth, state_of_birth, immediate_family)>
<!ELEMENT  year_recorded (#PCDATA)>
<!ATTLIST year_recorded id CDATA #FIXED "2006">
<!ELEMENT name (#PCDATA)>
<!ATTLIST name id ID #REQUIRED>
```

```
<!ELEMENT birthday (#PCDATA)>
<!ELEMENT city_of_birth (#PCDATA)>
<!ELEMENT state_of_birth (#PCDATA)>
<!ELEMENT immediate_family (number_of_immediate_family_members,
spouse, daughter+, son+)>
<!ELEMENT number_of_immediate_family_members (#PCDATA)>
<!ATTLIST number_of_immediate_family_members id CDATA
#REQUIRED >
<!ELEMENT year_recorded (#PCDATA)>
<!ATTLIST year_recorded id CDATA #FIXED "2006">
<!ELEMENT spouse (#PCDATA)>
<!ATTLIST spouse id ( husband | wife ) #IMPLIED >
<!ATTLIST spouse id ID #REQUIRED>
<!ELEMENT daughter (#PCDATA)>
<!ATTLIST daughter id ID #REQUIRED>
<!ELEMENT son (#PCDATA)>
<!ATTLIST son id ID #REQUIRED>
]>
<person>
      <year_recorded>2006</year_recorded>
      <name ID="ra01">John Raymond</name>
      <birthday>July 27, 1970</birthday>
      <city_of_birth>Des Moines</city_of_birth>
      <state_of_birth>Iowa</state_of_birth>
<immediate_family>
<number_of_immediate_family_members>5</number_of_immediate_
family_members>
      <wife ID="ra02">Donna Raymond</wife>
      <daughter ID="ra03">Jennifer Raymond</daughter>
      <daughter ID="ra04">Ashley Raymond</daughter>
      <son ID="ra05">Jeffrey Raymond</son>
      <son ID="ra06">Casey Raymond</son>
</immediate_family>
</person>
```

Save this file as "immediatefamilyattributesid.xml" and review it in Microsoft Internet Explorer®.

IDREF Attributes

IDREF attributes work, to some extent, like the links we find in HTML documents. They illustrate in the <!ELEMENT> tags specific relationships between some or all of the elements you have defined. We might want to provide information that links sets of information together. In the example we have been using, we could use IDREFs to link up the entire Raymond family. The IDREFs must refer to the ID attributes that are already defined in the same DTD. You can have some of the IDREFs refer to more than one ID attribute.

IDREF attributes allow us to see the metadata of some of the relationships that might not show up in the tree structure when we view the document through XML software. The relationships can be viewed in the source code that you see using Notepad and in a browser. Below is the coding with IDREFs integrated into the <!ELEMENT> tags:

```
<!ATTLIST name id ID #REQUIRED>
<!ELEMENT birthday (#PCDATA)>
<!ELEMENT city_of_birth (#PCDATA)>
<!ELEMENT state_of_birth (#PCDATA)>
<!ELEMENT immediate_family (number_of_immediate_family_members,
spouse, daughter+, son+)>
<!ELEMENT number_of_immediate_family_members (#PCDATA)>
<!ATTLIST number_of_immediate_family_members id CDATA
#REQUIRED >
<!ELEMENT year_recorded (#PCDATA)>
<!ATTLIST year_recorded id CDATA #FIXED "2006">
<!ELEMENT spouse (#PCDATA)>
<!ATTLIST spouse id ( husband | wife ) #IMPLIED >
<!ATTLIST spouse id ID #REQUIRED>
<!ELEMENT daughter (#PCDATA)>
<!ATTLIST daughter id ID #REQUIRED mother IDREF #IMPLIED
father IDREF #IMPLIED>
<!ELEMENT son (#PCDATA)>
<!ATTLIST son id ID #REQUIRED mother IDREF #IMPLIED father
IDREF #IMPLIED>
]>
<person>
      <year_recorded>2006</year_recorded>
      <name ID="ra01">John Raymond</name>
      <birthday>July 27, 1970</birthday>
      <city_of_birth>Des Moines</city_of_birth>
      <state_of_birth>Iowa</state_of_birth>
      <immediate_family>
      <number_of_immediate_family_members>5</number_of_
      immediate_family_members>
            <wife ID="ra02">Donna Raymond</wife>
            <daughter ID="ra03" mother="ra02" father="ra01">
            Jennifer Raymond</daughter>
            <daughter ID="ra04" mother="ra02" father="ra01">
            Ashley Raymond</daughter>
            <son ID="ra05" mother="ra02" father="ra01">
            Jeffrey Raymond</son>
```

```
            <son ID="ra06" mother="ra02" father="ra01">Casey
            Raymond</son>
      </immediate_family>
</person>
```

Save this file as "immediatefamilyattributesidref.xml" and review it in Microsoft Internet Explorer®. You can have some of the IDREFs refer to more than one ID attribute. As you can see, we have designated who the mother and father is in the son and daughter elements.

Thus far in the "person" XML coding, we have structured our code so it can only store information about either a father or a mother and their sons and daughters if in fact they have any. If we wanted to be able to compile information for anyone in the Raymond family, we would have to restructure our code. Below is the DTD with the appropriate IDREF designations for Ashley Raymond:

```
<?xml version="1.0" standalone="yes"?>
<!DOCTYPE person [
<!ELEMENT  person (year_recorded, name, birthday,
city_of_birth, state_of_birth, immediate_family)>
<!ELEMENT  year_recorded (#PCDATA)>
<!ATTLIST year_recorded id CDATA #FIXED "2006">
<!ELEMENT name  (#PCDATA)>
<!ATTLIST name id ID #REQUIRED>
<!ELEMENT birthday (#PCDATA)>
<!ELEMENT city_of_birth  (#PCDATA)>
<!ELEMENT state_of_birth (#PCDATA)>
<!ELEMENT immediate_family (number_of_immediate_family_members,
father?, mother?, spouse?, daughter*, son*)>
<!ELEMENT number_of_immediate_family_members (#PCDATA)>
<!ATTLIST number_of_immediate_family_members id CDATA
#REQUIRED >
<!ELEMENT father (#PCDATA)>
<!ATTLIST father id ID  #REQUIRED>
<!ELEMENT mother (#PCDATA)>
<!ATTLIST mother id ID  #REQUIRED>
<!ELEMENT sister (#PCDATA)>
<!ATTLIST spouse id ID  #REQUIRED>
<!ELEMENT brother (#PCDATA)>
<!ATTLIST brother id ID  #REQUIRED>
<!ELEMENT spouse (#PCDATA)>
<!ATTLIST spouse id ( husband | wife ) #IMPLIED >
<!ATTLIST spouse id ID  #REQUIRED>
<!ELEMENT daughter (#PCDATA)>
<!ATTLIST daughter id ID #REQUIRED mother IDREF #IMPLIED
father IDREF #IMPLIED>
```

```
<!ELEMENT son (#PCDATA)>
<!ATTLIST son id ID #REQUIRED mother IDREF #IMPLIED father
IDREF #IMPLIED>
]>
<person>
    <year_recorded>2006</year_recorded>
    <name ID="ra03">Ashley Raymond</name>
    <birthday>February 26, 2001</birthday>
    <city_of_birth>Minneapolis</city_of_birth>
    <state_of_birth>Minnesota</state_of_birth>
    <immediate_family>
        <number_of_immediate_family_members>5</number_of_
        immediate_family_members>
            <father ID="ra01">John Raymond</father>
            <mother ID="ra02">Donna Raymond</mother>
            <sister ID="ra04">Jennifer Raymond</sister>
            <brother ID="ra05">Jeffrey Raymond</brother>
            <brother ID="ra06">Casey Raymond</brother>
    </immediate_family>
</person>
```

Save this as "immediatefamilyattributesidref1.xml". It should look like the screenshot shown in Figure 2.6.

Using the same DTD that we used above, but changing the XML information values, we could store information for Donna Raymond, Jeffrey Raymond, or Casey Raymond. That we put question marks after "father"

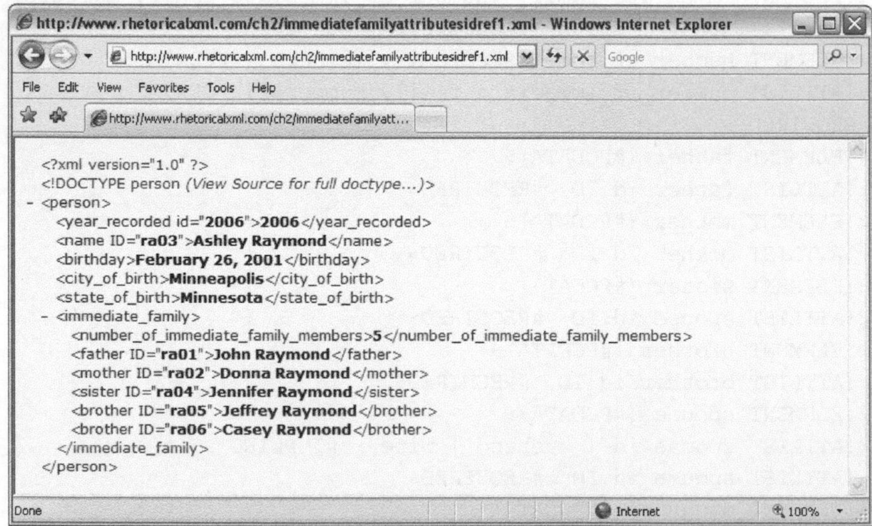

Figure 2.6 "Immediatefamilyattributesidref1.xml" in Microsoft Internet Explorer®

and "mother" signals that this coding would work for Donna Raymond; "father" or "mother" would not show up in the XML coding as we have designed this database for the immediate family, but her husband, sons, and daughters would show up with their attendant IDREFs.

```
<!ATTLIST year_recorded id CDATA #FIXED "2006">
<!ELEMENT name (#PCDATA)>
<!ATTLIST name id ID #REQUIRED>
<!ELEMENT birthday (#PCDATA)>
<!ELEMENT city_of_birth  (#PCDATA)>
<!ELEMENT state_of_birth (#PCDATA)>
<!ELEMENT immediate_family (number_of_immediate_family_members,
father?, mother?, spouse?, daughter*, son*)>
<!ELEMENT number_of_immediate_family_members (#PCDATA)>
<!ATTLIST number_of_immediate_family_members id CDATA
#REQUIRED >
<!ELEMENT father (#PCDATA)>
<!ATTLIST father id ID #REQUIRED>
<!ELEMENT mother (#PCDATA)>
<!ATTLIST mother id ID #REQUIRED>
<!ELEMENT sister (#PCDATA)>
<!ATTLIST spouse id ID #REQUIRED>
<!ELEMENT brother (#PCDATA)>
<!ATTLIST brother id ID #REQUIRED>
<!ELEMENT spouse (#PCDATA)>
<!ATTLIST spouse id ( husband | wife ) #IMPLIED >
<!ATTLIST spouse id ID #REQUIRED>
<!ELEMENT daughter (#PCDATA)>
<!ATTLIST daughter id ID #REQUIRED mother IDREF #IMPLIED
father IDREF #IMPLIED>
<!ELEMENT son (#PCDATA)>
<!ATTLIST son id ID #REQUIRED mother IDREF #IMPLIED father
IDREF #IMPLIED>
]>
<person>
      <year_recorded>2006</year_recorded>
      <name ID="ra02">Donna Raymond</name>
      <birthday>March 26, 2001</birthday>
      <city_of_birth>Des Moines</city_of_birth>
      <state_of_birth>Iowa</state_of_birth>
      <immediate_family>
         <number_of_immediate_family_members>5</number_of_
         immediate_family_members>
            <husband ID="ra01">John Raymond</husband>
            <daughter ID="ra03">Ashley Raymond</daughter>
            <daughter ID="ra04">Jennifer Raymond</daughter>
```

```
        <son ID="ra05">Jeffrey Raymond</son>
        <son ID="ra06">Casey Raymond</son>
    </immediate_family>
</person>
```

Label this "immediatefamilyattributesidref2.xml" and see how it looks in Internet Explorer®.

Namespaces

As we have already noted, XML allows us to exchange or combine data with others with some ease. In some cases, we might want to combine XML coded data from two or more databases devoted to the storage of similar or complementary types of information. The problem with this is that some of the elements from the different databases might have the same name, but these elements could mean different things and this would make the information we were trying to compile useless.

In a hypothetical example, what if professionals who were compiling XML coded data from the World Health Organization (WHO) and the U.S. Department of Health and Human Services (HHS) were using information from both sources and there were some "collisions" between some identical element tags because these tags had different semantic meanings? For example, perhaps the concept of "death by natural causes" has different meanings in the two databases but they both used the tags <natural_causes> to describe their meanings. One organization might determine that if a person dies of a heart attack at age sixty-five in a given country in a given year that this person died of a natural cause. Another organization might assume that this should not be included in "death by natural cause" data; it might attribute this to "death by heart disease."

To declare a namespace, we need to start with the xmlns attribute which is easy to remember as the "ns" stands for "namespace." Then we give it a value with a Uniform Resource Indicator (URI). The convention is to use the Uniform Resource Locator (URL) of the organization for this as it is convenient and we can be sure that there is only one of them.

For example, to indicate a declaration for namespace associated with the WHO, we would use,

```
xmlns="www.who.int/en/"
```

For the HHS, we would use,

```
xmlns="www.hhs.gov/"
```

While the address that we use to identify namespaces looks like we are using some external website and trying to extract some information from it, we are just using this convention to give the namespace a unique name.

Our hypothetical element tags would look like this for the HHS data:

```
<natural_causes xmlns="www.hhs.gov/">4.65 million
</natural_causes>
```

If the WHO compiled data that did not include death by heart disease, we might use the following tags:

```
<natural_causes xmlns="www.who.int/en/">3.44 million
</natural_causes>
```

We are using elements with the same name, "natural causes," but with different namespace declarations. We might imagine that it would be easier to change all of the tags from one of the databases to better reflect a different meaning instead of acknowledging the different sources of information with namespace declarations. For example, we could use <natural_causes_and_heart_failure> for the HHS data, even though this might be a bit unwieldy. However, this might not be easy because there could be so many different documents in some institutional databases and this would prove impractical. Also, it is to our advantage to indicate the source of different sets of data, and namespaces allow us to do this.

In the following XML code we illustrate how a namespace can be used to indicate that the data on causes of mortality is from the WHO. The WHO namespace is in the

```
<cause_specific_per_one_hundred_thousand_population>
```

tag and this indicates that all of the data such as mortality by cause of "non-communicable diseases" and "injuries" is encapsulated within this element and is from the WHO source.

```
<?xml version="1.0" standalone="yes"?>
<!DOCTYPE health_status_mortality [
<!ELEMENT cause_specific_per_one_hundred_thousand_population
( HIV-AIDS, TB_HIV_negative_people, TB_HIV_positive_people ) >
<!ELEMENT HIV-AIDS (#PCDATA)>
<!ELEMENT TB_HIV_negative_people (#PCDATA)>
<!ELEMENT TB_HIV_positive_people (#PCDATA)>
<!ELEMENT age-standardized_mortality_rate_per_one_hundred_
thousand_population (non-communicable_disease, cardiovascular_
disease, cancer, injuries)>
<!ELEMENT non-communicable_disease (#PCDATA)>
<!ELEMENT cardiovascular_disease (#PCDATA)>
<!ELEMENT cancer (#PCDATA)>
<!ELEMENT injuries (#PCDATA)>
```

```
]>
<health_status_mortality>
      <cause_specific_per_one_hundred_thousand_population
       xmlns="www.who.int/en/">
            <HIV-AIDS>&lt;10</HIV-AIDS>
      <TB_HIV_negative_people>&lt;1</TB_HIV_negative_people>
      <TB_HIV_positive_people>&lt;1</TB_HIV_positive_people>
      </cause_specific_per_one_hundred_thousand_population>
      <age-standardized_mortality_rate_per_one_hundred_
      thousand_population>
      <non-communicable_disease>460</non-communicable_disease>
            <cardiovascular_disease>188</cardiovascular_
            disease>
            <cancer>134</cancer>
            <injuries>47</injuries>
      </age-standardized_mortality_rate_per_one_hundred_
      thousand_population>
</health_status_mortality>
```

Save this file as "whomortality1.xml" and review it in a browser. You can see that we used the "<" special symbol to indicate the less than "<" sign. In Chapter 5 we will come back to this example and further develop namespaces to show how they can be used with other technologies.

Entities

Internal General Entities

The XML documents we have been working on so far have been your basic standalone XML files where the information can be described in one complete XML coded file with the option of having an internal or external DTD to validate it against. One of the simplest uses of entities is to attach boilerplate kinds of information to XML documents so you would not have to keep repeating the same information for every file you compiled in XML. These are called internal general entities. In the upcoming "person" example, you will see that we added the following line of code:

```
<!ENTITY contact "This data compiled and copyrighted by State
Census Bureau.  To contact, phone at (555) 555-5555 or e-mail
at sctstats@sct.gov.">
```

We then indicated where we wanted this "contact" information that is within the quotes placed in our <!ENTITY> tags above by writing "contact" between a "&" and a ";". This shows up as "&contact;" in the code listing, right before the "</person>" tag.

```
<?xml version="1.0" standalone="yes"?>
<!DOCTYPE person [
<!ELEMENT person (name, birthday, city_of_birth,
state_of_birth, immediate_family)>
<!ELEMENT name (#PCDATA)>
<!ELEMENT birthday (#PCDATA)>
<!ELEMENT city_of_birth  (#PCDATA)>
<!ELEMENT state_of_birth (#PCDATA)>
<!ELEMENT immediate_family (spouse, daughter+, son+)>
<!ELEMENT spouse (husband | wife)>
<!ELEMENT husband (#PCDATA)>
<!ELEMENT wife (#PCDATA)>
<!ELEMENT daughter (#PCDATA)>
<!ELEMENT son (#PCDATA)>
<!ENTITY contact "This data compiled and copyrighted by State
Census Bureau.  To contact, phone at (555) 555-5555 or e-mail
at sctstats@sct.gov.">
]>
<person>
      <name>John Raymond</name>
      <birthday>July 27, 1970</birthday>
      <city_of_birth>Des Moines</city_of_birth>
      <state_of_birth>Iowa</state_of_birth>
      <immediate_family>
            <spouse>
                  <wife>Donna Raymond</wife>
            </spouse>
            <daughter>Jennifer Raymond</daughter>
            <daughter>Ashley Raymond</daughter>
            <son>Jeffrey Raymond</son>
            <son>Casey Raymond</son>
      </immediate_family>
      &contact;
</person>
```

Save this file as "immediatefamilyentity.xml" and then review it in Microsoft Internet Explorer®. You should see the image captured in Figure 2.7.

As you can see in Figure 2.7, the contact information does not show up in quotes, nor is it surrounded by tags as the other information is. The value of this technique (using entities) is that it allows us to more easily represent information that needs to show up on every document. Once the DTD is established, we would only need to write "&contact;" in the same place as we begin to add other "persons" to this document. Also, if any of the contact information changed—the phone number, the e-mail address—we would only have to make changes in the DTD above and not in every XML document coded with the information for every "person" in our database.

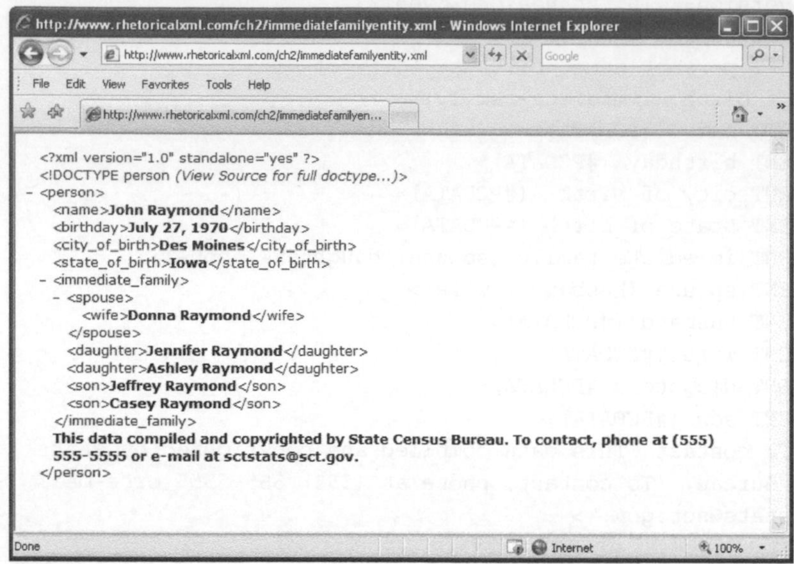

Figure 2.7 "Immediatefamilyentity.xml" in Microsoft Internet Explorer®

External General Entities

The power of XML in part derives from the fact that XML documents can be coded in such a way where we can select from many different files that might be in one or more databases to produce a larger document composed of these files. This is one of the real values of XML; we can build from external general entities.

In the following extended example, we are going to breakdown a longer document into a series of units or modules and label them as entities. The document is published by the U.S. Department of Energy (DOE) and is entitled "Advisory Committee on Human Radiation Experiments—Executive Summary." We have a print copy of this document included in Appendix A at the end of this book.

Type in or access from our website the following separate modules. Save them as instructed at the end of each module. Each of the separate modules starts with an opening <module> tag, and ends with a closing </module> tag.

To start with our first entity, type in the following and save it as "achre_publication_information.ent." ACHRE is our acronym for "Advisory Committee on Human Radiation Experiments." To better differentiate between each module, we have added "begin" and "end" markers surrounding each module. Be sure to start each module with <module> and end with </module>.

Begin Publication Information Entity.

```
<module>
      <title>PUBLICATION INFORMATION</title>
      <para>The Final Report of the Advisory Committee on Human
      Radiation Experiments (stock number 061-000-00-848-9),
      the supplemental volumes to the Final Report
      (stock numbers 061-000-00850-1, 061-000-00851-9, and
      061-000-00852-7), and additional copies of this Executive
      Summary (stock number 061-000-00849-7) may be purchased
      from the Superintendent of Documents, U.S. Government
      Printing Office.
      </para>
</module>
```

End of Publication Information Entity.

Save as "achre_publication_information.ent".

Begin Telephone Orders Entity.

```
<module>
      <line>All telephone orders should be directed to:</line>
      <line>Superintendent of Documents</line>
      <line>U.S. Government Printing Office</line>
      <line>Washington, D.C. 20402</line>
      <line>(202) 512-1800</line>
      <line>FAX (202) 512-2250</line>
</module>
```

End of Telephone Orders Entity.

Save as "usgpo_telephone_orders.ent." USGPO is our acronym for U.S. Government.

Begin Mail Orders Entity.

```
<module>
      <line>All mail orders should be directed to: </line>
      <line>U.S. Government Printing Office</line>
      <line>P.O. Box 371954</line>
      <line>Pittsburgh, PA 15250-7954</line>
</module>
```

End Mail Orders Entity.

Save as "usgpo_mail_orders.ent".

Begin Internet Archives Entity.

```
<module>
      <para>An Internet site containing ACHRE information
      (replicating the Advisory Committee's original gopher)
      will be available at George Washington University.
      The site contains complete records of Advisory Committee
      actions as approved; complete descriptions of the
      primary research materials discovered and analyzed;
      complete descriptions of the print and non-print
      secondary resources used by the Advisory Committee;
      a copy of the Interim Report of October 21, 1994,
      and other information. The address is www.seas.gwu.edu/
      nsarchive/radiation. The site will be maintained by the
      National Security Archive at GWU.</para>
</module>
```

End Internet Archives Entity.

Save as "achre_internet_archives.ent".

Begin Printing Location Entity.

```
<module>
      <line>Printed in the United States of America</line>
</module>
```

End Printing Location Entity.

Save as "usgpo_printing_location.ent".

Begin Creation of Advisory Committee Entity.

```
<module>
      <title>THE CREATION OF THE ADVISORY COMMITTEE</title>
      <para>On January 15, 1994, President Clinton appointed
      the Advisory Committee on Human Radiation Experiments.
      The President created the Committee to investigate
      reports of possibly unethical experiments funded by the
      government decades ago.</para>
      <para>The members of the Advisory Committee were fourteen
      private citizens from around the country: a
      representative of the general public and thirteen experts
      in bioethics, radiation oncology and biology, nuclear
      medicine, epidemiology and biostatistics, public health,
      history of science and medicine, and law.</para>
```

```
    <para>President Clinton asked us to deliver our
    recommendations to a Cabinet-level group, the Human
    Radiation Interagency Working Group, whose members are
    the Secretaries of Defense, Energy, Health and Human
    Services, and Veterans Affairs; the Attorney General;
    the Administrator of the National Aeronautics and Space
    Administration; the Director of Central Intelligence;
    and the Director of the Office of Management and Budget.
    Some of the experiments the Committee was asked to
    investigate, and particularly a series that included
    the injection of plutonium into unsuspecting hospital
    patients, were of special concern to Secretary of Energy
    Hazel O&apos;Leary. Her department had its origins in
    the federal agencies that had sponsored the plutonium
    experiments. These agencies were responsible for the
    development of nuclear weapons and during the Cold War
    their activities had been shrouded in secrecy. But now
    the Cold War was over.</para>
    <para>The controversy surrounding the plutonium
    experiments and others like them brought basic questions
    to the fore: How many experiments were conducted or
    sponsored by the government, and why? How many were
    secret? Was anyone harmed? What was disclosed to those
    subjected to risk, and what opportunity did they have for
    consent? By what rules should the past be judged? What
    remedies are due those who were wronged or harmed by the
    government in the past? How well do federal rules that
    today govern human experimentation work? What lessons
    can be learned for application to the future? Our Final
    Report provides the details of the Committee&apos;s
    answers to these questions. This Executive Summary
    presents an overview of the work done by the Committee,
    our findings and recommendations, and the contents of the
    Final Report.</para>
</module>
```

End Creation of Advisory Committee Entity.

Save as "achre_creation_of_advisory_committee.ent".

Begin President's Charge Entity.

```
<module>
    <title>THE PRESIDENT&apos;S CHARGE</title>
    <para>The President directed the Advisory Committee to
    uncover the history of human radiation experiments during
```

```
        the period 1944 through 1974. It was in 1944 that the
        first known human radiation experiment of interest was
        planned, and in 1974 that the Department of Health,
        Education and Welfare adopted regulations governing the
        conduct of human research, a watershed event in the
        history of federal protections for human subjects.</para>
        <para>In addition to asking us to investigate human
        radiation experiments, the President directed us to
        examine cases in which the government had intentionally
        released radiation into the environment for research
        purposes. He further charged us with identifying the
        ethical and scientific standards for evaluating these
        events, and with making recommendations to ensure that
        whatever wrongdoing may have occurred in the past cannot
        be repeated. We were asked to address human experiments
        and intentional releases that involved radiation. The
        ethical issues we addressed and the moral framework we
        developed are, however, applicable to all research
        involving human subjects.</para>
        <para>The breadth of the Committee&apos;s charge was
        remarkable. We were called on to review government
        programs that spanned administrations from Franklin
        Roosevelt to Gerald Ford. As an independent advisory
        committee, we were free to pursue our charge as we saw
        fit. The decisions we reached regarding the course of
        our inquiry and the nature of our findings and
        recommendations were entirely our own.</para>
</module>
```

End President's Charge Entity.

Save as "presidents_charge.ent".

Begin Committee's Approach Entity

```
<module>
        <title>THE COMMITTEE&apos;S APPROACH</title>
        <para>At our first meeting, we immediately realized that
        we were embarking on an intense and challenging
        investigation of an important aspect of our nation&apos;s
        past and present, a task that required new insights and
        difficult judgments about ethical questions that persist
        even today. </para>
        <para>Between April 1994 and July 1995, the Advisory
        Committee held sixteen public meetings, most in
        Washington, D.C. In addition, subsets of Committee
```

```
members presided over public forums in cities throughout
the country. The Committee heard from more than 200
witnesses and interviewed dozens of professionals who
were familiar with experiments involving radiation.
A special effort, called the Ethics Oral History Project,
was undertaken to learn from eminent physicians about
how research with human subjects was conducted in the
1940s and 1950s.</para>
<para>We were granted unprecedented access to government
documents. The President directed all the federal
agencies involved to make available to the Committee any
documents that might further our inquiry, wherever they
might be located and whether or not they were still
secret.</para>
<para>As we began our search into the past, we quickly
discovered that it was going to be extremely difficult
to piece together a coherent picture. Many critical
documents had long since been forgotten and were stored
in obscure locations throughout the country. Often they
were buried in collections that bore no obvious
connection to human radiation experiments. There was no
easy way to identify how many experiments had been
conducted, where they took place, and which government
agencies had sponsored them.</para>
<para>Nor was there a quick way to learn what rules
applied to these experiments for the period prior to the
mid-1960s. With the assistance of hundreds of federal
officials and agency staff, the Committee retrieved and
reviewed hundreds of thousands of government documents.
Some of the most important documents were secret and
were declassified at our request. Even after this
extraordinary effort, the historical record remains
incomplete. Some potentially important collections
could not be located and were evidently lost or
destroyed years ago.</para>
<para>Nevertheless, the documents that were recovered
enabled us to identify nearly 4,000 human radiation
experiments sponsored by the federal government between
1944 and 1974. In the great majority of cases, only
fragmentary data was locatable; the identity of subjects
and the specific radiation exposures involved were
typically unavailable. Given the constraints of
information, even more so than time, it was impossible
for the Committee to review all these experiments, nor
could we evaluate the experiences of countless individual
subjects. We thus decided to focus our investigation on
```

```
    representative case studies reflecting eight different
    categories of experiments that together addressed our
    charge and priorities. These case studies included:
    </para>
<list>
    <line>experiments with plutonium and other atomic bomb
    materials</line>
    <line>the Atomic Energy Commission&apos;s program of
    radioisotope distribution</line>
    <line>nontherapeutic research on children</line>
    <line>total body irradiation</line>
    <line>research on prisoners</line>
    <line>human experimentation in connection with nuclear
    weapons testing</line>
    <line>intentional environmental releases of
    radiation</line>
    <line>observational research involving uranium miners and
    residents of the Marshall Islands</line>
</list>
<para>In addition to assessing the ethics of human radiation
experiments conducted decades ago, it was also important to
explore the current conduct of human radiation research.
Insofar as wrongdoing may have occurred in the past, we needed
to examine the likelihood that such things could happen today.
We therefore undertook three projects:</para>
<list>
    <para>A review of how each agency of the federal
    government that currently conducts or funds research
    involving human subjects regulates this activity and
    oversees it.</para>
    <para>An examination of the documents and consent forms
    of research projects that are today sponsored by the
    federal government in order to develop insight into the
    current status of protections for the rights and
    interests of human subjects.</para>
    <para>Interviews of nearly 1,900 patients receiving out-
    patient medical care in private hospitals and federal
    facilities throughout the country. We asked them whether
    they were currently, or had been, subjects of research,
    and why they had agreed to participate in research or had
    refused.</para>
    </list>
</module>
```

End Committee's Approach Entity.

Save as "achre_committees_approach.ent".

DTDs for Documents with General Entities

Now we have to write a DTD that is a little different than the ones we have seen before. This will include the list of entities that we describe above. Note that the entity tags are similar to element tags, but we start with "<!ENTITY" and close with ">". We are also using the "SYSTEM" marker that we have used before.

```
<!ELEMENT achre_executive_summary_radiation_experiments
(title?, module+)>
<!ELEMENT title (#PCDATA)>
<!ENTITY achre_publication_information SYSTEM "achre_
publication_information.ent">
<!ENTITY usgpo_telephone_orders SYSTEM "usgpo_telephone_
orders.ent">
<!ENTITY usgpo_mail_orders SYSTEM "usgpo_mail_orders.
ent">
<!ENTITY achre_internet_archives SYSTEM "achre_internet_
archives.ent">
<!ENTITY usgpo_printing_location SYSTEM "usgpo_printing_
location.ent">
<!ENTITY achre_creation_of_advisory_committee SYSTEM "achre_
creation_of_advisory_committee.ent">
<!ENTITY achre_presidents_charge SYSTEM "achre_presidents_
charge.ent">
<!ENTITY achre_committees_approach SYSTEM "achre_committees_
approach.ent">
<!ELEMENT module (title?, heading?, line?, para?, line?,
list?)>
<!ELEMENT title (#PCDATA)>
<!ELEMENT heading (#PCDATA)>
<!ELEMENT line (#PCDATA)>
<!ELEMENT para (#PCDATA)>
<!ELEMENT list (para?, line?)>
<!ELEMENT para (#PCDATA)>
<!ELEMENT line (#PCDATA)>
```

Note that we have also put in several <!ELEMENT> tags as we usually do in a DTD for a "title" and "module." Within the modules, we have title, heading, para, line, and list to better allow us to granularize the information in the ACHRE—Executive Summary. We put a "?" sign after all of the elements as we might or might not have include them in any of the modules as they all are unique. Save this as "achre1.dtd".

Document Type Declaration

Here is our Document Type Declaration for our ACHRE—Executive Summary document coded in XML:

```
<?xml version="1.0" standalone="no"?>
<!DOCTYPE achre_executive_summary_radiation_experiments SYSTEM
"achre1.dtd">
<achre_executive_summary_radiation_experiments>
<title>Advisory Committee on Human Radiation Experiments:
Executive Summary</title>
&achre_publication_information;
&usgpo_telephone_orders;
&usgpo_mail_orders;
&achre_internet_archives;
&usgpo_printing_location;
&achre_creation_of_advisory_committee;
&achre_presidents_charge;
&achre_committees_approach;
</achre_executive_summary_radiation_experiments>
```

Save this as "achre1.xml".

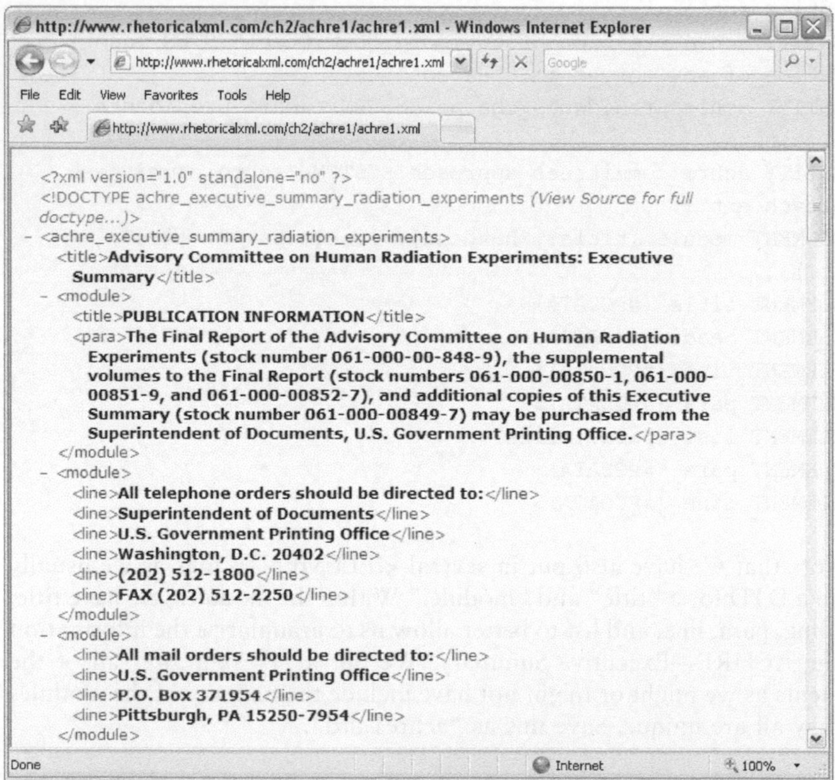
http://www.rhetoricalxml.com/ch2/achre1/achre1.xml - Windows Internet Explorer
http://www.rhetoricalxml.com/ch2/achre1/achre1.xml
Google
File Edit View Favorites Tools Help
http://www.rhetoricalxml.com/ch2/achre1/achre1.xml
<?xml version="1.0" standalone="no" ?>
<!DOCTYPE achre_executive_summary_radiation_experiments (View Source for full doctype...)>
- <achre_executive_summary_radiation_experiments>
<title>Advisory Committee on Human Radiation Experiments: Executive Summary</title>
- <module>
<title>PUBLICATION INFORMATION</title>
<para>The Final Report of the Advisory Committee on Human Radiation Experiments (stock number 061-000-00-848-9), the supplemental volumes to the Final Report (stock numbers 061-000-00850-1, 061-000-00851-9, and 061-000-00852-7), and additional copies of this Executive Summary (stock number 061-000-00849-7) may be purchased from the Superintendent of Documents, U.S. Government Printing Office.</para>
</module>
- <module>
<line>All telephone orders should be directed to:</line>
<line>Superintendent of Documents</line>
<line>U.S. Government Printing Office</line>
<line>Washington, D.C. 20402</line>
<line>(202) 512-1800</line>
<line>FAX (202) 512-2250</line>
</module>
- <module>
<line>All mail orders should be directed to:</line>
<line>U.S. Government Printing Office</line>
<line>P.O. Box 371954</line>
<line>Pittsburgh, PA 15250-7954</line>
</module>
Done
Internet
100%

Figure 2.8 Screenshot of "achre1.xml" in Microsoft Internet Explorer®

As you can see, this is a not a standalone document as we are referring to an external DTD, "achre1.dtd", and to external entities that are referenced within it. In this case, all of these coded documents—the entities, DTD, and XML coded files—need to be placed in one file on your computer.

Note that we have placed all of the external entities we have described in a listing below the title tags. When we do list these entities, we always place an ampersand "&" before them and a semicolon ";" after them. For example, the first one is,

```
&achre_publication_information;
```

We can see that the root element in the document type declaration is,

```
<achre_executive_summary_radiation_experiments>
```

If all of these elements are in the same file, you can pull up the entire document in Internet Explorer® by clicking on "achre1.xml". A screenshot of the "achre1.xml" file is found in Figure 2.8.

Using an External DTD and External Entities on a Website

If we placed our DTD for this document and all of the entity modules on a website as opposed to a single file on our computer, we would have to change both our DTD and XML files. The DTD would look like this:

```
<!ELEMENT achre_executive_summary_radiation_experiments
(title?, module+)>
<!ELEMENT title (#PCDATA)>
<!ENTITY achre_publication_information SYSTEM "www.
rhetoricalxml.com/ch2/achre1/achre_publication_information.
ent">
<!ENTITY usgpo_telephone_orders SYSTEM "www.rhetoricalxml.com/
ch2/achre1/usgpo_telephone_orders.ent">
<!ENTITY usgpo_mail_orders SYSTEM "www.rhetoricalxml.com/ch2/
achre1/usgpo_mail_orders.ent">
<!ENTITY achre_internet_archives SYSTEM "www.rhetoricalxml.com/
ch2/achre1/achre_internet_archives.ent">
<!ENTITY usgpo_printing_location SYSTEM "www.rhetoricalxml.com/
ch2/achre1/usgpo_printing_location.ent">
<!ENTITY achre_creation_of_advisory_committee SYSTEM
"www.rhetoricalxml.com/ch2/achre1/achre_creation_of_advisory_
committee.ent">
```

```
<!ENTITY achre_presidents_charge SYSTEM "www.rhetoricalxml.com/
ch2/achre1/achre_presidents_charge.ent">
<!ENTITY achre_committees_approach SYSTEM "www.rhetoricalxml.
com/ch2/achre1/achre_committees_approach.ent">
<!ENTITY achre_historical_context SYSTEM "www.rhetoricalxml.
com/ch2/achre1/achre_historical_context.ent">
<!ELEMENT module (title?, heading?, line?, para?, line?,
list?)>
<!ELEMENT title (#PCDATA)>
<!ELEMENT heading (#PCDATA)>
<!ELEMENT line (#PCDATA)>
<!ELEMENT para (#PCDATA)>
<!ELEMENT list (para?, line?)>
<!ELEMENT para (#PCDATA)>
<!ELEMENT line (#PCDATA)>
```

Save this file as "achre1website.dtd". Note that to reference the external entities that now reside on an Internet Web server, we have to include the URL and the name of the entity after SYSTEM. A new XML file looks like this:

```
<?xml version="1.0" standalone="no"?>
<!DOCTYPE achre_executive_summary_radiation_experiments SYSTEM
"www.rhetoricalxml.com/ch2/achre1/achre1website.dtd">
<achre_executive_summary_radiation_experiments>
<title>Advisory Committee on Human Radiation Experiments:
Executive Summary</title>
&achre_publication_information;
&usgpo_telephone_orders;
&usgpo_mail_orders;
&achre_internet_archives;
&usgpo_printing_location;
&achre_creation_of_advisory_committee;
&achre_presidents_charge;
&achre_committees_approach;
&achre_historical_context;
</achre_executive_summary_radiation_experiments>
```

Save this file as "achre1website.xml". Observe that instead of just referencing the DTD file, "achre1.dtd", that was stored in a file on our computer, we are referencing this DTD file that is now on the Internet Web server that supports *The Rhetorical Nature of XML* website. Thus, we now have SYSTEM "www.rhetoricalxml.com/ch2/achre1/achre1website.dtd". The advantage of this practice is that it allows members of a large organization to share and extract or point to DTDs and entities with greater ease.

In order to demonstrate how facile this system is, we can exclude some of the entities in our XML document below and add one or more entities. If we only wanted to extract "the committee's approach" and we also wanted to add "the historical context" documents in our document, we would have the following entities listed:

```
<?xml version="1.0" standalone="no"?>
<!DOCTYPE achre_executive_summary_radiation_experiments SYSTEM
"www.rhetoricalxml.com/ch2/achre1/achre1website.dtd">
<achre_executive_summary_radiation_experiments>
<title>Advisory Committee on Human Radiation Experiments:
Executive Summary</title>
&achre_committees_approach;
&achre historical context;
</achre_executive_summary_radiation_experiments>
```

Save as "achre2website.xml". A screenshot of the "achre2website.xml" file is shown in Figure 2.9.

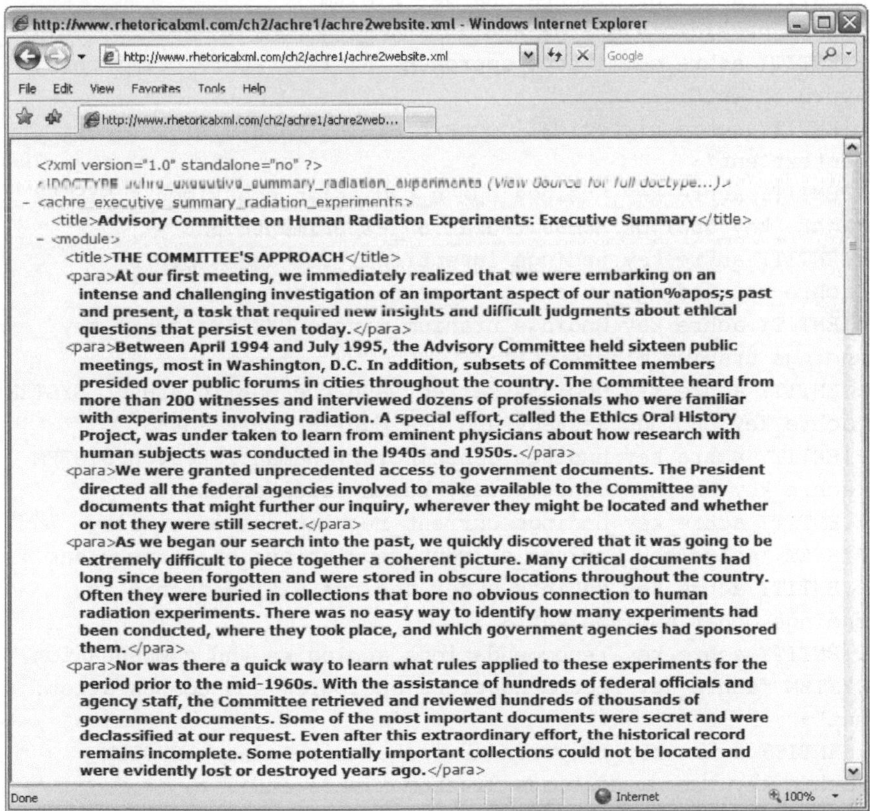

Figure 2.9 Screenshot of "achre2website.xml" in Microsoft Internet Explorer®

If we had a long list of entities and DTDs that were stored in our hard drives or on an external server, we could pick and choose just what modules we would need to produce a specific document for our needs. This shows how entities can be employed by technical communicators who are drawing from previously written documents, a practice known as single sourcing. We will cover this practice in some detail in Chapter 3 and again in Chapter 6. Below is the DTD for an expansion of the file's Document Type Declaration for our ACHRE—Executive Summary. You will need to go to our website, www.rhetoricalxml.com, to extract all of the additional "key recommendations," "key findings," and "advisory committee's legacy" entities. There are thirteen of them:

```
<!ENTITY achre_internet_archives SYSTEM "achre_internet_
archives.ent">
<!ENTITY usgpo_printing_location SYSTEM "usgpo_printing_
location.ent">
<!ENTITY achre_creation_of_advisory_committee SYSTEM "achre_
creation_of_advisory_committee.ent">
<!ENTITY achre_presidents_charge  SYSTEM
"achre_presidents_charge.ent">
<!ENTITY achre_committees_approach  SYSTEM "achre_committees_
approach.ent">
<!ENTITY achre_historical_context SYSTEM "achre_historical_
context.ent">
<!ENTITY achre_key_findings_human_radiation_experiments SYSTEM
"achre_key_findings_human_radiation_experiments.ent">
<!ENTITY achre_key_findings_intentional_releases  SYSTEM
"achre_key_findings_intentional_releases.ent">
<!ENTITY achre_key_findings_uranium_miners SYSTEM "achre_key_
findings_uranium_miners.ent">
<!ENTITY achre_key_findings_secrecy_and_the_public_trust  SYSTEM
"achre_key_findings_secrecy_and_the_public_trust.ent">
<!ENTITY achre_key_findings_contemporary_human_research SYSTEM
"achre_key_findings_contemporary_human_research.ent">
<!ENTITY achre_key_findings_current_regulations_on_secrecy
SYSTEM "achre_key_findings_current_regulations_on_secrecy.ent">
<!ENTITY achre_key_findings_other_findings SYSTEM "achre_key_
findings_other_findings.ent">
<!ENTITY achre_key_recommendations_apologies_and_compensation
SYSTEM "achre_key_recommendations_apologies_and_compensation.
ent">
<!ENTITY achre_key_recommendations_uranium_miners SYSTEM
"achre_key_recommendations_uranium_miners.ent">
<!ENTITY achre_key_recommendations_improved_protection SYSTEM
"achre_key_recommendations_improved_protection.ent">
```

```
<!ENTITY achre_key_recommendations_secrecy_and_national_
security SYSTEM "achre_key_recommendations_secrecy_and_
national_security.ent">
<!ENTITY achre_key_recommendations_other_recommendations SYSTEM
"achre_key_recommendations_other_recommendations.ent">
<!ENTITY achre_advisory_committees_legacy SYSTEM "achre_
advisory_committees_legacy.ent">
<!ELEMENT module (title?, heading?, line?, para?, line?,
list?)>
<!ELEMENT title (#PCDATA)>
<!ELEMENT heading (#PCDATA)>
<!ELEMENT line (#PCDATA)>
<!ELEMENT para (#PCDATA)>
<!ELEMENT list (para?, line?)>
<!ELEMENT para (#PCDATA)>
<!ELEMENT line (#PCDATA)>
```

Save this file as "achre.dtd". Below is the expanded XML file:

```
<?xml version="1.0" standalone="no"?>
<!DOCTYPE achre_executive_summary_radiation_experiments SYSTEM
"achre.dtd">
<achre_executive_summary_radiation_experiments>
<title>Advisory Committee on Human Radiation Experiments:
Executive Summary</title>
&achre_publication_information;
&usgpo_telephone_orders;
&usgpo_mail_orders;
&achre_internet_archives;
&usgpo_printing_location;
&achre_creation_of_advisory_committee;
&achre_presidents_charge;
&achre_committees_approach;
&achre_historical_context;
<title>Key Findings</title>
&achre_key_findings_human_radiation_experiments;
&achre_key_findings_intentional_releases;
&achre_key_findings_uranium_miners;
&achre_key_findings_secrecy_and_the_public_trust;
&achre_key_findings_contemporary_human_research;
&achre_key_findings_current_regulations_on_secrecy;
&achre_key_findings_other_findings;
<title>Key Recommendations</title>
&achre_key_recommendations_apologies_and_compensation;
&achre_key_recommendations_uranium_miners;
&achre_key_recommendations_improved_protection;
```

```
&achre_key_recommendations_secrecy_and_national_security;
&achre_key_recommendations_other_recommendations;
&achre_advisory_committees_legacy;
</achre_executive_summary_radiation_experiments>
```

Save this file as "achre.xml". If you open this in Microsoft Internet Explorer®, you will see how easy it is to expand by pointing to or referencing different DTDs and entities. Note that this file is not based on files that were extracted from a website. Instead, it is based on files that are on a hard drive. You can download all of these files and put them on your computer by going to our website, www.rhetoricalxml.com.

Parsed and Unparsed Entities

So far, we have been talking about parsed data, which is information that is analyzed by XML parsers to determine if it is valid. Parsed data is usually text, and all the examples we have been using have been examples of text. Unparsed data can also be text, but usually it refers to data like graphics saved as .jpegs and .gifs.

Remember the parsed entity we used in the "person" form earlier in the chapter:

```
<!ENTITY contact "This data compiled and copyrighted by State
Census Bureau.  To contact, phone at (555) 555-5555 or e-mail
at sctstats@sct.gov.">
```

We start with "<!ENTITY" and end with a ">" and we first type in an abbreviation such as "contact" that will be referenced later in the XML coded information. Then we follow with the actual text that is the information in the entity. If the entity information is not included in the DTD of the standalone document, we would use SYSTEM to indicate this.

In unparsed entities, we use a similar method, but instead we have to add "NDATA" and follow this with some notation that describes the information we are going to be using.

For example, if we wanted to include the seal of the ACHRE, we would use the following entity in our DTD:

```
<!ENTITY achre_seal SYSTEM  achre_seal.gif NDATA GIF>
```

If you go to our website, www.rhetoricalxml.com, you will find this graphic entitled "achre_seal.gif". When we place this .gif graphic in the body of XML data, we cannot use the code that we typically use for entities, which would be "&achre_seal;". Instead we would use, in this example, the following code:

```
<GRAPHIC SOURCE=achre_seal_gif/>.
```

We will also have to add information in the DTD to tell the browser just what kind of tool it will need to use to allow the graphic to show up. This is because this information is unparsed or not presented as regular XML code. To do this, we have to provide another line of code in the DTD that indicates the notation for the graphic.

```
<!NOTATION gif SYSTEM "image/gif">
```

Note that we start with a <!NOTATION tag here, then describe the kind of extension that will be needed to allow the graphic to show up. In this case, as in the <!ENTITY declaration above, we will need to use "gif." Then we put SYSTEM in as this entity was not described in its entirety in the DTD. Following this, we have to describe the content information, which in this case is "image/gif."

Next, we have to place an empty element in the DTD code to mark where in the structure of the document our graphic will show up. Remember that empty elements contain no information; they just serve as markers. Below is the empty element we will use:

```
<!ELEMENT graphic EMPTY/>
```

It is important to put a forward slash at the end of an empty element so the software does not look for a closing tag.

To finish our unparsed entity coding in this example, we need declare an attribute for this graphic in our DTD:

```
<!ATTLIST graphic source ENTITY #REQUIRED>
```

Putting it all together as a single document, here is what we now have:

```
<?xml version="1.0" standalone="no"?>
<!DOCTYPE achre_executive_summary_radiation_experiments  [
<!ELEMENT achre_executive_summary_radiation_experiments
(title?, module+)>
<!ELEMENT title (#PCDATA)>
<!ENTITY achre_publication_information SYSTEM "achre_
publication_information.ent">
<!ENTITY usgpo_telephone_orders SYSTEM "usgpo_telephone_
orders.ent">
<!ENTITY usgpo_mail_orders SYSTEM "usgpo_mail_orders.ent">
<!ENTITY achre_internet_archives SYSTEM "achre_internet_
archives.ent">
<!ENTITY usgpo_printing_location SYSTEM "usgpo_printing_
location.ent">
```

```
<!ENTITY achre_creation_of_advisory_committee SYSTEM "achre_
creation_of_advisory_committee.ent">
<!ENTITY achre_presidents_charge  SYSTEM "achre_presidents_
charge.ent">
<!ENTITY achre_committees_approach  SYSTEM "achre_committees_
approach.ent">
<!ENTITY achre_historical_context SYSTEM "achre_historical_
context.ent">
<!ELEMENT module (title?, heading?, line?, para?, line?,
list?)>
<!ELEMENT title (#PCDATA)>
<!ELEMENT heading (#PCDATA)>
<!ELEMENT line (#PCDATA)>
<!ELEMENT para (#PCDATA)>
<!ELEMENT list (para?, line?)>
<!ELEMENT para (#PCDATA)>
<!ELEMENT line (#PCDATA)>
<!ENTITY achre_seal SYSTEM "achre_seal.gif" NDATA GIF>
<!NOTATION gif SYSTEM "image/gif">
<!ELEMENT GRAPHIC EMPTY>
<!ATTLIST GRAPHIC SOURCE ENTITY #REQUIRED>
]>
<achre_executive_summary_radiation_experiments>
<GRAPHIC SOURCE="achre_seal"/>
<title>Advisory Committee on Human Radiation Experiments:
Executive Summary</title>
&achre_publication_information;
&usgpo_telephone_orders;
&usgpo_mail_orders;
&achre_internet_archives;
&usgpo_printing_location;
&achre_creation_of_advisory_committee;
&achre_presidents_charge;
&achre_committees_approach;
&achre_historical_context;
</achre_executive_summary_radiation_experiments>
```

Save this file as "achre_seal.xml". The graphic will show up if you have coded your style sheet properly and whether or not the browser you are using is set to display unparsed entities. We will discuss style sheets in greater detail in Chapter 4. In Figure 2.10, you can see that the graphic is placed at the top of the document, above the title.

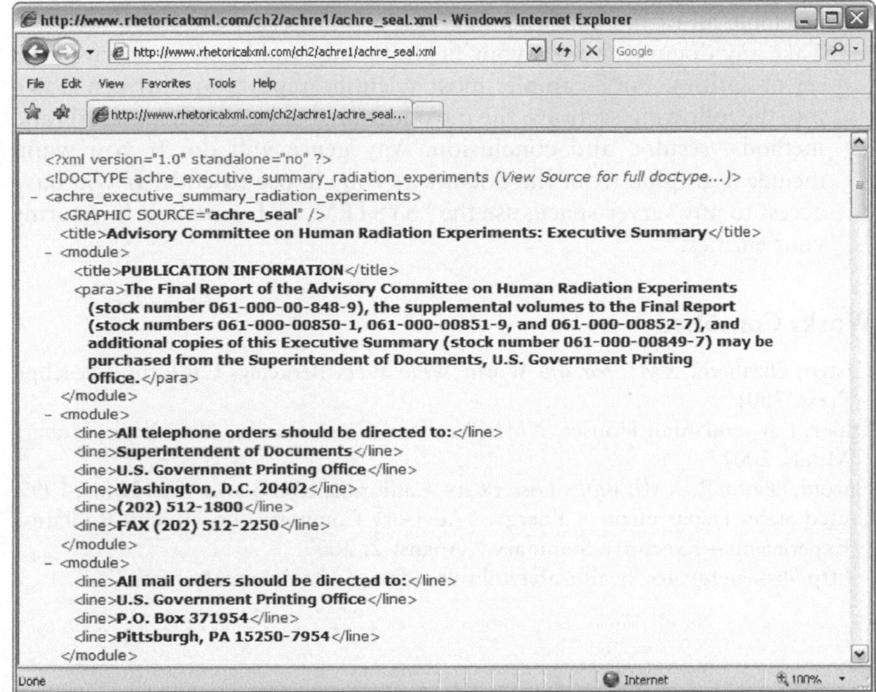

http://www.rhetoricalxml.com/ch2/achre1/achre_seal.xml - Windows Internet Explorer
http://www.rhetoricalxml.com/ch2/achre1/achre_seal.xml
Google
File Edit View Favorites Tools Help
http://www.rhetoricalxml.com/ch2/achre1/achre_seal...
<?xml version="1.0" standalone="no" ?>
<!DOCTYPE achre_executive_summary_radiation_experiments (View Source for full doctype...)>
- <achre_executive_summary_radiation_experiments>
<GRAPHIC SOURCE="achre_seal" />
<title>Advisory Committee on Human Radiation Experiments: Executive Summary</title>
- <module>
<title>PUBLICATION INFORMATION</title>
<para>The Final Report of the Advisory Committee on Human Radiation Experiments (stock number 061-000-00-848-9), the supplemental volumes to the Final Report (stock numbers 061-000-00850-1, 061-000-00851-9, and 061-000-00852-7), and additional copies of this Executive Summary (stock number 061-000-00849-7) may be purchased from the Superintendent of Documents, U.S. Government Printing Office.</para>
</module>
- <module>
<line>All telephone orders should be directed to:</line>
<line>Superintendent of Documents</line>
<line>U.S. Government Printing Office</line>
<line>Washington, D.C. 20402</line>
<line>(202) 512-1800</line>
<line>FAX (202) 512-2250</line>
</module>
- <module>
<line>All mail orders should be directed to:</line>
<line>U.S. Government Printing Office</line>
<line>P.O. Box 371954</line>
<line>Pittsburgh, PA 15250-7954</line>
</module>
Done
Internet
100%

Figure 2.10 Screenshot of "achre_seal.xml" in Microsoft Internet Explorer®

Activities

1. As an alternative to the extended "person" examples we have gone through in this chapter, design your own set of element tags for college students at the university you attend. For example, one element might serve to define a student's major. What might be the specifics of a database that would serve both the students and a university's administration professionals? Make sure that the DTD that defines your tagging system is sophisticated enough so that some tags are embedded within other tags. Also, assign each student an ID attribute. After you have done this, check out your code using Internet Explorer®.
2. Reconstitute the IDREF example of the Raymond family that we have been working with, but make Casey Raymond the "person" in your example.
3. Breakdown the U.S. Declaration of Independence into at least six separate entities—there are plenty of copies of it online. Name the entities as you see fit. For example, you might name the first section the "usdipreamble.ent". Be specific in your naming of each section; let the names you choose for the tags accurately reflect the content. After

designing these entities, construct a DTD for this document, and then assemble all of the entities and check it on Internet Explorer®.

4. Take any document from your field of study and break it down into a set of entities. For example, most scientific reports are broken down into the following sections: the introduction, background, materials and methods, results, and conclusion. Any genre will do. If you want, include a graphic from the document you choose to code. If you have access to any server space, use the "SYSTEM URL" method for storing your entities.

Works Consulted

Castro, Elizabeth. *XML for the World Wide Web*. Berkeley, California: Peachpit Press, 2001.

Ethier, Kay, and Alan Houser. *XML Weekend Crash Course*. New York: Hungry Minds, 2001.

Harold, Elliotte R. *XML Bible*. Foster City, California: IDG Books Worldwide, 1999.

United States Department of Energy. "Advisory Committee on Human Radiation Experiments—Executive Summary." August 2, 2007. http://hss.energy.gov/healthsafety/ohre/roadmap/achre/summary.html

3 Semantics and Classification Systems

Single Sourcing and Methods for Knowledge Managers

Chapter Overview

In the first two chapters, we discussed the fundamentals of knowledge management and XML coding. In this chapter, we will explain how communication professionals can apply theories of semantics to better describe how we can name and arrange XML elements relative to one another. We will accomplish this by illustrating how objects have been identified, named, and classified in some historical and contemporary contexts to reveal the benefits of properly arranging complex bodies of information. Equally important, we also describe the potential hurdles we might encounter as knowledge managers when we engage in classifying practices because complex bodies of information are rhetorical constructs that need to address the needs of different audiences. To further explicate how to contextualize knowledge using XML technologies, we will review a series of concrete organizing methods and provide specific examples drawn from the fields of technical communication, knowledge management, and library science that will allow us to meet the rhetorical needs of our audiences.

The Semantic Web

In *Weaving the Web*, Tim Berners-Lee writes about his fascination with the way words are defined by other words and that information is really defined by other information that it is related to. He takes it a little further when he writes "There is little else to meaning. The structure is everything" (12). What he means by structure is that things derive their meaning by how they are situated in regards to other things. He provides the example of how the cells in our brain are, by themselves, just cells, but because they are connected to other things, or other cells, they form the basis of all that we know.

Often we take for granted just how much we know, especially when it comes to language. For example, early in his career, Berners-Lee tried to design a computer program that would understand language the way all of the connected cells in our brains allow us to understand language. He "asked" the computer the question "How much wood would a woodchuck chuck if a woodchuck could chuck wood?" The computer "replied" with "How much wood would a woodchuck chuck if a wood chuck chuck wood wood chuck chuck chuck . . ." (Berners-Lee 13). Of course, we know as humans that this is just a clever tongue twister with no real answer, but the computer was not sophisticated enough to see it as such as it really "tried" to provide an answer that really looked, in part, like the question it was asked. We can see that "would" and "wood" sound the same, but we can distinguish their meaning by the syntax, or order of the words in the question, and the spelling of each word. When we hear someone say this, we know that the word "chuck" can stand alone as a verb, but it can also be part of a word, "woodchuck," and mean something entirely different. We can discern the semantic differentiation of these words by the way they are connected to each other in this sentence.

Berners-Lee's interest in connections led him to develop the early technical elements that served as the basis for the WWW. When he was working at CERN, a particle physics laboratory in Switzerland, he developed a program he called ENQUIRE, which allowed physicists who were working on different projects to share information with one another. Later, he worked with others to establish the standards that would allow computers to connect to one another across the world and form the basis for the Internet. What Berners-Lee now advocates is that computers and programs that run on them become more sophisticated so they create more than just an "information space" where people can send e-mail and look at websites. The next step for the Internet is to create a "Semantic Web" where databases in computers connect better to other databases. This vision has been partially realized as we have seen ever more sophisticated search engines evolve such as the ones that support Google. But we could have an even more robust Semantic Web if we included specific metadata that better allowed for more precise and far reaching searches into the databases of organizations that were willing to make their information accessible.

Usually, databases are relational, which means that they have bodies of information set in different columns that are tied together. For example, the information in the columns could be rainfall measurements, barometric pressure, and temperatures on certain dates in certain cities or geographic regions. This makes sense to us that these columns would be tied together as rainfall varies by factors such as geography, date, and barometric pressure (Berners-Lee 180). They are called relational databases as the information is semantically tied to other information, and this is what provides their meaning. For this to happen, we need information that is coded in such a way where we can readily access this information. HTML can direct one to a basic website with lots of information, but it is not very good at directing us to specific parts of large websites, or, even more promising, databases that usually warehouse much more information. Berners-Lee sees XML as the universal coding language that allows us to search for XML-tagged information based on the semantics used to tag the information, tags such as <barometric_pressure> (Berners-Lee 181). HTML has been successful because anyone can use it and no one owns it. This can be also said for XML. If we can all agree on one language for one purpose, whether it is HTML for layout or XML for storing data, the greater incentive will exist for companies to build software technologies that can recognize and decode this language. Thus more people will use a specific language to build their own websites or databases, and more of us will have access to the information they encode in these sources.

This leads us back to the idea of words being defined by other words and words or specific designations being used to define information. Seeking data from a number of databases would be more successful if the Semantic Web was facile enough to do things like equate "mean-diurnal-temperature" with "daily-average-temperature" (Berners-Lee 186). A meteorologist knows that these two phrases indicate the same thing and might be interested in acquiring such data. Over time, the Semantic Web could "learn" to understand that these two phrases mean the same thing through "inference languages." XML namespaces can also be used to allow people to combine data that might be tagged or named differently (Berners-Lee 188). We talked about namespaces briefly in Chapter 2, and we will return to that topic in more detail in Chapter 5.

While XML-related technologies will eventually enable us to realize Berners-Lee's concept of the Semantic Web, we still have to remember that what information is made available in databases and the relationship between pieces of information will serve as the basis for this phenomenon. It is not just about the technologies; it is about how humans make rhetorical choices about naming and arranging the things they named between each other. In Chapter 2, we described how we can use XML tags to name pieces of data and how these pieces of data were structurally related to each other in DTDs. In the sections that follow, we will illustrate how the practice of identifying, classifying, and arranging objects and concepts relative to each other in other forms of information storage and representation is by its very nature a rhetorical endeavor.

Classification Systems

A conventional, Aristotelian classification scheme sets up binary, either-or relationships between objects; if something does not belong in one category, we can put it in an opposing category as it is not characteristic of the first category (Bowker and Star 62). Things are often defined by negation—they present features or qualities when examined that other things do not and this determines how they are different from other objects.

Cicero, who followed Aristotle and benefited from his work, described this method of classification in *Topics*: "when you have taken the feature which the thing you want to define shares with other things, continue along this a route until a peculiar property is established which cannot be applied to any other thing" (Topica 29). To define a word using Cicero's method, we designate the class or genus of things that the word can be part of, and then determine how it differs from all the members of the class that it has been grouped in (Crowley and Hawhee 216). The differences, one discovers, allow for a more specific definition of the word; it designates the term in its most particular form or species.

If we were going to use this method to define what a hybrid car is, we would first group it as a general member of the car class. Hybrids have similarities with all other cars such as they have four wheels, are built for transporting humans, and have steering wheels. Like some other cars, hybrids get very good gas mileage, but even here, there are a few hybrids that cannot claim this as a distinguishing feature. However, hybrids clearly differ from standard internal combustion powered cars in that they are powered, in part, from an electric engine that works in concert with the gas engine. Some hybrids can be distinguished from other hybrids because they employ regenerative brakes that capture and then transfer the kinetic energy that is normally lost when the driver puts on the brakes back to the battery.

However, the methods by which we distinguish differences between things are often a bit fuzzier than the binary method. In prototype theory, we already have a broad understanding of what something is and we do not have to whittle away at it to discover the similarities and differences that the thing possesses relative to other things as we do in the hybrid example above. Cognitive scientists and psychologists interested in computer science and, in particular, artificial intelligence, have long acknowledged the capacity of the brain to function in this fashion. They have created their own term—schema—to apply to this type of template-based model of knowledge storage (Schank and Abelson). Prototype theory follows this same idea. By sitting in a chair, we know what it is without asking specific questions to determine how it differs from everything else in the universe. In effect, we have a prototypical sense for what a chair is, and this sense allows us to identify a variety of different looking objects as chairs. The decisions we make when we design classification systems are often based on prototypes. Michel Foucault makes light of this practice by quoting a passage from a short story by Jorge Luis Borges where a "certain Chinese encyclopedia" has a classification scheme where:

> animals are divided into: (a) belonging to the Emperor, (b) embalmed, (c) tame, (d) sucking pigs, (e) sirens, (f) fabulous, (g) stray dogs, (h) included in the present classification, (i) frenzied, (j) innumerable, (k) drawn with a very fine camelhair brush, (1) *et cetera,* (m) having just broken the water pitcher, (n) that from a long way off look like flies. (xv)

Borges often wrote short stories that describe the way we become trapped in our own fictions or arrangements of the things in the world. While most prototypical systems are based on thinking that is not as whimsical as the example from Borges, this passage does stretch our imagination a bit and demonstrates different patterns of subjectivity. Perhaps there are or have been people from some cultures who feel it is important to distinguish between animals that belonged to the leader or "Emperor" of a country from other animals. It is easy to see that animals that are "tame" or "frenzied" might be of special importance to humans as they are ones we could or could not domesticate, but this classification is based on human needs, not the intrinsic nature of the animal. The phrase "included in the present classification" at first sounds appropriately objective, but it perhaps suggests how we can get caught up in absurd bureaucratese. Is this phrase really suggesting any possible division that is useful?

Classification systems affect the way we view the world. When we review a classification system that neatly arranges objects, a pattern emerges and this gives us the feeling that nature is stable and ordered and manageable. To be considered valid, classification systems must appear objective and Aristotelian.

In what Foucault calls the Classical Age (the seventeenth and eighteenth centuries, which most historians call the Age of Reason), he tells us that scholars believed that language was transparent, that language could describe the differences and similarities among things so people could perfectly arrange and thus understand the order of things.

Language neatly parses or arranges the universe in a way that gives us the feeling of control over it. Foucault thus sees language as a repressive institution that keeps us from understanding all the things that it cannot adequately convey given its present net of words and the syntactical structures that arrange these words into identifiable constructs of meaning. In Foucault's "archaeology," he attempts to establish a theoretical machinery that exposes this phenomenon "primarily [through] the play of analogies and differences" (160), or, as we described above, the manner in which we think of naming and classifying things using binary thinking.

All structures of thought—myth, religion, science, philosophy—rely on the ability of language to name and order things. These structures of thought are based on what Foucault calls an "episteme," the prevailing foundation of language or discourse that propels the "thinking" of an age and extends this foundation into other systems of order. For example, the history of living forms in the natural history of the Age of Reason was based on the visual ordering in tables used to show the relationship between different species in the study of natural history. This was based on the belief that language was the temporal ordering of thought. Thought itself was not believed to be a temporal entity; thought was organic, mushed together, something that existed behind the eyes of a subject as a distinct, instantaneous, happening-all-at-once entity. Language was believed to be the analysis of thought, the placing of words into an order that produced sentences that were eventually placed into paragraphs, essays, and books that explained the ideas of the writer. This "profound establishment of order in space" (Foucault 83) emphasized the visual "order of things," thus instituting in the field of natural history the practice of similarly ordering plants and animals along the continuum of the table. Here, the morphological structures that could be *seen* on these plants and animals—the shape of the neck, the relative size of the leaf compared to the stem that attaches it to the stalk—were examined and used to show how species were similar or different from one another.

On a visual and systematic level, there were strains in the Classical episteme that could be seen as arising out of the same concerns brought out above at the linguistic level. As the practice of nomination (the naming of objects) became more complex, it became more difficult to ascertain the specific characteristics that were needed to place an organism on a table in a contiguous relation to other organisms. The character of animals became more amorphous and incomplete when it became obvious that for every physical similarity that described the relationships between samples that were to be studied and classified, there were many more differences that became "visible" because of the attention to visual analysis. Animals with similar morphological structures might be able to fit neatly into a natural historian's scheme of things, but there were glaring differences that could be seen when two animals were placed side by side on a table—a table that approximates the periodic tables chemists use today—that read left to right and top to bottom like the pages of the books that tried to capture the reasoning of scholars. Natural history could not be described as neatly as the syntactical arrangement of words in sentences the grammarians of the Classical Age were trying to prescribe.

In conventional terms, an ideal classification is a "spatial, temporal, or spatio-temporal segmentation of the world" that arranges information according to the following criteria (Bowker and Star 10, 11):

1. The rules of classification are consistent. For consistency, there must be a clear ordering rule or set of rules that can be used to place every constituent part in the system in its place. Using the criteria of origin and descent allows for a consistent ordering of a genealogical record. All one needs is to determine when someone was born and who her parents were. Sorting our e-mail correspondences can be determined by the time we received them, thus creating a consistent classification scheme.
2. There is no overlap between separate categories. All entities "neatly and uniquely fit" into one place in the classification scheme; they are "mutually exclusive." For example, there can be only one first-born son from one set of parents, and, with this distinction, one place in a classification chart that this first-born son alone will fit in. We can also see in the periodic table of elements that there is only one element that can fit in one place that is designated for the one element with only two valence shells that contain two electrons in the first valence shell and six electrons in the second valence shell. This element is oxygen.
3. A system of classification is designed to cover everything. Each classification scheme completely encompasses the world it purports to describe. In the periodic table of elements, all known elements have a

position or place, and all elements that will be discovered in the future or have been described theoretically and are slated to be produced in a laboratory have a place set aside for them. This would be true of a system used to describe how all of the botanical systems that exist; all non-animal living species need to have a "place" where can be seen on a chart relative to all other botanical species.

While in theory it seems logical that these three criteria should be adhered to when we design classification systems, in practice it is not always easy to do this. In the next section, we will demonstrate some of the difficulties in classifying objects in the real world.

The ICD

The International Classification of Diseases (ICD) has been employed by the world's public health officials to chart the nature, frequency, and geographic origins of diseases and causes of death in human populations since the late nineteenth century. The ICD has been modified regularly since the 1890s, and the latest version, ICD-10, came out in 1990. A detailed study of these changes by Geoffrey Bowker and Susan Star, in concert with the work of others on the practices employed in information mapping, can be used to better understand the technology of storage and retrieval systems in computer technologies: "the history of the ICD attaches directly to the development of information processing technology this century" (126).

Their study of the ICD also illustrates how classification systems arise and change to fit multiple social worlds and the manner in which diseases are characterized. In many cases, this has never been easy. For example, some medical professionals see cancer as a disease that is localized in one part of the body that is sufficiently dealt with by cutting it out, while others see it as a cause of a patient's entire immune system being weakened. If the "seat" of the disease—where it presents or is found on the body—is the determining factor for where we place it in a classification scheme, we would have two different ideas about how this disease is to be classified.

Additionally, Bowker and Star demonstrate how classification systems can become so entrenched in our thinking that they become "invisible," thus undermining our ability to adapt them as future needs or insights arise. When physicians fill out death certificates that become the sources of the ICD data, the classification system steers them towards associations between diseases and their causes, but keeps them from imagining other possible causes of diseases. For example, if there are only three causes listed for a particular disease, this implicitly suggests that there could be no other causes than the three listed. This phenomenon is analogous to the patterns of thinking scientists fall into when they practice what Kuhn refers to as "normal science."

Even if classification systems are invisible and keep us from thinking outside of them, there is a real need for complete and well-organized classification systems. This is evident in the case of the ICD and data collection from death certificates as it allows public health officials to better invest resources to healing the afflicted. However, just listing every possible cause of death on a form is not enough. If every known disease is listed as a potential cause of mortality, the system would be unwieldy and would increase in "randomness" as physicians would not take the time to review and properly fill out the forms or really understand the nature of every disease category. Conversely, if we include too few categories in a classification scheme, the information is not useful. If doctors were only allowed to indicate that death was from "natural causes" or "accident," we would not be able to track disease trends that affect mortality. We need some middle ground where we can have enough identifiable diseases in the ICD data banks to yield specific enough information that would allow for better policies and practices by medical organizations and physicians (Bowker and Star 159).

Even if we have a reasonable amount of known diseases listed on the death certificate, allowing the many hard working physicians of the world to document the diseases they see, there can be some problems with classification schemes. These classification systems need to be open and dynamic so they can accommodate new diseases or record more about existing diseases if they become more accurately diagnosed and characterized. When new ways of thinking come along, we can either fit them into the "other" category that we so often find on forms, which is not really helpful, or we can alter our meaning systems to better include these ideas or objects in our classifications (Lemke 177).

One way of structuring the ICD to make it more dynamic and useful would be to include a category for the material surroundings of the patient who might acquire a certain disease—e.g., poor housing or economic circumstances as contributing elements of tuberculosis. This would validate the epidemiological value of this association by suggesting that there might be a connection or connections between poor ventilation systems, or older and potentially toxic building materials, and tuberculosis. However, even this potential opening up of the classification scheme might implicitly exclude, as traditional understandings of classification systems go, the potential connection between newer building materials and tuberculosis if these variables are not related in the meaning system of the ICD. If there is no way to connect rates of tuberculosis and recently built living quarters, the physicians who fill out the forms or researchers who analyze the information might not make a connection between these elements. This would make it harder for professionals to imagine and/or win support for funding that would support a study of these elements (Bowker and Star 82). Likewise, if classification schemes did not allow for the connection of sudden infant death syndrome (SIDS) to both social and environmental causes, researchers would not be able to use the data to construct research studies (Bowker and Star 69). Classification schemes that have this effect are invisible systems that do not draw attention to their subjective nature because their description of reality is more likely to become uncritically accepted as "true," but it might exclude ideas that are also of value.

There are many socially constructed variations of disease classification schemes and they can be seen as the product of the needs and compromises of a variety of stakeholders, those who have a personal or professional investment in the way the ICD is designed:

- Statisticians who are concerned with epidemiological trends are not concerned with isolated occurrences of diseases, while public health officials concerned with the potential outbreak of the bubonic plague or Ebola would want to be aware of any one documented case (Bowker and Star 70).
- For practicing physicians, carefully filling out a cause of death form cuts into their time that they can be spending with other patients, but statisticians feel that careful data collection eventually yields important results (Bowker and Star 144).
- Physicians are more likely to spend more time filling out a form of a younger person who has passed away than an older one (Bowker and Star 145).
- Medical specialists want to see an emphasis on different strains of diseases, whereas public health statisticians want to see "broader, action-oriented" categories on nutritional deficiencies and environmental factors that would allow them to propose new policies (Bowker and Star 145).
- When the ICD was first established, statisticians wanted it to only contain 200 categories that would not change. Additionally, the diseases were to be classified by their seat—or where in the body they are found—as opposed to what caused the diseases as the later distinction is not always easy to determine. This simplicity would insure the stability of the table's categories so the occurrences of specific diseases could be compared and studied across the decades (Bowker and Star 145–6).
- Spanish authorities wanted diseases to be classified in a fashion that would better allow public officials to react to them. The categories they distinguished were, 1) general and sporadic, 2) epidemic, 3) imported, 4) common to people and animals, and 5) professional intoxications (Bowker and Star 146).
- Some statisticians wanted social-biological factors such as violent death be given prominence on the list so they could more readily distinguish what deaths were caused by external causes (Bowker and Star 146).
- Insurance companies wanted disease data to be correlated with the ages of patients and when their "compulsory" medical insurance began, but this was difficult as there is no standard across countries for this (Bowker and Star 147).
- Late nineteenth-century German chemical companies were the first to document causes of death. One of the categories relevant to them was based on whether or not the patient had worked inside or outside.

Another asked physicians to determine if the patient had handled certain substances (Bowker and Star 147).

- For pharmaceutical companies, the list of diseases is important because it dictates just what their products are good for and therefore, what will sell. Because birth control pills have the side effect of increasing blood pressure, they can be prescribed as a drug that can be used for low blood pressure, or hypotension, with the "side effect" of keeping a patient from getting pregnant. Because hypotension is listed as a disease, physicians can prescribe contraceptives to patients so they can practice birth control. In Catholic countries like Spain where there are religious restrictions against birth control, it is easier to gain the benefit of this side effect (Bowker and Star 147).

In contrast to the examples above where one group of stakeholders insist on one way of classifying or naming something for their epidemiological or political needs, the creation of boundary objects, objects classified under different categorical headings, enables different communities of stakeholders to use the same piece of data for their own specific purposes (Star "Structure" 50). For example, hospital attendants might be asked to record certain symptoms that are commonly associated with epilepsy to aid the physicians so they can adjust a patient's medication. This data might also be compiled in a larger database that could offer insights to researchers who are working to construct general theories about brain function. The same data recorded by the attendants would have different meanings to different communities, and as boundary objects, would still be of value to both groups.

Similarly, the "dagger and asterisk" system of footnoting has been introduced in recent versions of the ICD to cross-reference particular diseases so it can be noted that they reside in more than one category. This allows for more dynamic histories of a disease, one that would allow different communities to more readily understand the disease in terms of their own experience with it (Bowker and Star 73). For example, tuberculosis does not have a single cause, nor does it originally show up in the same part of the body. To properly reflect its ever changing or protean nature, it needs to exist in more than one category (Bowker and Star 73). A dagger or asterisk by the name of the disease as it is listed can show up at the bottom of the page before a footnote to indicate that this disease is also listed on another page in the ICD.

Designation or naming practices also contribute to the spin inherent in invisible systems. All qualifiers such as "presumably" and "possibly" were once to be removed from disease descriptions in death certificates even if they accurately reflected the equivocal or uncertain nature of a physician's diagnosis (Bowker and Star 81). Additionally, in early versions of the ICD, the geographical site of a disease's first documentation became part of its name. In later versions, this was often replaced by a generic place and then a cause; we would go from "Baghdad boil" to "urban cutaneous" to a general agent that might bring about the disease. Similarly, gone is the practice of using the name of the researcher who first identifed, or patient(s) who first suffered with, the disease. In general, we have moved from "places visited," "heroic sufferers," and "great doctors," to abstractions (Bowker and Star 80).

When Gay-Related Immune Disorder (GRID) was later designated as Acquired Immune Deficiency Syndrome (AIDS), the potential for different cultural associations of the disease was made possible. No longer does the name of this condition specifically call to mind one group of people and their lifestyle, and researchers and physicians will not always be asked to associate the causes of this disease by just saying or thinking about the name of the disease. The acronym AIDS also stands for a more informative description of the disease; the effect of the disease is that it attacks the immune system, and people with this disease are not born with or are genetically predisposed to it. They "acquire" it.

In later recent versions of the ICD, the context of the disease is also deemed to be important. In the ICD-10, a series of criteria that described the patient's housing situation was added to better reflect the social condition of the afflicted and the disease. Criteria such as the quality of heating and ventilation, whether or not the person lived alone, was impoverished, or was homeless added a sociological dimension. Also, the kind of work performed in the medical laboratory in the diagnosis of a disease also became part of the classification scheme. For example, there were separate categories for tuberculosis confirmed "histologically," "sputum microscopy," and "culture only." In the ICD-9, if a patient mentions that she has epilepsy, the classification designation is "with intractable epilepsy." However, if a patient states that she has migraines, the designation is "with intractable migraines, so stated." The "so stated" suggests that for migraines, patients are viewed with some "suspicion" and cannot always be counted on for completely accurate diagnoses. In these last two context-based classifications, the patients are removed from the "ownership of their conditions"(Bowker and Star 84).

Legal distinctions are also factors in determining certain classifications. The general definitions of blindness in the ICD-9 range from "profound impairment" to "unqualified visual loss" to "legal blindness, as defined in the United States." In the United States, legal blindness is described as "severe impairment," whereas for the WHO, which determines international definitions of blindness, "profound impairment" serves as the standard (Bowker and Star 84).

Bowker and Star refer to these sorting and classifying phenomena as the principle of convergence, the "double process by which information artifacts and social worlds are fitted to each other and come together," or converge (82). Information artifacts constitute the social world of those who use them to exchange ideas as they are the bricks and mortar of discourse, the language they use to communicate in their field, and through formal and informal discussion of their work, professionals also create a language that augments these artifacts as they fulfill their need to make sense of what they are interested in. This concept of convergence can be likened to Thomas Kuhn's theories about how the language of science comes into being and shapes the discourse scientists use to explain their ideas and understand nature.

Below are three overarching ideas that Bowker and Star ask us to be mindful of when designing classification schemes (324–5):

1. We need to understand that classifying is a practice where we have more than one audience. To be effective we need to allow for some "ambiguity" in our designations. For example, we will have some boundary objects, objects that are understood to have different meaning to various constituencies. To assume that every element in a classification must reside in only one specific space and cannot show up in several spaces would undermine this practice.
2. It is important to be aware of "residual" categories, categories that are often referred to in classification schemes as "other." Residual categories have much to do with the categories that do show up in a classification scheme as they are like the "silences in a symphony" that makes a pattern of the music (Bowker and Star 325). In the example above regarding the Nursing Interventions of Classification (NIC), there are many practices that nursing professionals engage in that are not represented. Perhaps explaining the change in condition of a patient to a visiting family member when the primary physician is not around could be considered an example of this kind of work. It is not necessarily an intervention that a nurse can get credit for in her daily routine, and it is the kind of task that professionals perform in their day that fit into this residual category —an important, yet undesignated task that falls between the cracks.
3. We need to remember who initially contributed to the system and what might have been some of the initial uncertainties, political tensions, and tradeoffs that went in to them; usually, classification schemes are difficult to design, and there are competing interests and compromises. When these systems have an "invisible infrastructure" or are "black boxed," we are not cognizant of the process by which they came into being and we are less likely to question them. For example, when the NIC was first devised, it failed to designate nursing tasks that were thought to be invisible. Some work "just gets done," but it is not really noticed. Work such as "inserting oral or nasal airways, as appropriate" (Bowker and Star 234) might be considered an essential procedure on a patient during the course of a nurse's shift, something that can be checked off on a chart and be visible. Being "available to listen to a patient's feelings" (Bowker and Star 235) might be considered as something that should not be checked off on a chart; it is an invisible practice, but an important practice. Knowing this history and that there were varying opinions and debate about what should be in the NIC allows us to understand that it was designed by humans; that while useful, it does

> not capture everything that a nurse does that is of value; and that it can be challenged and redesigned. When classification systems are not understood in this way and are seen as "invisible infrastructures," those who work with or are affected by them feel left out of any participation in policy making (Bowker and Star 325).

The way we organize and work with each other is closely tied to the technical aspects of storage and retrieval. The physical repositories of information and data entry artifacts of infrastructures—the hardware, software, GUIs, ledgers, and forms—channel the values and philosophies of organizations. For the ICD, they provide the "large-scale organizational memory" that supported decisions regarding the control of diseases and the allocation of resources (Bowker and Star 36).

The determining dimensions of information infrastructures that shape and are shaped by classification systems are based on the relationships of language, technology, and work culture (Star and Ruhleder 4). Infrastructures "occur" when they can be readily employed by those for their own purposes—they are more than just substrates that are set out for use. As we have become accustomed to using telephones, faxes, and e-mail, our language as it relates to time and distance has changed as have our relationships with each other. Formal infrastructures often give way to local knowledge and practices that existed before they were imposed.

The ICD has employed such local knowledge and practices via the workarounds employed by health care professionals to meet the needs of changing populations. In earlier forms of the ICD, a single cause of death was preferred on death certificates as it made it easier to compile statistically. As longevity increased in some populations, the underlying cause of death was more likely to consist of multiple disease processes, and therefore attending physicians began listing several of these processes in the category "Disease or condition directly leading to death" on death certificates. Because of this workaround practice, an "antecedent causes" category was added to the form, thus allowing for the designation of "Morbid conditions, if any, giving rise to the above cause" (Bowker and Star 73). Renal failure could be listed as the disease relating directly to death and diabetes could be the antecedent cause, thus allowing for a more accurate representation.

For an infrastructure to work, it must foster a sense of a democratic community of producers and consumers of information. For this to happen, the minimum technical threshold for its use must be acquired and within the abilities of the audiences who will be asked to use them (Star and Ruhleder 125).

A classification strategy tells a story that identifies its architect(s). Likewise the genres of organization communication indicate the organizing practices of the individuals who use them. The purpose and rhetorical situation that define a type of discourse (a memo, a list, a letter of recommendation) serve as the basis for the rules that define these genres. These rules both constrain the classifying nature of these documents, and are also altered by the social practices of the individuals who use them (Yates and Orlikowski "Genres" 306). Additionally, the genre repertoire—the array and relative proportion of various discursive forms utilized by the community—tells us much about the social structure of a community, the way it defines itself relative to outsiders, the way in which the community evolves, and how the genres help to stabilize a community's practices (Yates and Orlikowski "Repertoire" 546).

In the DOE's "Advisory Committee on Human Radiation Experiments—Executive Summary" that we coded in XML in Chapter 2 and will be discussing in the coming sections, we can imagine how the general arrangement of sections in the executive summary genre of organizational communication could have influenced the writers. Executive summaries should be clear and concise to meet the needs of a busy executive or citizen who desires to understand the substance and scope of a much longer report. That there is a "Publication Information Section" in this executive summary reinforces in rhetorical fashion the idea that the public has a right to this information and they can acquire it if its members so choose. This can be thought of as a standard and stabilizing element in this genre. However, executive summaries of larger reports might not necessarily have a substantial "Creation of the Advisory Committee Section" in them, but for this report, it was deemed necessary by the writers because it rhetorically underscores the idea that private citizens with wide ranging expertise were given the authority to go to great lengths to honestly assess and possibly expose a violation of trust between a government and its citizens in decades past. To counter this violation of trust, this section indicates that great care was taken in fairly establishing and composing the committee. The specific names of each XML entity in the summary could also reflect the rhetorical choices made by its writers. In the next section we discuss single sourcing because it establishes a methodology for breaking down information into discrete units and then reassembling them depending on the informational needs of specific audiences.

Single Sourcing

Single sourcing is the practice of using one document for different outputs. For example, a legal disclaimer that warns that a product is only to be used within the accordance of the directions that are included can be stored in a file and used in all of a company's documentation for all of its products. A description of general safety precautions that one should always be mindful of when using a company's products can be stored as a single sourced file. When software products are upgraded to another version, many of the same operations of the newest version of the software are the same as those in the previous version, such as directions that describe how to open up a file. If this task has been single sourced, it does not have to be rewritten and can be reused in the documentation for the latest version of the software. The XML modules used in the DOE's "Advisory Committee on Human Radiation Experiments—Executive Summary" in Chapter 2 can be thought of as documents that could be reused in future documentation efforts.

Perhaps one of the greatest values of single sourcing is that some writing that has already been done by one writer or group of writers in one part of an organization does not have to be redone by another writer or group of writers in the same organization. There is already some existing documentation that they can draw on for their own needs. All too often, professionals in organizations write in isolation from others, and this is referred to as the "silo trap" (Rockley "Managing" 5). It is as if people are isolated in silos and cannot leverage the hard work of others that work in different departments of an organization.

To enable professionals to be able to find documentation that has already been produced by others, the information has to be written with reuse in mind and stored in a CMS that would allow relatively easy access to this information. Later, in Chapter 6, we will spend more time discussing CMSs and building a simple CMS and parser using XML. For now, it is only necessary to understand that a CMS is a system for transporting modular data from one location to another and for facilitating search and access to the content stored within.

In order for the information in a CMS to be searchable, members of different branches within an organization have to be able to agree on the kinds of information they can use that would benefit everyone. Equally important, the organization's members have to be aware of the semantic differences of the elements of information that might be in play between the different branches.

Single sourcing can mean using the same documents or parts of documents that have already been written and are stored on a technical communicator's hard drive. Sometimes the content is modified for a particular audience or medium such as a manual or a website (Williams 321). The phrase "write once, publish many times" is often invoked when discussing single sourcing.

Today, this phrase still holds, but single sourcing is now supported by the application of XML coding and CMSs. Ann Rockley asserts that technical communicators are going through a paradigm shift because the advent of this practice and technology are asking us to think of information as "objects" that are "referenced" in and then "drawn" from a database to become part of a document (Rockley "Impact" 189). This paradigm shift means that we are getting away from thinking about single sourcing as cutting and pasting from legacy documents that have already been in use before. Instead, we are moving from a document-centered to an object-oriented way of thinking about information (Williams 321). Now, information can be coded in XML and thus packaged in modular elements that reside in databases, like those we discussed in Chapter 2. These elements can be drawn and arranged as needed with the use of DTDs.

According to Rockley, there are four levels of single sourcing that help illustrate this paradigm shift. Level one single sourcing, "Identical Content, Multimedia," describes the practice of using content from one medium, say a printed manual, and then taking the same content and putting it in another medium such as a PDF file or an online Web page. This raises the concern that perhaps the content that was in the original form is not necessarily suited for the other. If a large paper-based document is just made into a PDF file, this does not really work as well as it might in an online environment where information is better broken down into smaller elements and linked. When users in the online environment see a large PDF file, they cannot maneuver well within it as they have to scroll up and down through a large document and often end up just printing out the entire PDF document. If any updates need to be made on content, this means that the updates have to be made both in the print manual and the online PDF file, which makes for more work. Also, in the original print manual, the graphics that are included might be embedded appropriately in the written text, whereas in an online environment, it might be best to have the written text presented with links to the attendant graphics, thus giving the viewer the option of reviewing the graphics or not. If the online presentation is just a PDF document, this does not take advantage of the way links can make access to the graphics an option.

Level two single sourcing, "Static Customized Content," is the practice of taking single sourced content and then modifying it for different platforms or audiences. Improving on the practice described in the example above, this would mean that the traditional paper-based manual would be transferred to a website in a way that best takes advantage of this electronic medium. The original document would be broken down into a number of links to specific parts, and these parts would be linked accordingly, thus giving the viewer the navigational choices on the website that would make it easier to view on the screen.

Some companies manufacture suites of products that use some of the same documentation for similar functions. Microsoft® Office, for example, contains Excel®, PowerPoint®, and Word®, and some of the documentation for some operations in each of these software programs can be reused as they are the same. However, new specific directions on the different software programs would have to be added to ensure that they are properly documented. Software programs are often upgraded, but most of the original descriptions for using their functions remain the same and these can be single sourced when documentation for later versions is updated.

There are multiple information products for products such as training materials, online help, reference cards, and user guides that can use the same core material, but then be modified for each of these mediums. Additionally, single-sourced information can be employed and then tailored for different audiences, ranging from novice to expert.

What is most significant about level two single sourcing is that technical communicators are writing with the idea that they will be asked to write and be able to later extract modular units or objects for different uses. This is in contrast to the level one example where an entire document is transferred in total to another medium with little thought to changing its form to best meet the need of the medium.

"Dynamic Customized Content" describes level three single sourcing. In this scheme, each user has the content customized for her own needs. Users log in and are presented with information that serves their specific profile. Oftentimes this information is generated because each user is asked to fill out a form that describes her needs. User needs are also identified by the use selection patterns they engage in when viewing a site and additional material is sent to them based on relevance. In Chapter 6, we will also build a simple level three single sourcing project to go along with our CMS project. In this content, we ask users to self-select their own individual skill levels based on "beginning," "intermediate," or "advanced" categories.

This practice of single sourcing coupled with customer profiles, metadata, and databases has been most notably implemented by e-commerce; perhaps Amazon.com is the most widely known commercial example of this technology. But Rockley holds that this can also be used in industrial settings. Here "information is drawn from a database, not from static, pre-built files of information" that assumes all patrons are the same (Rockley "Impact" 191).

Level four single sourcing is built on an electronic performance support system (EPSS) that builds on level three technologies by providing users with support material such as usage questions and training manuals that are tailored to their needs "before they know they need it" (191). EPSS technology generates profiles of each individual so that her needs can be anticipated. This technology can be used by online vendors or an organization's intranets.

Rockley contends that the team approach to producing documentation in a single sourcing environment will be the likely outcome of this technology. Because writers will not be relegated to rewriting old documentation, but will instead be asked to produce new single source elements for features of new products or new versions of products, their work will not be "boring." Also, when a single sourced document is changed in a database, it means that this element is changed in different media or "wherever it is used" (192).

Primary and Secondary Modules

Kurt Ament further articulates how we can use single sourcing methods to breakdown information into modular units that can be readily used and reused as needed. While Ament is not writing about the use of XML and CMS systems exclusively, we can see how parts of his method can aid us as we employ these technologies.

Ament distinguishes between two different levels of content: the document level and the element level. At the document level, we have full bodied productions such as print manuals that are linear in arrangement. They consist of many smaller parts arranged in a specific order.

At the element level, he asks us to think of modular building blocks such as lists, paragraphs, or sections of documents that can stand alone and be used as needed in the production of larger documents. While Ament does not use the term "element" like we did in Chapter 2 when we were describing the basic XML encapsulating units, this is not necessarily an inaccurate comparison as we can use XML to tag lists and paragraphs. Though Ament is not talking specifically about XML, his ideas are certainly relevant to this discussion.

The first step in producing a single-sourced array of content is to identify the specific modules, both primary and secondary, that will constitute the larger document. Primary modules are the most common building blocks of larger documents, and secondary modules are usually contained in primary modules (Ament 26). Below we illustrate the basic primary modules and list the secondary modules that Ament identifies:

Definition lists. Definition lists can be broken down into the following kinds of lists:

- Component lists consist of the hardware that go into making up a product such as monitor, CPU, and printer.
- Terminology lists consist of proprietary terms, terms that an organization has exclusive rights over such Network Node Manager, or public domain terms that are in use to describe terms such as File Transfer Protocol (FTP) (Ament 63).

Glossaries. Glossaries are master lists that include all of the component lists and terminology lists for a given document and are usually found at the end of a traditional document or as a separate link in hypertext page. They may include all items from all definition lists that can be found within a larger document.

Procedures. Procedures are descriptions of the ordered steps one takes to perform a task. Ament describes four kinds of procedures (118):

- Single-step procedures involve just one task. It might be something as simple as creating a password for an organization's intranet. They consist of imperative commands that tell someone precisely what to do.
- Multi-step procedures are a set of single-step procedures and should be arranged in an ordered list. Such a procedure might be how to locate the login page of a corporate intranet, create a user name, and then create a password so one can log in successfully.

- Superprocedures are compilations of other procedures. Ament suggests that if any procedure contains more than nine steps, it should be converted into a superprocedure.
- Subprocedures explain steps in superprocedures.

Processes. Processes are similar to procedures in that they describe a method for performing a task, but they are usually narrative in structure. As opposed to procedures that tell the user the specific text that needs to be entered in a software program, they describe in general terms how one might solve a problem. For example, a process could be an extended example of how to use a search engine on an intranet. If a basic search does not work, then perhaps the user needs to reconsider the word choices she has made. Research has shown that the best time to educate a user about the search process is after an initial search has failed or otherwise frustrated the user (Morville and Rosenfeld 182). Perhaps an advanced search option is available in the intranet's architecture and this can be utilized to better educate the user and advance their needs.

Topics. Topics are written to describe the "who, what, when, where, or why" (Ament 26, 140). Topics could serve to answer the following questions. Below are some questions we generated using XML as the basic subject:

- Who should learn XML?
- What is XML?
- When should an organization employ XML technology?
- Where is XML technology best implemented in an organization?
- Why should an organization employ XML?

Topics can answer these questions by employing forms such as "argument, description, exposition, or narration" (Ament 26, 141). Here are some examples we came up with of how these forms might be used using XML technology as the basic subject:

Argument. Persuading executives and co-workers that in the long term, employing XML technology will save the organization money by making it more efficient.

Description. Illustrating how XML elements look. One could present some simple XML coded elements of existing documentation such as product descriptions.

Exposition. Explaining the logic behind the building of a DTD. Exposition could also include the best practices that were posted by the Xerox service personnel who contributed their ideas to the Eureka project that we described in Chapter 1. Tables, itemized lists, or definition lists are often used in exposition.

Narration. Explaining how to perform a task in "sequential or chronological order" (Ament 141). For example, one could explain how to go about gathering information and deciding how many XML elements needed to be included in a DTD in paragraph form or an ordered list.

Troubleshooting Scenarios. Presenting problems that come up and their solutions. The problems can be explained in topic form and the solutions can be presented as the procedure one uses to rectify the problem. Examples could include the many ways an XML document is declared invalid and the solutions could include how one can rewrite the XML code to match the requirements of the DTD. As we described in Chapter 1, problems with copy machine repair and their solutions can also be thought of as troubleshooting scenarios, as can problems that come up when claims adjusters for insurance companies process a claim.

Ament tells us that primary modules usually include secondary modules, and these secondary modules include figures, itemized lists, notes, and tables (27).

Information Product and Element Models

Rockley uses the phrase "information product model" to describe the basic features of an organization's document. These features differ from the rhetorical forms that Ament describes in his primary and secondary modules. For example, an information product model for a press release would contain a subject, date, contact, body, and website address. The basic features of an information product model can be broken down further into "element models." For the press release element model, we would add the following features: a corporate description (short), announcement, product description, features, benefits, quote, corporate description (long), and availability ("Managing" 168).

While Rockley does use the term "element" in element model and does discuss XML elements and how they are tagged, she is not necessarily equating the two concepts. For example, the PCDATA that would make up the XML elements that described the actual month, day, year, and the first and last name of the contact for a press release are referred to as base elements. But the same information for these pieces of information that could be stored in Microsoft Word® and Adobe FrameMaker® files are also referred to as base elements in Rockley's organizing strategy (171). Rockley does not use the term element like we have been using it; we have been using this element only to describe XML elements that follow the rules of XML syntax.

"Semantic information," according to Rockley, is information that describes the "specific meaning" of information such as " website address" or "product description." This use of the word "semantic" is somewhat different than Tim Berners-Lee's use of the word in his Semantic Web; Berners-Lee wants to be sure that certain words or phrases used to describe an object in one computer's database can be equated to the same thing that might have a different word or phrase to describe it, even though both objects are the same thing. In the example we used before, "mean-diurnal-temperature" can be recognized as the same thing as "daily-average-temperature."

For Rockley, "generic information" refers to descriptions of information that do not tell us anything about the content of the information like "semantic information" does. Rather, generic information tells us about the information's basic form. For example, "para," "line," "list," and "title" would be generic tags. We know that some information will be presented in paragraph form when we see "para," but have no idea what the information might be about ("Managing" 170).

In Table 3.1 we organize our information using Rockley's method to illustrate how we can distinguish between the semantic and generic information for the DOE's "Advisory Committee on Human Radiation Experiments—Executive Summary." Additionally, we have added a column that tells us the name of the XML entity that we have coded because it more specifically illustrates how we could store a number of modules in a database and extract or point to these single-sourced elements for our own purposes. The information product model for this document is in all caps form in the left hand column and consists of the following:

- Publication Information
- Creation of the Advisory Committee
- President's Charge
- Committee's Approach
- Historical Context
- Key Findings
- Key Recommendations
- What's Next: The Advisory Committee's Legacy.

The basic features in this information product model could be used to describe other committees that were asked to perform specific tasks by a sitting president or government official.

We can see from Table 3.1 that the element model descriptions such as "Current Regulations on Secrecy in Human Research and Environmental Releases" present a further granularization and specificity of meaning that might distinguish this executive summary from others. In the right hand column we have the corresponding XML entity, "achre_key_findings_current_regulations_on_secrecy.ent", and its content is easy to discern because of the way we named it; it has something to do with ACHRE, it is in the "key findings" section of this committee's executive summary, and it is further distinguished because it contains information on "current regulations on secrecy." We can also see from the generic column in Table 3.1 that this entity contains a heading, list, and paragraph(s).

Table 3.1 Semantic, Generic, and XML Entity Descriptions for the DOE's "Achre—Executive Summary."

Semantic	*Generic*	*XML Entity Name*
ACHRE: Executive Summary		
PUBLICATION INFORMATION	module title para(s)	achre_publication_information.ent
USGPO Telephone Orders	module line(s)	usgpo_telephone_orders.ent
USGPO Mail Orders	module line(s)	usgpo_mail_orders.ent
ACHRE Internet Archives	module para(s)	achre_internet_archives.ent
USGPO Printing Location	module line(s)	usgpo_printing_location.ent
CREATION OF THE ADVISORY COMMITTEE	module title para(s)	achre_creation_of_advisory_committee.ent
PRESIDENT'S CHARGE	module title para(s)	achre_presidents_charge.ent
COMMITTEE'S APPROACH	module title para(s) list lines(s) para(s) list para(s)	achre_committees_approach.ent
HISTORICAL CONTEXT	module title para(s)	achre_historical_context.ent
KEY FINDINGS	title	
Human Radiation Experiments	module title list para(s)	achre_key_findings_human_radiation_experiments.ent
Intentional Releases	module list heading para(s)	achre_key_findings_intentional_releases.ent
Uranium Miners	module heading list para(s)	achre_key_findings_uranium_miners.ent

Table 3.1 continued

Semantic	*Generic*	*XML Entity Name*
Secrecy and the Public Trust	module	achre_key_findings_secrecy_and_ the_public_trust.ent
	heading	
	list	
	para(s)	
Contemporary Human Subjects Research	module	achre_key_findings_contemporary _human_research.ent
	heading	
	list	
	para(s)	
Current Regulations on Secrecy in Human Research and Environmental Releases	module	achre_key_findings_current_ regulations_on_secrecy.ent
	heading	
	list	
	para(s)	
Other Findings	module	achre_key_findings_other_ findings.ent
	heading	
	para(s)	
KEY RECOMMENDATIONS	title	
Apologies and Compensation	module	achre_key_recommendations_ apologies_and_compensation.ent
	heading	
	para(s)	
	list	
	para(s)	
Uranium Miners	module	achre_key_recommendations_ uranium_miners.ent
	heading	
	list	
	para(s)	
Improved Protection for Human Subjects	module	achre_key_recommendations_ improved_protection.ent
	heading	
	list	
	para(s)	
Secrecy: Balancing National Security and the Public Trust	module	achre_key_recommendations_ secrecy_and_national_security.ent
	heading	
	list	
	para(s)	
Other Recommendations	module	achre_key_recommendations_ other_recommendations.ent
	heading	
	para(s)	
WHAT'S NEXT: THE ADVISORY COMMITTEE'S LEGACY	module	achre_advisory_committees_ legacy.ent
	title	
Interagency Working Group Review	heading	
	para(s)	
Continued Public Right To Know	heading	
	para(s)	

Adding the name of the entity that would reside in a database in the same row as the semantic information used to describe it illustrates how we could make available in table form the single source modules that could be used and reused to produce other documents and form other XML documents with their own DTDs for different audiences. We can also accomplish this process without using separate entities for each module. We take this alternate approach in Chapter 6 when we develop our single sourcing parser project.

In the next section, we will discuss how we can employ these single sourcing methods as knowledge managers.

Organizing Knowledge as Knowledge Managers

As we have seen in our discussions in Chapter 1, professionals in organizations now realize that the knowledge of their employees is their greatest asset, but because this intellectual capital is catalogued in the minds of people, it can be challenging to direct and leverage. To better take advantage of this, organizations are instituting knowledge management systems so that this intellectual capital, like traditional forms of capital, is more readily available to all members of the organization. Repositories of knowledge in areas such as product design, technical support, employee skills, customer relations, and problem diagnosis can be analyzed and better utilized to decrease costs and increase revenues. This only occurs, however, if the knowledge management systems can be accessed by everyone.

For example, Arthur Andersen (AA), an international business auditing and consulting firm, employs a Proposal Toolbox, an online repository of proposals its members have submitted to their business clients (Dutta and De Meyer 390). This tool enables their consultants to reutilize parts of these proposals and also allows them to collaborate with the primary writers of each proposal as needed. Additionally, AA stores the general correspondence and intermediate work-in-progress documents in a file system dedicated to each client being served and all Arthur Andersen employees are granted access to this file. This supports the one-firm concept where teams of consultants work "across practice, across offices, and across countries if necessary." This allows them to better leverage their human capital as it reinforces informal networks of people exchanging their insights and skills.

The knowledge that a healthy organization possesses can be divided into four categories (Zack 25–7):

- Declarative knowledge is knowledge *about* things—the empirical data, terminology, and distinctions between ideas and objects important to the business of the firm.
- Procedural knowledge describes *how* things are done in an organization—the bureaucratic, industrial, or legal steps that need to be performed in a prescribed sequence.

- Causal knowledge includes the *why* of things, and might include factors that influence product quality or customer relations. These factors are often best described by stories or narratives of a corporation's employees.
- Relational knowledge is the awareness of how the declarative, procedural, and causal elements of knowledge above relate to one another. Evolving new products or establishing relationships with new clients is often most readily accomplished by a fusing of existing resources and skills that the company possesses.

The last three forms of knowledge in this list—procedural, causal, and relational knowledge—can include elements of tacit knowledge that we discussed in the first chapter. The most complex form is relational knowledge and it is often the knowledge that we acquire by articulating or describing and documenting our activities within and across the organizations we work in.

Knowing the kinds of knowledge is one thing, but to collect, store, and transmit knowledge is another. Additionally, organizations need to show workers how they can share their knowledge, and this can be done by designating professionals as knowledge managers. For example, each local office of Arthur Anderson has its own knowledge manager who oversees the collection and dissemination of archived knowledge (Dutta and De Meyer 392). When a project has been completed, project team members are asked to write reports for the knowledge manager summarizing what they have learned and to identify the best practices that they have come across in their client's organization. This knowledge is then passed on to AA global knowledge managers who filter, distill, and then apply this to firm-wide knowledge bases.

In an attempt to bring together employees in large organizations with specific skill sets, Microsoft® has employed a "knowledge map" (Davenport and Prusak 75–7). This is a complex undertaking because an individual's knowledge is multifaceted and it changes over time. The five major components for building and maintaining a knowledge map include:

1. Building an organization-wide structure of varying knowledge competencies and levels.
2. Describing the skill sets needed for different tasks.
3. Gauging the performance of different employees after a project is completed.
4. Integrating these knowledge competencies in an online information system.
5. Connecting this information to training programs.

Explicit competencies are those that describe an employee's ability to use certain tools such as Excel® or SQL 9.0. Higher order competencies, implicit competencies, characterize the abstract reasoning skills that an employee might possess. Some of these implicit skills included "Knowledge of Data

Warehousing" and "Network Administration" (Conway). This knowledge mapping system describes 137 implicit competencies and 200 explicit competencies.

We discussed in Chapter 1 that there is a strategic difference between information and knowledge, and human attention is what is all too often left out of the equation that allows us to distinguish between information and knowledge (Malhotra "Knowledge" 8). Below, Davenport and Prusak (113–14) describe the tasks that knowledge managers will need to undertake to accomplish this:

- Advocate the value of knowledge management systems so that they become integrated into the organization's culture.
- Oversee the implementation of the firm's knowledge management infrastructure.
- Find and negotiate with the appropriate external providers of information that would best suit the needs of their corporation. For example, Monsanto provides its scientists with external market data that allows them to take the initiative in developing new products (Davenport and Prusak 129).
- Suggest and critique methods that describe how knowledge creation can be undertaken in endeavors such as market research and business strategy development.
- Implement ways to measure the value of the information stored in the organization's knowledge repository.
- Create standards for knowledge managers within organizations.
- Identify the kinds of knowledge bases the firm can best utilize and the types of knowledge in which the firm is in short supply.

To better leverage an organization's human capital, Malhotra asks that we fine tune our ideas regarding knowledge management (Malhotra "Deciphering" 60). Like the community of professionals who worked to produce and evolve the ICD, organizations are communities of humans that can provide diverse meanings to the outputs generated by existing technological systems. Diverse interpretations allow the potential for "constructive conflict mode(s) of inquiry" (60). For this to happen, we need an information architecture that includes categories and metaphors that allow us to better identify skills and competencies of an organization's employees (60). For example, the AA Proposal Toolbox is a metaphorical description that allows people to understand that there are "tools" in existence, and if they are shared and utilized thoughtfully, they might better allow someone to accomplish the task of writing an effective proposal. On a metaphorical level, Microsoft's "knowledge map" suggests that there is knowledge "out there" in the organization, and if it can be located on the "map," it can be utilized effectively. It also suggests that a corporation's knowledge belongs to everyone, not just one individual (Davenport and Prusak 76).

Malhotra ("Deciphering") also asks professionals to recognize the value of tacit knowledge, human creativity, and imagination. To do this, technical professionals need to implement technical architectures that allow them to be more social, open, flexible, and respectful of individual users. Additionally, they need application architectures that serve their problem-solving needs as opposed to merely allowing for the (re)generation of output transactions of simple archived data. Because workers at AA are able to access the past proposals that have proven successful and then make contact with their colleagues in different AA corporate offices who generated these proposals, this enables these workers to exchange and leverage, for example, their tacit ideas about the different clients/audiences for whom they were working. A proposal for a defense contractor in Denmark might have to amplify certain concerns that are important to people working in this particular European business culture. Someone who wrote a proposal for an American dot.com corporation that specialized in disseminating information about agricultural markets might be able to suggest rhetorically effective techniques that she intuitively understood worked best for the needs of this unique corporate culture. The environments that workers find themselves situated within are often complex social constructions that cannot be learned from a book. When hard-to-come-by knowledge is not articulated and shared between an organization's members, it means that others will have to "relearn" this knowledge on their own, which can be costly and inefficient. Finding someone with the know-how one needs can be facilitated with a well-integrated knowledge management infrastructure.

Prevailing practices that implicitly suggest that "this is the way things are done" need to be de-emphasized (Malhotra "Deciphering" 60). To plan for the future, technical communicators need to free themselves from the idea that they use a limited number of ideas that, while perhaps effective in the past, might not be as effective in contemporary markets that demand continual innovation. We believe that if technical writers and other communication professionals take on the role of knowledge managers, they need to be mindful of the way explicit and tacit knowledge is identified and named within and between different divisions of organizations. In other words, if they are aware of Kuhn's four socially constructed elements—the symbolic generalizations, shared commitments, shared values, and shared exemplars that shape their work environments—they can be more cognizant of how the brand of knowledge management that Malhotra advocates might be implemented to affect change.

Knowledge Management and Information Science

Now that we know the rudiments of XML, we can see how it can be used to organize specific elements of information and thus produce useful knowledge. One of the best ways to understand the nature of the knowledge that knowledge managers are trying to manage is by finding out how it is articulated across different branches of an organization, closely examining it, and then breaking it down into its essential components. These practices are what one is compelled to do when producing XML code, and communication professionals can expand their territory into the realm of knowledge management by learning how to model knowledge via XML. Using XML offers them a way to provide not just data, but data with context. XML can then be used to directly support knowledge management practices, since we can think of knowledge as data with context.

XML allows professionals to connect with each other, or more precisely, with each other's databases. XML compels them to reexamine just what information is of value to their organizations. As Berners-Lee and others have pointed out, it also makes it easier to search within different databases for specific information they might need as opposed to information that is encoded in HTML or embedded in a traditional database.

Previous work on the required skills for data reporting using widely-employed spreadsheet tools see it as much a rhetorical enterprise as a procuring of discrete facts. To support the transmission and production of knowledge, information gathered from databases needs to be presented in meaningful patterns that meet the audience's needs, especially when tabular data is the sole content of the message. To do this, we need to know, in terms of rhetorical invention (Mirel),

- the value of the electronic data,
- the meaning of the data relationships, and
- the appropriate level of detail required to meet the needs of the audience.

Problems that could result from a lack of attention to these ideas include information overload, overly narrow content, random data, unprocessed data, and unintelligible data (Mirel 99–100). Procuring appropriate data so one can best meet the needs of an audience "is only feasible if the data are set up in a special way to allow writers to retrieve data from different databases" (Mirel 104).

Information scientists and archivists have understood for some time that the best way to organize data and concepts in information storage and retrieval systems (ISARs) is by indexing and abstracting them. Traditional metadata indices and summaries can be effectively employed in the syntax of the object-oriented languages like XML, and this practice will demand a more sophisticated ordering and labeling of objects because they will serve as the single source of data that will support more than one body of text.

If text indices and the rules of extraction utilized by an electronic indexing system are not apparent, the researcher can be similarly unaware of the system's limits. In this case, a user might be led to the false belief that there is no information available that might meet her unique research needs. When a researcher is trying to make new connections to other fields or ideas by tapping existing but unknown documents and does not have any concrete notion of what search terms that lead to potentially useful texts, a search system cannot aid in this endeavor if there is 1) an indeterminacy (or uncertainty) in the representation or words chosen to indicate the objects in a document by indexers or, 2) an indeterminacy of chosen search terms employed by the researcher who is performing the research (Blair 237). It is also important to remember that unless a system's users are librarians, researchers, or specialized professionals such as attorneys performing patent searches, they are not going to spend the time learning how to craft perfect queries and complex Boolean expressions (Morville and Rosenfeld 181). We must therefore invest even more effort into the indexing of our data.

A simple description of the physical object is not enough. As Stam has stated, what is important is "the *significance* of the piece—a concept representing a perceiver's judgment—based on any one of several criteria" (6). The groups of indices that need to be evolved include the signs, signification, and social context of, in Stam's example, an art object. In her depiction of the difficulty of cataloging fiber art, an emerging style of sculpture that curators had yet to evolve a descriptive vocabulary, Lunin posits that to capture the essence of any artform one must understand and describe the breakthrough that the artist experiences that precipitates the unprecedented artform. Additionally, new terminology needs to be evolved and disseminated that best describes, in this case, just what materials are being used in this artform. Even though art that was constructed from "art fabric" came into play soon after World War II, it was only in the seventies and eighties that terms such as "textiles," "fiber," and "fiberwork" appeared in art journals as accepted indicators of an artform. From this insight, one can decide on the set of indices that could account for the relevance of the artistic intent, materials, techniques, and the aesthetic elements such as line, form, unity, balance, and emphasis.

We might say that communication professionals/knowledge managers should devise a search system that includes a wider array of indexical elements. This would allow for a greater array of accepted terms that would point to a useful document. But in practical terms, we should be also be aware of going too far the other way when we are extracting indexical elements—to index every possible nuance of a document might bury some of the key elements in a text and also lead prospective researchers to believe that there are sources in existence that really do not amount to much. We all have put a few search terms into Google or a university library search system that we thought would be useful, but have come up with thousands of documents that we had to wade through that have nothing to do with our research needs.

Morville and Rosenfeld remind us that recall, or the total number of relevant documents retrieved over the total number of *relevant* documents in the collection, is inversely related to precision, or the number of relevant documents retrieved over the *total* documents in the collection (159). In other words, we cannot often return lots and lots of extremely high-quality results. We either sacrifice recall to gain precision or return more total documents and have less precise results. When building searchable collections for the World Wide Web, this is another reason that speaks to the importance of building metadata indices with the appropriate level of granularity but without obscuring useful data at the expense of being overly detailed.

As we have shown above in the extended description of how diseases are classified by Bowker and Star and how indexing professionals need to be selective when they choose the semantic elements that name objects so as to better enable researchers to find relevant information, judging what constitutes a relevant indicator of an object/text is a function of the needs of our audience. The list below summarizes the basic questions professionals need to consider when indexing complex documents (O'Connor 56), and it can be modified and implemented by knowledge managers who are working within an organization to decide how different elements of knowledge can be tagged using XML:

1. How many elements should be extracted from various bodies of information?
2. Which elements should be extracted?
3. Should the elements be extracted in their natural form or translated?
4. Should elements be in their natural order or constructed order?
5. Should generalizations of individual concepts take place?
6. What are the rules that guide extraction?

Answering these questions challenges knowledge managers to identify the many small parts that make up the whole of their material, but it also calls upon them to think about how the material needs to be organized in the hierarchy of a DTD. Moreover, breaking information down into its elementary parts and then asking themselves if they are in fact "objects" also challenges professionals to more critically frame their use of object-oriented languages such as XML (Price 71). This parallels Kuhn's concern that scientists thoughtfully examine the essential nature of the socially constructed elements they are working with before they deploy them; if they ask if something is in fact an object, what they are doing is asking themselves if a "symbolic generalization" that is commonly believed to be of value in a socially constructed environment really meets their needs.

For example, there is a small but vocal minority of economists who challenge the prevailing socially constructed notion of how they should gauge the economic health of our nation. It is traditionally thought that the value of all goods and services in one year should be added together to determine our gross domestic product (GDP). The larger the GDP, the healthier the economy. However, economists with environmentalist leanings believe that certain goods and services should be subtracted from the GDP if they are in fact employed to clean up the environment. If a large portion of the economy is devoted to cleaning up the mess people generate in the production of other goods and services, our standard of living is actually diminished; we have fewer goods such as homes and services such as medical care available to us that offer comfort. While this idea did not come about because the Department of Commerce decided to employ XML technology, it describes how knowledge managers might always be thinking about the potential for including new objects into such a complex matrix, objects such as goods and services devoted to environmental restoration. In the context of O'Connor's heuristics directly above, this example asks that knowledge managers working with economists consider whether or not environmental costs should actually be made into an XML object or element (whether or not it should be "extracted") and actually used to calculate the GDP formula. These professionals are also asking how objects relate to one another hierarchically (one object would be subtracted from the added value of the other objects), and it challenges the very idea that there is a set natural order of elements that they can all agree on to determine what we know as our standard of living.

Additionally, it would be naïve to assume that one professional could code data in such a way that all potential users would find the information meaningful. The problem with this is that different organizations (and branches within organizations) have their own take on just what the information in their databases means, and the more databases that are linked, the greater the potential for confusion.

To better manage the differences in an organization's setting, information professionals need to collaborate by articulating their needs with the needs of others across different branches of their organizations. If people are willing to work out the formatting of data for each application with each other, there would be no problem, but this takes a considerable amount of time. The greater the number of people seeking to connect with each other, the greater the effort of negotiating their differences in perspective.

Knowledge managers also need to understand that the strength of XML as a tool can also be its weakness. While XML allows them the ability to more ably store and transfer information, if they do not implement an integrated design method in the early stages of their development phase, they run the risk of creating a system that is too complex and trouble-ridden to use effectively.

One method for using XML efficiently is to create an architecture that represents how an organization does business. To do this, knowledge managers need to see an XML Document Design Architecture (XDA) as something that is set up in three layers (Simon 130):

- a conceptual layer,
- a logical layer,
- and a physical layer.

To develop a *conceptual layer*, knowledge managers need to acquire all of the documents that are currently in use throughout their organization. By doing so, they might be able to identify and combine certain documents, and also come to realize that there are other kinds of documents not in use from which they could benefit. This would yield a better sense for how their organization is represented via texts and data.

To understand the *logical layer*, knowledge managers need to determine the data that their documents commonly contain, and define how these pieces of information can be set in document data element types. For example, a memo would contain the elements that described the author(s), person(s) addressed, subject, date, and text. A spreadsheet that describes the annual economic activity of our nation would contain elements such as GDP, durable goods, nondurable goods, services, and perhaps even goods and services devoted to cleaning up the environment.

The *physical layer* would consist of the DTDs that the knowledge managers develop. These DTDs would designate which elements are to be used for each application and how they relate to one another.

What is key here is that each layer is kept separate so problems that might arise from creating too complex a database can be mitigated. As work is done on each layer, information should be analyzed for comprehensiveness, consistency, and redundancy. If this development process is adhered to, problems will be detected at each level or layer before they are sent on to the next level, thus eliminating wasted time spent on refurbishing the entire information architecture at a later date (Simon 131).

Chapter Summary

Once organizational information is available, teaching the organization's members how they can access it for their own articulation needs can also be part of the responsibility of technical communicators/knowledge managers and other communication professionals. Making sure this information is understandable is also important, and this is something that technical communicators are well practiced in. Historically, technical communicators have produced documentation for disparate audiences, and part of their work has always been to ensure that documents they turn out are used by others. Additionally, technical communicators working as knowledge managers can also keep people updated on new information as it is made available on databases that could potentially assist them with their work (Dick), and as technical communicators have been traditionally assigned the task of updating and disseminating documentation, this task would mirror their traditional responsibilities.

Creating a culture is one thing. To employ knowledge management systems in organizations also requires significant costs as some organization's members will have to devote their workdays to these endeavors. Other organization members not officially designated as members of knowledge management teams will also be asked to input their ideas into databases, thus drawing them away from their other responsibilities. However, there are significant long term benefits:

- Knowledge management allows communication professionals to be more aware of the differences between branches of a large organization and challenges them to redouble their efforts to articulate their ideas across different sectors of an organization.

- Knowledge management, as has been said before, better allows professionals to leverage the knowledge capital of an organization. New people do not have to be hired or work does not have to be farmed out to expensive consulting firms to solve a problem that an existing organization member already has experience dealing with, and new ideas and perspectives can be exchanged by members within an organization. This practice in general allows communication professionals to understand the value of seeking help from others in their organizations and would facilitate collaborative efforts in general, thus reducing the phenomenon of what Rockley refers to as "silo traps."("Managing" 5)
- If used wisely, XML also allows professionals to reexamine previous technologies that they might have taken for granted. As stated above, XML technology demands that professionals who use it think about the very nature of data, how data is often embedded within other data, and it demonstrates the weaknesses of other technologies that would not allow them to do this with such ease. XML technology also allows them to understand, relative to previous technologies, that knowledge bases can be added to and/or reconfigured, as Johnson-Eilola, Selber, and Selfe might suggest ("Interfacing"); this allows professionals to be more critical of technology because they do not always have to accept the output of information gathering and representing technologies to which they are presently tied to in the workplace. Communication professionals working as knowledge managers can work with others in their organizations to rewrite the XML code.
- The very nature of XML allows technical communicators to think critically about knowledge. It demands that they break information down and reconsider its value; they are more than just translators of information as Slack, Miller, and Doak would hope. As many of the critics in the first section of the first chapter would want, it also challenges all members of an organization to become more critical about the way they organize information; they can become "symbolic-analysts" (Johnson-Eilola "Relocating"). They can also become more aware of the socially constructed elements that constitute the workplace and that govern what they think of as successful work (Kuhn 47).

Because successful technical communicators/knowledge managers possess the collaborative and interpersonal skills that their field has traditionally demanded, they would do well in helping to establish a culture that encourages employees to deposit information and take advantage of the ideas of their co-workers as it would indicate to all that everyone's ideas are valued. In their role as knowledge managers, communication professionals could help facilitate this environment.

Discussion Questions

1. Consider a relatively complex PDF file or document that has been saved in Word®. After reviewing what Rockley describes as "level two single sourcing, 'static customized content'" that we describe in this chapter, decide how you would break it down into its essential components and reconstitute it as a document that would be easily navigable in a website environment. How many parts of the document would exist as separate links? How would you link them together? How might you breakdown the separate components of this document into single source documents that could be stored in a database for reuse?
2. Using the questions we ask in the Discussion Questions 1 exercise directly above, how might Ament's description of primary and secondary modules better allow us to understand how we can breakdown or granularize parts of a larger document so it could be reconfigured for a Web environment or a set of single source modules?
3. Review the six rules of extraction that O'Connor provides and then apply them to some of the efforts illustrated in the evolution of the ICD. How do these rules allow us to better understand this process? For example, when O'Connor asks that professionals make a decision about what elements should be extracted from a body of data, what does he mean by elements in the context of disease classification efforts? Symptoms? Cause of death? The circumstances that describe the environment in which a patient lives? How many of these elements should be named and classified? How do O'Connor's rules allow us to imagine the many factors that go into designing classification systems?

References

Ament, Kurt. *Single Sourcing: Building Modular Documentation*. Norwich, NY: William Andrew Publishing, 2003.

Berners-Lee, Tim. *Weaving the Web: The Original Design and Ultimate Destiny of the World Wide Web*. New York: HarperCollins, 2000.

Blair, David. "Indeterminacy in the Subject Access to Documents." *Information Processing and Management*. 22.3 (1986): 229–41.

Bowker, Geoffrey, and Susan Star. *Sorting Things Out: Classification and Its Consequences*. Cambridge, MA: MIT Press, 2000.

Conway, S. Personal E-mail Communication. GPM-Knowledge Management, Microsoft Co. January 13, 2002, 8:48 pm.

Crowley, Sharon, and Debra Hawhee. *Ancient Rhetorics for Contemporary Students, Second Edition*. Boston: Allyn and Bacon, 1999.

Davenport, Thomas, and Laurence Prusak. *Working Knowledge*. Boston, MA: Harvard Business School, 1998.

Dick, Kevin. *XML: A Manager's Guide*. Reading, MA: Addison Wesley. 2000.

Dutta, Soumitra, and Arnoud De Meyer. "Knowledge Management at Arthur Andersen (Denmark): Building Assets in Real Time and Virtual Space." Yogesh Malhotra, ed. *Knowledge Management and Business Model Innovation*, Hershey, PA: Idea Group Publishing, 2001. 284–401.

Foucault, Michel. *The Order of Things*. Trans. Alan Sheridan. New York: Vintage, 1973.

Johnson-Eilola, Johndan. "Relocating the Value of Work: Technical Communication in a Post-Industrial Age." *Technical Communication Quarterly* 5. 3.2 (1996): 245–70.

Johnson-Eilola, Johndan, and Stuart Selber. "After Automation: Hypertext and Corporate Structures." *Electronic Literacies in the Workplace: Technologies of Writing*. P. Sullivan and J. Dautermann, eds. Urbana, IL and Houghton, MI: National Council of Teachers of English and Computers and Composition Press, 1996. 115–41.

Johnson-Eilola, Johndan, Stuart Selber, and Cynthia Selfe. "Interfacing: Multiple Visions of Computer Use in Technical Communication." *Three Keys to the Past: The History of Technical Communication*. T. C. Kynell and T. Moran, eds. Stamford, CT: Ablex, 1999. 197–226.

Kuhn, Thomas. *The Structure of Scientific Revolutions*. 2nd ed. Chicago: University of Chicago Press, 1970.

Lemke, Jay. *Textual Politics: Discourse and Social Dynamics*. London: Taylor and Francis, 1995.

Lunin, Lois. F. "Analyzing Art Objects for an Image Database." *Challenges in Indexing Electronic Text and Images*. R. Fidel, T. Bellardo Hahn, E. M. Rasmussen, and P. J. Smith, eds. Medford: American Society for Information Science, 1994. 57–72.

Malhotra, Yogesh. "Deciphering the Knowledge Management Hype." *The Journal for Quality and Participation*. 21.4 (1998). 58–60.

—— "Knowledge Management for E-Business Performance: Advancing Information Strategy to 'Internet Time.'" *Knowledge Management and Business Model Innovation*. Yogesh Malhotra, ed. Hershey, PA: Idea Group Publishing, 2001. 2–15.

Mirel, Barbara. "Writing and Database Technology: Extending the Definition of Writing in the Workplace." *Electronic Literacies in the Workplace: Technologies of Writing*. Patricia Sullivan and Jennie Dautermann, eds. Urbana, IL and Houghton, MI: National Council of Teachers of English and Computers and Composition Press, 1996. 91–114.

Morville, Peter, and Louis Rosenfeld. *Information Architecture for the World Wide Web*. Sebastopol, CA: O'Reilly, 2007.

O'Connor, Brian C. *Explorations in Indexing and Abstracting: Pointing, Virtue, and Power*. Englewood: Libraries Unlimited, 1996.

Price, Jonathan. "Introduction: Special Issue on Structuring Complex Information for Electronic Publication." *IEEE Transactions on Professional Communication*. 40.2 (1997): 69–77.

Reinhart, Tobias. *Cicero's Topica*. Oxford: Oxford University Press, 2003.

Rockley, Ann. "The Impact of Single Sourcing and Technology." *Technical Communication*. 28.2 (2001): 189–93.

Rockley, Ann. *Managing Enterprise Content*. Indianapolis: New Riders, 2003.

Schank, Roger C., and Robert P. Abelson. *Scripts, Plans, Goals, and Understanding: An Inquiry into Human Knowledge Structures*. Hillsdale, NJ: Lawrence Erlbaum, 1977.

Simon, Solomon H. *XML: eCommerce Solutions for Business and IT Managers*. New York, NY: McGraw-Hill, 2001.

Slack, Jennifer, David J. Miller, and Jeffrey Doak. "The Technical Communicator as Author: Meaning, Power, Authority." *Journal of Business and Technical Communication* 7.1 (1993): 12–36.

Stam, Deirdre C. "The Quest for a Code, or a Brief History of the Computerized Cataloging of Art Objects." *Art Documentation* 8.1 (1989): 7–15.

Star, Susan. "The Structure of Ill-Structured Solutions: Heterogenous Problem-Solving, Boundary Objects and Distributed Artificial Intelligence." M. Huhns and L. Gasser, eds. *Distributed Artificial Intelligence* 2. Menlo Park, CA: Morgan Kaufman, 1989. 37–54.

Star, Susan and Karen Ruhleder. "Steps Toward an Ecology of Infrastructure: Design and Access for Large Information Spaces." *Information Systems Research* 7.1 (1996): 111–34.

Wick, Corey. "Knowledge Management and Leadership Opportunities for Technical Communicators." *Technical Communication* 47.4 (2000): 515–29.

Williams, Joe D. "The Implications of Single Sourcing for Technical Communicators." *Technical Communication* 50.3 (2003): 321–7.

Yates, Joanne, and Wanda Orlikowski. "Genre Repertoire: The Structuring of Communicative Practices in Organizations." *Administrative Science Quarterly*. 39.4 (1994): 541–74.

—— "Genres of Organizational Communication: A Structurational Approach to Studying Communication and Media." *Academy of Management Review* 17.2 (1992): 299–326.

Zack, Michael. "If Managing Knowledge Is the Solution, Then What Is the Problem?" *Knowledge Management and Business Model Innovation*. Yogesh Malhotra, ed. Hershey, PA: Idea Group Publishing, 2001. 16–36.

4 The Visual Rhetoric of XML

Using CSS and XSL to Format and Display XML Projects

Chapter Overview

In this chapter we consider the visual dimension of XML data manipulation by examining technologies such as CSS and the XSL. As document authors or knowledge managers, it is important that we understand both how to organize and classify data in a careful way and how to enable access to that data in an aesthetically pleasing fashion. The visual dimension of information presentation is especially important considering the massive amounts of unformatted raw data destined for the World Wide Web that is available in organizations. Some organizations also have historical data, or "legacy data," that needs to be shaped into a presentable form for a corporate website, brochure, or newsletter. As we discussed in Chapter 2, creating accessible knowledge management systems involves selecting the appropriate amount of detail for our particular audiences. Visualization techniques are useful when trying to focus user attention on particular facets of data or when trying to highlight salient parts of an XML database for a particular community.

CSS is a style sheet language that is compatible with both HTML and XML, while XSL is a family of style sheet and transformation languages used specifically and exclusively with XML data. These languages are important to understand because they allow one to separate metadata describing the data itself (or its *content*) and the shape that data should take (or its *form*) into two different logical files or locations. Subsequently, the task of writing semantic XML tags can be separated from the task of writing visual XML style sheets, which is helpful since these procedures require different skills and abilities. In many companies, graphic designers will handle the CSS or XSL and technical communicators or programmers will handle the construction of XML code. It is useful, however, to understand the basics of both techniques as these individuals must work together as a team in order to produce rhetorically effective documents.

Fortunately, the mechanics behind style sheets and their associated data files are straightforward. Since semantic coding and visual coding tasks are separated, one updates the original XML document to change or add new semantic tags and uses external CSS or XSL files to change display information using formatting tags. In other words, when the data inside an XML or HTML file needs to be changed, it is only necessary to alter the HTML or XML file containing that data. This could be done by adding new tags into the file or by removing or changing existing tags or the data residing within them. When the visual formatting or layout of this data needs changing, one then alters the CSS or XSL file containing the instructions relevant to the data's formatting and visualization. The concept is simple, but bears repeating as it is critically important: style sheet languages make it possible to consider the visual formatting and display instructions independently from the document's data content. Rhetorically speaking, this separates *what* is being communicated in an XML document from *how* it is communicated. The semantic XML tags describe *what* and the CSS and XSL technologies describe *how*.

Under this divided model, HTML or XML files contain the data and the CSS or XSL style sheets contain instructions specifying how to present or re-present that data. These instructions are stored in external files that are linked from XML documents; this is done in the same way an author would link a DTD or declare a namespace. While it is possible in CSS to use both semantic and display tags in the same file, this defeats the purpose of entirely separating content from form and can be both technically and rhetorically confusing for the document author. Also, this is only possible in HTML documents and not in XML documents.

In order to ease into a description of the new syntax required by CSS and XSL, we first consider some of the rhetorical facets of imagery and visual communication. Next, we outline some simple examples of how to use technologies such as CSS with a simple HTML document and then with one of the XML documents from Chapter 2. Finally, we discuss the more sophisticated XSL, which allows a document creator to translate one XML document into another—a process that is useful for a variety of tasks such as when one is using data across multiple contexts or presenting for multiple audiences.

Rhetoric, Imagery, and XML

While the "division of content and form" strategy of CSS has proven wildly successful in the maintenance and upkeep of large websites, it also poses some interesting rhetorical questions. For example, what does it mean to have purely descriptive information that is abstracted entirely away from its presentation? How is the interpretation of a message changed when that message is presented in a visual style that is very much in contrast with the data? Style is one of the original rhetorical canons, and for good reason. Classical rhetoricians used the term *elocutio* to refer to the stylistic components of persuasive speech and rhetorical style has since been broadened to include textual and electronic styles of shaping language. What classic and modern rhetoricians have discovered is not surprising: the style of a message, whether delivered orally, in writing, or electronically, can be as important as, if not more important than, the message itself. At the very minimum, style is an element of electronic rhetoric that we cannot forget about as it influences everything from how corporate ethos is presented to how trust is established and maintained with online consumers.

Stylistic concerns have been part of the rhetorical tradition for thousands of years, dating back to the Romans and beyond. For example, in addition to the historical notoriety garnered from his military conquests and his financial support of a young Julius Caesar, the Roman general and politician Marcus Crassus (115–53BC) was notable for criticizing the division between philosophy, or what he characterized as "wise thinking," from rhetoric, or his term for "elegant speaking" (Whitburn 45). There has always been a great deal of interest in the differences and similarities between philosophy and rhetoric, but it is certain that one does not always need to be a wise person and a careful thinker in order to be a persuasive communicator. Particularly in the electronic age, there are a variety of ways to shape one's message to seem more appealing than it might be if considered on its intellectual merit alone.

If we consider rhetoric as a means of persuasion, then it is obvious that the visual style of a message is a very important part of its rhetorical constitution. It is sometimes unfortunate, but by separating a wise thought from its elegant packaging, it is true that we remove some of the persuasive veneer that often prompts the recipient of a message to receive it favorably (or, in rhetorical situations that strive for agency from an audience, to move from passive observer to active participant).

Think of trying to purchase an automobile based purely on the quantitative (numerical) parameters of its mechanical and electrical components. Buying one Sport Utility Vehicle over another might not seem to be such a difficult choice when comparing raw data such as gas mileage and storage capacity. The aesthetics of the vehicles, though, cannot be overlooked or ignored. Furthermore, when this quantitative information is filtered through a carefully crafted rhetorical presentation such as a television commercial or commercial website, a consumer's decision to act may be much more based on emotion (one's reaction to the beauty of the vehicles) than logic (one's analysis of the superiority of one vehicle over another in terms of performance or maintenance data). In general, consumers want the vehicle that is filled with adventurous and beautiful people, or the one that reliably tows a railroad trailer up a mountain in the middle of a tornado, or the one that accelerates down a roadway racing a computer generated cheetah. These are all visual expressions which are carefully crafted to link positive consumer emotions such as happiness, sexual desire, or excitement with the vehicles being advertised.

We can think of this vehicle purchasing example from the context of XML and databases. The professionals in charge of such advertising presentations carefully use different data sets like advertising demographics and statistically arranged consumer buying habits in order to select the best times and channels for television commercials and the optimal websites on which to place their promotions. The vehicle's statistics and specifications are carefully screened for target demographics (young adults are especially targeted as they have the potential to be long term and active consumers) and only the most impressive facts and figures are likely to show up in an advertisement. The data is then given an entirely new rhetorical life when surrounded by visual spectacles such as those produced by mass media on television or in film.

The same is true for an advertisement on the World Wide Web. When we store the vehicle's data in an XML file and that data's formatting instructions in a CSS or XSL file, we can give full attention to the data when structuring knowledge representations and then turn our attention to the display of that data when we construct the style sheet. For instance, a product specialist or engineer could choose to create a special attribute named advertisingReady and set that attribute's value to true for all facts or figures that are useful in promotional materials. A graphic designer could then choose to focus his or her attention on creating a visual aesthetic that is pleasing to the eye and frames the vehicle in an appropriate way for the audience. This divided technique makes sense because these two domains often have entirely different sets of parameters. What something *is* is often quite different than how something *looks*.

Knowledge managers must be aware that visual content is another powerful layer that can be used (or misused) to communicate with audiences and facilitate information exchange within an organization. Not surprisingly, much of our knowledge about the visual communication process comes from the field of advertising. In television or print advertising, images are effectively used to set the tone or mood of a message without the need for written annotation or dialogue. Think of a dark and gloomy castle and the emotions and thoughts that this environment conveys. Now, think of a busy city park on a summer day. By varying a few pictorial elements and aesthetic approaches, an author can predispose her audience to a certain manner of thinking or feeling before the primary message is even presented. Then, when actors appear on the television screen or text is emblazoned across a magazine spread, the message—which is generally to purchase a given product or endorse a particular brand, individual, or community—is that much more powerful and memorable.

Recent research into consumers' experience with imagery examines how the images used in advertising communicate through coded representational systems. Many of our earliest "writing" systems, such as cave paintings, were formed using pictographic representations of reality. An interesting finding from contemporary advertising research is that certain types of imagery may convey uniform messages, though these messages are not always predictable. In one experiment, Linda Scott from Oxford University and Patrick Vargas from the University of Illinois at Urbana-Champaign varied images of a cat, a sunset, and a set of abstract paintings in order to gauge consumers' reactions to these pictures as they applied to a particular product (facial tissues). Using a sample of seventy-seven undergraduate students, they found that the differing images communicated simple, but consistent, rhetorical messages to a variety of participants. For instance, the image of a fluffy cat communicated colorlessness, softness, absorbency, high price, and fragility. A sunset communicated colorfulness, softness, and absorbency. While some of these outcomes were to be expected, others, such as the ability of the fluffy cat to communicate high price and fragility, were less predictable. In this study, Vargas and Scott demonstrated that the statements generated by participants went "well beyond resemblance to an object or the sensory effects of formal features" (353).

What is particularly fascinating about the Scott and Vargas study is that even abstract images were found to communicate specific features and elicit certain emotional responses from participants. When students were asked to record their observations of the abstract O' Keeffe painting *Black Spot No. 2*, respondents sometimes read soft colors as communicating softness, but almost always read the curvature of the lines as communicating this quality. In addition, they noted that the black object at the bottom of the painting was an indicator of strength and that the shading technique was used to communicate absorbency.

Another interesting rhetorical dimension of visual design is found when considering the cultural differences of one's audience. Certain colors, images, texts, and patterns may work very well for one culture and at the same time be off-putting or even offensive to another group of individuals. For example, in August of 2007, the BBC News reported a story in which the U.S. military distributed soccer balls to the Khost province of Afghanistan (Leithead online). While the soccer balls were intended to be gifts for Afghan children to enjoy, they featured flags of various countries, including the flag of Saudi Arabia. This flag contains an image of the shahada, one of the five pillars of Islamic faith. To Islamic residents, kicking this soccer ball would be the equivalent of kicking a Bible to a Christian—a highly offensive and blasphemous act.

As the soccer ball example shows, even the act of embedding flags or other nationalistic materials into one's product, which seems at first to be rather innocuous, is complicated when internationalization is involved. When developing a product with an international audience, it is necessary to carefully research the ideological beliefs and values of individuals from each culture that will be accessing or viewing your materials. Fortunately, websites that take advantage of CSS or XSL can address issues of multiculturalism simply by providing a separate external style sheet for different geographic locations and then allowing the user to specify the region from which they are accessing your content. A single database can then be expressed through different templates to better serve disparate international audiences that need access to the same data source.

Addressing Visual Complexity

In order for a practitioner to address the many rhetorical challenges of visualization, she needs to be aware of text-based technologies like CSS and XSL that are used to represent XML data in a visually appealing fashion or to transform it into another form more suitable for humans to read. Unlike HTML, which does not force one to differentiate between descriptive and formatting markup, XML information is purely descriptive. In other words, XML describes what the data is and how it is arranged, but it does not prescribe how that data should be presented to a human. While a fundamental goal of XML is to make data machine-readable, there are many situations in which this content must be made human-readable as well.

The computers interpreting XML files have no need for sophisticated graphics and pleasing aesthetic themes, but this lack of visual sophistication is rhetorically damaging when XML information must be interpreted or examined by a human user rather than a computer. For instance, think about how important first impressions are when encountering something new—a new restaurant you are planning to eat in, a new person you are meeting for the first time, or a new home you are planning to rent or purchase. That first visual encounter leaves a lasting impression that directly influences how you interact with that new space or person in the future. A particularly bad impression might lead you to avoid this person or place altogether. This relates to prototype theory and cognitive schema, which we discussed in Chapter 3. We form mental models of the world which influence how we perceive our environment based on previous experiences and complex psychological interactions between emotions, social relationships, and cognition. These mental models are not necessarily true or representative of a person or location, but they will nonetheless influence and dictate how we think and act when faced with repeated encounters.

The same type of process holds true for our visits to websites, particularly in regards to aesthetics. Researchers have found that visual elements have a significant impact on how credible or reliable an online information source is perceived to be—the *online ethos* of a website (Warnick 262). Our prior experiences with well-designed and visually appealing websites have trained us to be more willing to engage with and see authority in new online resources with attractive visual styles. For this reason, it is important that we consider some of the supportive visual technologies of XML.

CSS

CSS are powerful textual documents used for specifying the layout and formatting of Internet documents. In contrast to XML, which describes only the data, CSS is used to describe how that data should appear when rendered in a Web browser. Since CSS is now the preferred method for formatting and displaying HTML content, we will begin this chapter by examining the ways in which CSS and HTML work together.

Using CSS with HTML

By combining CSS with HTML, one can mimic the separation between content and form that is enforced by XML documents. This is exactly what happens when professional designers construct large websites. However, this was not always the case. Historically, it was possible (and, in fact, is still possible) to use HTML documents without CSS files for websites and to simply include additional HTML tags and attributes for formatting within each content page. For instance, try typing the following code example into your text editor. Save the file as "ch4_ex1.html" and load this file into your Web browser.

```
<html>
<head>
<title>Welcome to Cascading Style Sheets</title>
</head>
      <body>
            <h1>This is an example of a first order heading.
            </h1>
            <h2>This is an example of a second order heading.
            </h2>
            <p>This is an example of a normal paragraph.
            <font color="green"><u>Here is some green,
            underlined text.</u></font>
            </p>
      </body>
</html>
```

When viewing this example in your browser, you will see something similar to the screenshot shown in Figure 4.1. (As the screenshots in this text are in black and white, you may wish to visit the text's accompanying website at www.rhetoricalxml.com to see the screen captures in full color. For now,

Figure 4.1 Formatting Using HTML

just recognize that in a Web browser the last underlined sentence shown in the screen capture would also be displayed as green text.)

Although it is certainly handy to be able to add formatting tags directly to HTML files, this technique is difficult to manage with large websites that have numerous content pages that are distributed over multiple file locations. The next step would be to include CSS code within the HTML document. Although this is still not as easy to manage as external style sheets, it does move us one step closer to separating content from form. Now, we will use CSS to create the same effect.

The same code rewritten to take advantage of CSS tags is shown here:

```
<html>
<head>
<title>Welcome to Cascading Style Sheets</title>
<style type="text/css">
    .greenunderline
        {
        color: green;
        text-decoration: underline;
        }
</style>
</head>
    <body>
        <h1>This is an example of a first order heading.
        </h1>
        <h2>This is an example of a second order heading.
        </h2>
        <p>This is an example of a normal paragraph.
        <p class="greenunderline">Here is some green,
        underlined text.</p>
    </body>
</html>
```

This new file can be saved as "ch4_ex2.html" if you are working along with the examples. The first thing to notice about this revised code is that we have added an additional element within the <head> portion of the document. This style definition is what a designer uses to specify visual and layout information that can then be used to arrange or format the information within the HTML document. In this case, we have chosen to create what is known as a class, or special selector reference, named .greenunderline. This class definition contains two CSS triplets that are each made up of a selector, property, and value. In fact, CSS documents are composed entirely of these triplets, which are collectively referred to as the "rules" of the document. Each rule specifies how a given unit of information, such as the color of the font, should be formatted upon encountering a selector of that type within the hypertext document.

In this example, .greenunderline is the selector, color, and text-decoration are the properties, and "green" and "underline" are the values. Since there is no .greenunderline element built into the standard HTML library, this means that the HTML page is expecting us to define a customized point of reference using our own named selector. In essence, this is precisely what a class is: a means of defining our own selector elements which can then be used either in addition to or in replacement of standard HTML elements.

The general structure for a collection of CSS rules is to first list the selector name, which may be an HTML element such as P, BODY, H1, or H2 or even a custom class like .greenunderline, then define a list of one or more custom values which can be applied to the properties of that selector. The list of values and properties will be encapsulated by curly brackets (braces) immediately following the named selector. So, the general structure of a CSS rule looks something like this:

```
SELECTOR
      {
      PROPERTY: VALUE 1, VALUE 2, ... , VALUE N;
      }
```

It is also possible to associate multiple selectors with a single rule. This is done by adding the additional selector elements after the first element using a comma to separate them. The multiple selector format looks like this:

```
SELECTOR 1, SELECTOR 2, ... , SELECTOR N
      {
      PROPERTY: VALUE 1, VALUE 2, ... , VALUE N;
      }
```

This syntax allows a content developer to apply a single CSS rule to multiple selectors. For example, to write a CSS rule that sets the font face of data within the P, H1, H2, H3, and TD elements to Arial, one could use the following syntax in order to save space within the CSS file:

```
P, H1, H2, H3, TD
      {
      font-family: Arial;
      }
```

In a situation like this, where elements share common properties and will be using the same values, it makes sense to combine multiple rules into a single CSS rule definition. Otherwise, we would be needlessly repeating the same instruction for these elements. Specialized rules for textual data within these elements can always be applied using a rule that is defined later in the document (this is the *cascading* part of CSS, discussed later in the chapter). Once these CSS rules have been defined, we will need a way of linking these rules to the tags in our HTML or XML document. Let us return to our original HTML example, reprinted here:

```
<html>
<head>
<title>Welcome to Cascading Style Sheets</title>
<style type="text/css">
    .greenunderline
            {
            color: green;
            text-decoration: underline;
            }
</style>
</head>
    <body>
            <h1>This is an example of a first order heading.
            </h1>
            <h2>This is an example of a second order heading.
            </h2>
            <p>This is an example of a normal paragraph.
            <p class="greenunderline">Here is some green,
            underlined text.</p>
    </body>
</html>
```

The next thing to note about this example is that we have slightly modified the <p> tag surrounding the sentence "Here is some green, underlined text." This tag has been replaced with <p class="greenunderline">. We have added the class="greenunderline" directive to instruct the browser that this particular paragraph should be formatted according to the rules assigned to that descriptor. We can define as many additional rules as we like. The rules can be found in one of three places:

1. In a special block of text found between the head opening and closing tags (<head> and </head>) of the HTML document (as shown in this example).
2. Inline with the HTML selector. An example of this sort would look something like: <p style="color: green">Here is some green text.</p>. While useful when documents need to be self-contained, this method can quickly become cumbersome as it requires the style directive to be applied to each and every tag in which a style sheet will be used.

3. In a separate, external CSS file linked from the HTML document using a special instruction in this format: <link rel="stylesheet" type="text/css" href="stylesheet_filename.css"> where "stylesheet_filename.css" is the name of the external file containing your CSS rules. This instruction would also appear between the <head></head> tags of the HTML document. For websites containing many CSS rules, this is the preferred method as it enables one to quickly locate, add, and revise rule triplets as needed. The external file means that one will not need to wade through embedded HTML data in order to find a CSS rule for a particular HTML element; all the CSS rules will be in one place.

Loading the revised document into your Web browser using any of these methods will yield the same exact results as shown previously in Figure 4.1. Nothing has changed except for the mechanism working behind the scenes.

At this point, we may want to further customize the visual presentation of our document. We can therefore rewrite the code once more as we see here:

```
<html>
<head>
<title>Welcome to Cascading Style Sheets</title>
<style type="text/css">
      h1
            {
            font-size: 18px;
            color: red;
            font-family: Helvetica, Arial, Sans-Serif;
            font-weight: bolder;
            text-decoration: underline;
            }
      h2
            {
            font-size: 15px;
            color: blue;
            font-family: Times New Roman, Times, Serif;
            font-weight: bold;
            text-decoration: italic;
            }
      p
            {
            font-size: 13px;
            color: black;
            }
      .greenunderline
            {
```

```
            color: green;
            text-decoration: underline;
            }
</style>
</head>
      <body>
            <h1>This is an example of a first order heading.
            </h1>
            <h2>This is an example of a second order heading.
            </h2>
            <p>This is an example of a normal paragraph.
            <p class="greenunderline">Here is some green,
            underlined text.</p>
      </body>
</html>
```

Save this file as "ch4_ex3.html". In this example, we see that CSS allows a designer to take advantage of built-in selectors such as <h1>, <h2>, and <p> as well as to create custom selectors such as the .greenunderline class we have been using. You can open this file in your browser to see how easily these styles are applied to default HTML selectors.

The CSS specification contains a large library of properties. Using the font-family property, a designer can specify the preferred font for rendering text on a Web page. Commas are used to specify additional font choices in descending order; if the first choice is not installed on the reader's computer, it will follow the series in order until it finds a selection that is installed. For this reason, the last choice in the series is usually a very general value such as Sans-Serif or Serif. These selections will apply any available Sans-Serif or Serif fonts to the text.

The color property is used to manipulate the color of the text surrounded by selected HTML tags indicated in the CSS rule. For instance, the CSS code shown previously specifies that all level 1 heading (<h1>) text should be red. Level 2 headings (<h2>) should be blue, paragraph (<p>) text should be black, and our special class named .greenunderline will produce green and underlined text in the browser.

Hexadecimal Color Codes and CSS Properties

Hexadecimal color codes such as #FFFFFF (white), #000000 (black), or #FF0000 (red) can be used with properties that accept color values. The RGB color model used by HTML allows a designer to specify precise combinations of reds, greens, and blues in order to generate millions of different colors (assuming the person viewing your site has a 32-bit video card). The pattern for generating a hexadecimal color is the pound symbol (#) followed by two digits representing the red intensity, two digits representing the green intensity, and two digits representing the blue intensity.

\# _ _ _ _ _ _ or # R R G G B B

Since the numbers are encoded in hexadecimal, the following chart can be used to find the highest and lowest intensities.

0, 1, 2, 3, 4, 5, 6, 7, 8, 9, A, B, C, D, E, F

lowest intensity ... highest intensity

In this numbering system, once A is reached, it simply continues counting forward so that A=10, B=11, C=12, D=13, E=14, and F=15. "Hexadecimal" means that each position in a hexadecimal number is a power of sixteen rather than a power of ten, as we see in the decimal system we use every day. So, just as the number 255 in decimal can be expressed in powers of ten from right to left ($10^0 \times 5 + 10^1 \times 5 + 10^2 \times 2 = 255$) so can this number be expressed in hexadecimal using only two digits ($16^1 \times 15 + 16^0 \times 15 =$ 255 decimal = FF hexadecimal).

Using this system, the full range of allowable color codes for the color value as used in CSS (or any other HTML color specification) is from #000000 (red=0, green=0, and blue=0) to #FFFFFF (red=255, green=255, and blue=255). Though this system allows for a very wide range of colors (approximately 16 million, or $255 \times 255 \times 255$) not all colors are considered to be "Web safe" or "browser safe" colors. For more information on good color values to use, try a Google search on "Web safe HTML colors." In the earlier days of Web development, 216 colors were identified as cross-operating system and cross-browser compatible. Today, most computers have at least a 16-bit video card and many more colors are safe to use.

Font Sizing

The font-size property accepts several different units of measurement in order to change the size of displayed text on a Web page. Values for this property can be either absolute, meaning the sizes are compared to a default size table stored in the user's browser, or relative, meaning that the size is specified relative to the font size of the parent element. Units for font-size values can be given in ems, points, and pixels. The example code shown so far uses pixels to specify the font-size property's value. This is an example of an absolute size.

One way to achieve relative sizing is to use the em unit to specify size values. Ems are textual units that use the size of the surrounding text as a reference point to adjust the property applied to a selector. The CSS code shown here adds a new class named .importantText that sets the font size to twice as large as the font size of the surrounding element. If we wanted to instead make this text one and a half times as large as its parent element we would use the code font size: 1.5em rather than font size: 2em. Consider the following CSS code:

```
<style type="text/css">
    h1
          {
          font-size: 18px;
          color: red;
          font-family: Helvetica, Arial, Sans-Serif;
          font-weight: bolder;
          text-decoration: underline;
          }
    h2
          {
          font-size: 15px;
          color: blue;
          font-family: Times New Roman, Times, Serif;
          font-weight: bold;
          text-decoration: italic;
          }
    p
          {
          font-size: 13px;
          color: black;
          }
    .importantText
          {
          font-size: 2em;
          }
    .greenunderline
          {
          color: green;
          text-decoration: underline;
          }
</style>
```

What is nice about this code is that the font's size will now be adjusted relative to *any* parent element. This means that regardless of where the text appears—in a heading, subheading, or paragraph—it will always appear at twice the height of its parent element's font size. We can then use this new class by adding the following HTML code:

```
<h1>This is an example of a <span class="importantText">
first order</span> heading.</h1>
<h2>This is an example of a <span class="importantText">
second order</span> heading.</h2>
<p>This is an example of a normal paragraph.
<p class="greenunderline">Here is some <span class=
"importantText">green, underlined</span> text</p>.
```

The full listing for this code now looks like this:

```
<html>
<head>
<style type="text/css">
     h1
          {
          font-size: 18px;
          color: red;
          font-family: Helvetica, Arial, Sans-Serif;
          font-weight: bolder;
          text-decoration: underline;
          }
     h2
          {
          font-size: 15px;
          color: blue;
          font-family: Times New Roman, Times, Serif;
          font-weight: bold;
          text-decoration: italic;
          }
     p
          {
          font-size: 13px;
          color: black;
          }
     .importantText
          {
          font-size: 2em;
          }
     .greenunderline
          {
          color: green;
          text-decoration: underline;
          }
</style>
</head>
<body>
<h1>This is an example of a <span class="importantText">
first order</span> heading.</h1>
<h2>This is an example of a <span class="importantText">
second order</span> heading.</h2>
<p>This is an example of a normal paragraph.
<p class="greenunderline">Here is some <span class=
"importantText">green, underlined</span> text.</p>
</body>
</html>
```

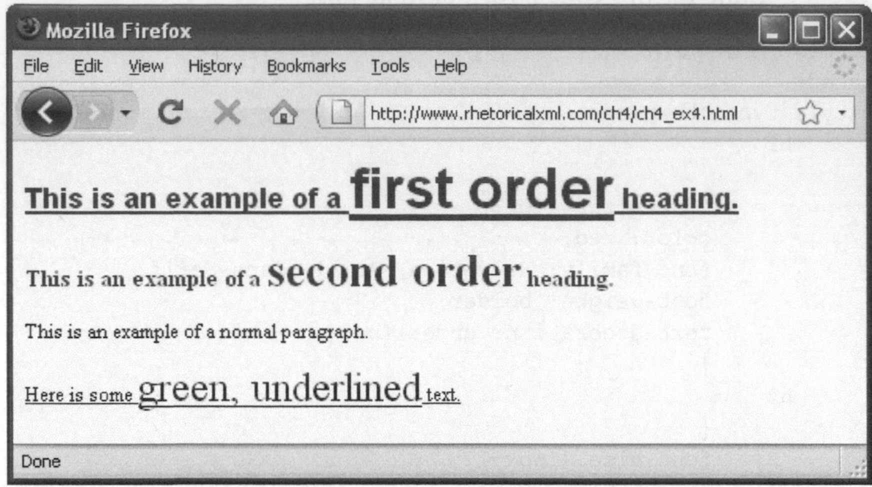

Figure 4.2 Browser Output of Relative Font Size

In this example, the .importantText class is applied to units of text within parent elements such as H1, H2, and P using a special HTML entity known as SPAN. We will learn more about SPAN and the similar tag DIV later in this chapter. See Figure 4.2 for the browser output of this file or type it in yourself (save as "ch4_ex4.html") and see how it looks in your browser.

Classes and Cascading

Returning briefly to our original example, there are some other things we should notice about how classes are used. First, note that the CSS rules sometimes use a slightly different syntax than traditional HTML code. For instance, to combine our new class with the CSS rule we have defined for the <h1> selector, we would need to type in the HTML code shown here:

```
<body>
      <h1 class="greenunderline">This is an example of a first
      order heading.</h1>
      <h2>This is an example of a second order heading.</h2>
      <p>This is an example of a normal paragraph.
      <p class="greenunderline">Here is some green, underlined
      text.</p>
</body>
```

A screen capture showing what this code looks like can be seen in Figure 4.3. You can see in this instance we are no longer using the .importantText class but that we are instead applying parts of both the H1 selector's CSS definition and of the .greenunderline selector's class definition.

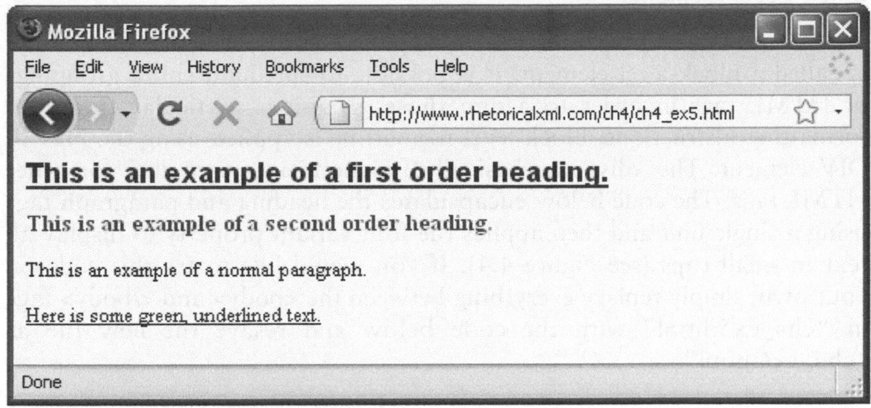

Figure 4.3 Cascading in Action

This combinatory feature is what makes CSS so powerful. It is the *cascading* part of CSS. When you type in this new example and load it into your Web browser (save as "ch4_ex5.html"), you should notice that the text encapsulated by the <h1> tags is now rendered exactly the same as before, only in green. In other words, it has retained the properties defined by our original <h1> rule (the 18 pixel font size, the Helvetica font face, the bold and underlined text), but it has replaced the original red font color with the green color specified by our class. We also saw this phenomenon at work in our .importantText class example when portions of text were rendered in the same color and font-face as their parent elements, but were doubled in size as specified by the 2em value for font-size.

We can think of the mechanism behind CSS as a virtual waterfall. At the top of the waterfall, or the beginning of the CSS definition, an author defines certain formatting characteristics and visual layout properties for elements. It is entirely possible for these elements to retain the exact same properties when they reach the bottom of the waterfall, or the end of the CSS definition. It is equally possible, however, that these elements might be changed as they make their way down the waterfall (perhaps by splashing into abutting rocks, if we continue the metaphor). The important thing to remember is this: any original properties that are not redefined further downstream will remain in place when the bottom of the waterfall is reached. Only subsequent CSS properties applied later in the CSS definition will alter the appearance and layout of elements already imbued with formatting instructions. In this way, document-wide consistency can be maintained at the beginning of the document and further customization is made possible as sub-selectors, classes, and special elements pick up their customized and individualized instructions further down the waterfall, or further down the CSS hierarchy.

Block-Level Elements

CSS can easily be applied to entire groups of HTML elements. Using what is called a block-level element, it is possible to surround entire groupings of HTML tags in order to assign those groupings particular layout or formatting instructions. Block-level formatting is applied using the HTML DIV element. The <div> tag basically functions as a container for other HTML tags. The code below encapsulates the heading and paragraph tags using a single unit and then applies the font variant property to display all text in small caps (see Figure 4.4). If you would like to try this code on your own, simply replace everything between the <body> and </body> tags in "ch4_ex5.html" with the code below and resave the new file as "ch4_ex6.html".

```
<div style="font-variant: small-caps;">
     <h1>This is an example of a <span class="importantText">
     first order</span> heading.</h1>
     <h2>This is an example of a <span class="importantText">
     second order</span> heading.</h2>
     <p>This is an example of a normal paragraph.
     <p class="greenunderline">Here is some <span class=
     "importantText">green, underlined</span> text</p>.
</div>
```

In certain situations, you may also want to group a number of elements within a block-level element in order to provide custom formatting for smaller units of information. For instance, you may want to apply CSS rules

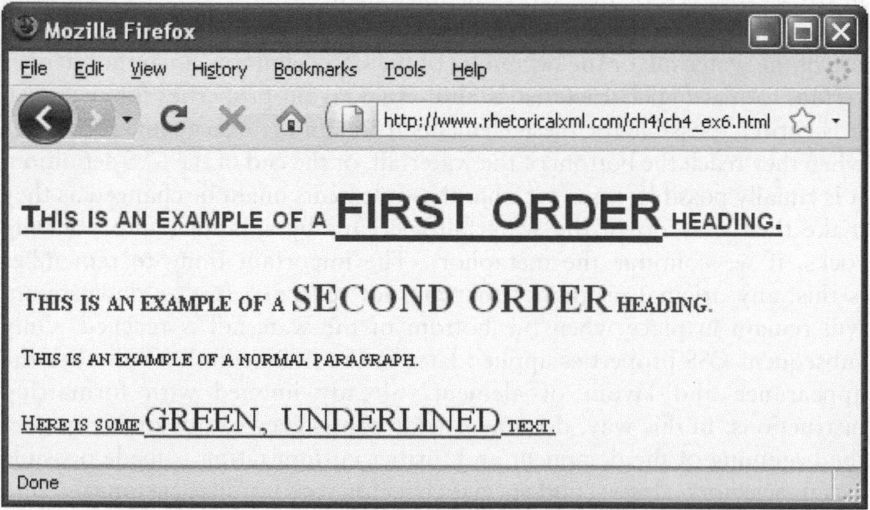

Figure 4.4 DIV Tag and Small Caps CSS

to several sentences within a large paragraph that has already been formatted according to the style bound to the <p> tag. This can be accomplished using the SPAN element in HTML.

Using DIV and SPAN for Web 2.0 Applications

DIV and SPAN elements are useful for designing "Web 2.0" XML applications that continuously refresh the browser in the background, without explicit instructions from the user. "Web 2.0" is a name given to sites that are highly interactive, depend largely on social and community interaction and contributions, and utilize cutting-edge technologies such as asynchronous XML to push the envelope of the online experience. Visualization is often an important part of Web 2.0, since the normal cues (such as the icon in the browser showing a refresh animation) no longer apply to the exchange of data between server and client computers. In other words, a user no longer needs to press a submit button and then wait for a result: the results are retrieved in the background and the user can continue browsing other areas of the website. A creative use of visual rhetoric is therefore necessary to guide the visitor through the interactive experience with signposts or to otherwise highlight salient points within a formatted listing of data.

The technologies enabling Web 2.0 interactivity have been around for many years and they often make use of traditional tags such as <div> and <span> as well as technologies such as client-side JavaScript. Client-side JavaScript is special programming code that is run on the client's computer rather than on the Web server which hosts the content. While a detailed explanation of how this process works is outside the scope of this text, a brief outline is provided here to demonstrate how important it is to understand the <div> and <span> CSS tags when working with HTML and XML for Web 2.0:

1. An HTML page is created and certain portions of the page are earmarked for dynamic content by surrounding them with <div> or <span> tags.
2. JavaScript is used along with an XML-encoded communication stream to update the content inside these earmarked locations by accessing a special property of the DIV or SPAN element known as innerHTML.
3. The innerHTML property allows JavaScript to dynamically add HTML content within the DIV or SPAN areas of the HTML page without reloading the page in the user's browser.

An example of this dynamic refreshing is a page that does not require the user to click the refresh button in order to see new content loaded from an XML file or other data source. *Google Maps* was a pioneer in this area and created a map that pulled new information from a data repository without requiring the user to continually refresh her browser. The updated data was pulled in asynchronously using JavaScript code and provided the user with a "seamless" map that could be panned and zoomed without the traditional lag-time associated with page reloading.

The framework which allows this process to happen is referred to as AJAX, or Asynchronous JavaScript and XML. AJAX was made famous in the Web development community by Jesse James Garrett (Garrett online) and further popularized by its use in popular websites such as *Google Maps*, *Google Suggest*, *Netflix*, and *Flickr*.

CSS Positioning

Thus far, we have been using CSS largely as a tool for formatting text. In addition to formatting, the CSS specification also contains a powerful set of properties that can be used for layout purposes. In other words, CSS can also be used for positioning and arranging content. The main positioning property used by CSS is aptly named position. Like the font-size property we discussed earlier, the position property can also be used in a relative or absolute fashion. Relative positioning will position content relative to its normal position, while absolute positioning will position an element anywhere on the page according to exact coordinates. A third value enables fixed positioning, which will position an element relative to the browser window. Relative and absolute positioning are nothing new to the experienced graphic designer or technical communicator; in fact, even popular word processing programs such as Microsoft Word® include options for relative or absolute positioning of images and other content.

In order to use positioning in CSS, one must pair the position property with offset values for left, right, top, or bottom margins. The position property will specify the mode of positioning used by the browser and the offset property will read a value that can then be used to line up and place the content. When working with positioning properties, it is helpful to visualize the browser window as a grid composed of pixels (as it actually is) and then calculate how far in one direction or another you may wish to move a given element. When content is unknown in length or quantity, it is helpful to use relative positioning as this can adapt to different situations depending on the amount of text already on the page. For example, if "footer" content such as a time modification stamp or contact information is embedded at the bottom of each page, relative positioning allows a designer to always position the footer relative to where the last block of text ends on a page. With absolute positioning, one runs the risk of having certain units of text overlapping or otherwise ending in unexpected positions.

The next code example includes a CSS property added to the h2 selector in order to define some relative positioning. This example includes an offset of 50 pixels from the left hand border of the parent element (or the leftmost browser boundary if the selector does not reside within a DIV or SPAN element or other HTML structure such as a table).

```
h2
    {
    font-size: 15px;
    color: blue;
    font-family: Times New Roman, Times, Serif;
    font-weight: bold;
    text-decoration: italic;
    position: relative;
    left: 50px;
    }
```

Let us return to our earlier HTML code example, which uses this h2 selector:

```
<body>
    <h1>This is an example of a <span
    class="importantText">first order</span> heading.</h1>
    <h2>This is an example of a <span
    class="importantText">second order</span> heading.</h2>
    <p>This is an example of a normal paragraph.
    <p class="greenunderline">Here is some <span
    class="importantText">green, underlined</span> text.</p>
</body>
```

Using this HTML, the Web browser will now display the h2 text with an offset of 50 pixels from the left hand border, as shown in Figure 4.5. If you

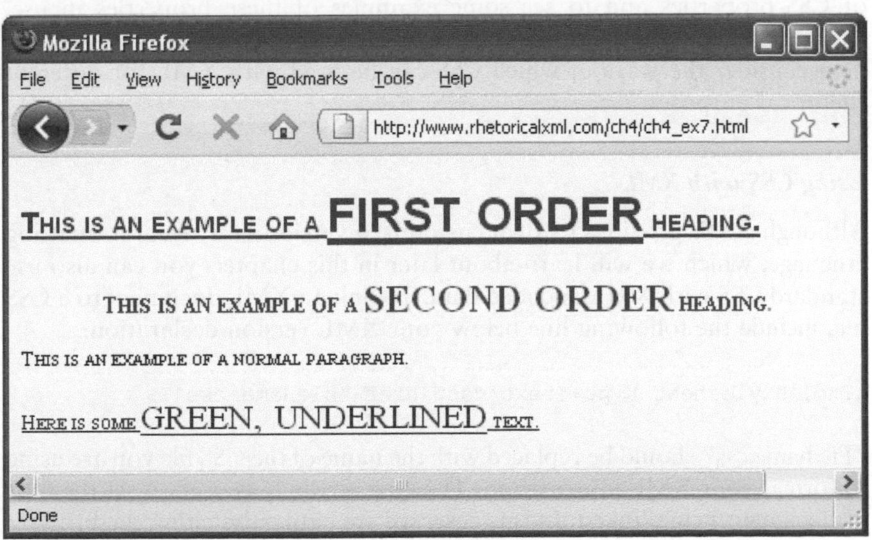

Figure 4.5 Relative Positioning in CSS

are working along on your own, replace the previous h2 CSS code from "ch4_ex6.html" with this new h2 definition and save the new file as "ch4_ex7.html".

Positioning can be tricky in CSS, particularly when working with browser idiosyncrasies. It is a good idea to gradually develop the CSS and test it frequently using various Web browser software. This is more helpful than developing the entire style sheet for a single browser such as Internet Explorer®, Firefox®, or Safari® since you can never fully predict the types of browsers that will be used by your visitors. You can, however, make some informed guesses and try to cater to the most widely used and popular software.

To view an example of "putting it all together" in terms of positioning, font size, and modularization of classes with HTML, you can visit the text's accompanying website in order to study its style sheet code or download the example CSS files from this chapter. It is helpful to see how the various elements of style sheets combine together in order to create a unified, consistent, and (we hope) rhetorically appealing aesthetic. This code demonstrates many of the properties we have discussed in this chapter as well as new properties such as margin, width, height, border, padding, and so forth. A dedicated CSS reference manual is a necessity for any information designer working with visualization on the World Wide Web.

The many properties of CSS and the ability of this technology to influence the rhetoric of a website is perhaps best demonstrated in "live" fashion through a website such as the *CSS Zen Garden*. A link to this website is provided in the Additional Online Resources at the end of this chapter and a visual rhetoric activity using *CSS Zen Garden* is also outlined in the end-of-chapter activities. You may wish to visit the sites listed in the Additional Online Resources section of this chapter to browse through the full listing of CSS properties and to see some examples of these properties in use. Additional resources are also available as links from this text's website. We next consider the ways in which CSS can be used with XML for different rhetorical purposes.

Using CSS with XML

Although XML provides its own customized version of a visual formatting language, which we will learn about later in this chapter, you can also use standard CSS with XML-encoded data. To link an XML document to a CSS file, include the following line below your XML version declaration:

```
<?xml-stylesheet type="text/css" href="filename.css"?>
```

"Filename.css" should be replaced with the name of the CSS file you are using to format your XML information. The type attribute is used to set the type of document being linked. In this case, we are explicitly telling the browser or parser that we are including a document with textual information—specifically, a CSS file.

Before illustrating the use of CSS with XML, let us return briefly to the issue of rhetorical significance in XML-encoded information. One of the easiest ways to quickly indicate importance or magnitude of data is to use big, bold colors or other visually distinctive techniques. XML data on its own, in an unformatted state, gives equal weight to each segment of data. We can, however, use CSS techniques to format and display data according to particular criteria. We might want to highlight a special-of-the-day on an online menu, indicate sale prices for a warehouse application, or highlight new titles in an online video game collection. CSS makes all of these tasks very simple.

We can follow the video game example further to see how XML and CSS integrate. Let us imagine a video game enthusiast with a passion for playing video games of all genres and platforms. Her collection has grown out of hand and she needs a robust information architecture to better manage her collection and preserve her sanity when she wants to find a particular game. If this video game enthusiast were to encode her video game collection using XML, she could develop a series of elements such as GAME, TITLE, DEVELOPER, and GENRE to better classify and organize her collection by building facets important to her own informational needs. She could then develop XML tags for these elements along with an external style sheet named "gamestyle.css" that she could gradually edit in order to influence how the information looked when it was retrieved from her XML database and displayed on her computer screen. After adding a few games, her initial XML file would look something like this:

```
<<?xml version="1.0" encoding="ISO-8859-1"?>
<?xml-stylesheet type="text/css" href="gamestyle.css"?>
<xbox360-collection>
      <game>
            <title>BioShock</title>
            <developer>2k Games</developer>
            <purchase_cost>59.99</purchase_cost>
            <genre>First Person Shooter</genre>
            <release_year>2007</release_year>
      </game>
      <game>
            <title>Nhl 2k9</title>
            <developer>Visual Concepts</developer>
            <purchase_cost>59.99</purchase_cost>
            <genre>Sports</genre>
            <release_year>2008</release_year>
      </game>
      <game>
            <title>Street Fighter IV</title>
            <developer>Capcom</developer>
            <purchase_cost>59.99</purchase_cost>
```

```
            <genre>Fighter</genre>
            <release_year>2009</release_year>
      </game>
</xbox360-collection>
```

Note that the gamer is using an encoding attribute of ISO-8859–1, which is a standard character encoding of the Latin alphabet. The rest of the code shown in this example should be familiar by now as it contains only elements, the XML version declaration, and the external style sheet instruc-

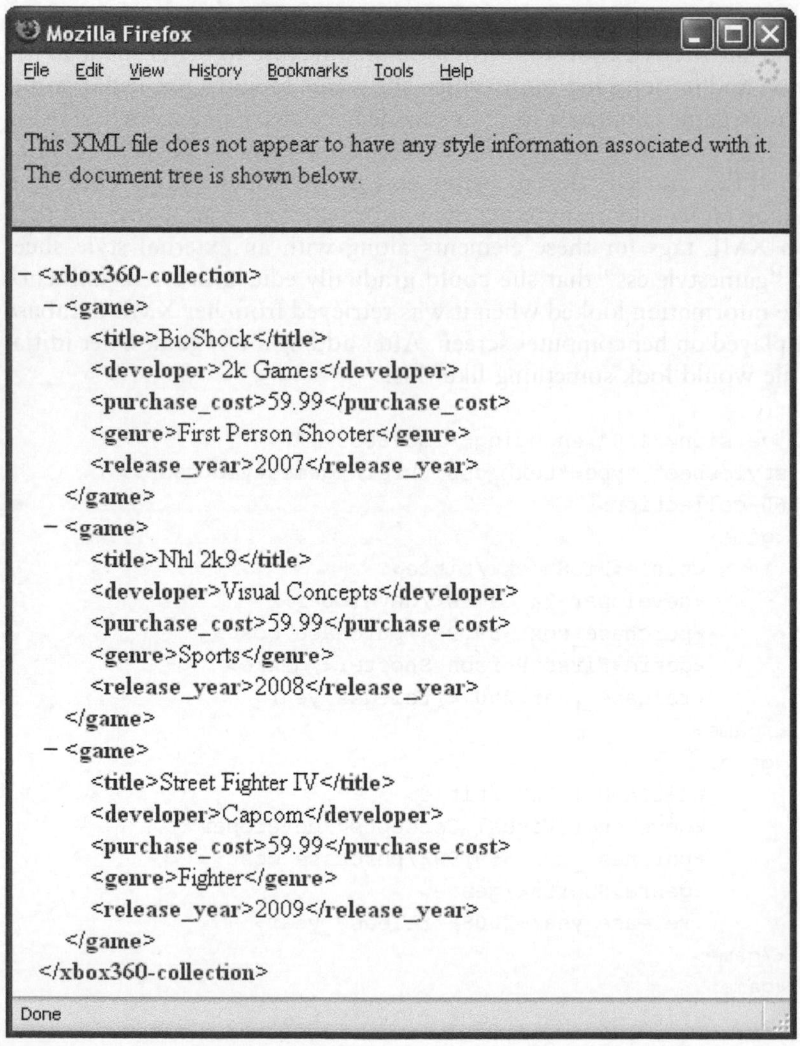

Figure 4.6 XML Content with No Style Sheet

tion. If you are working along with the example, save this XML file as "Ch4_gamelist.xml".

If our hypothetical gamer were to remove the external style sheet instruction found in the second line of code, she would see the output shown in Figure 4.6 when viewing her XML code in a Web browser.

Adding the missing line back in, though, produces a very different output, which is shown in Figure 4.7. So, what has happened here? Now that we have linked the style sheet to the XML document, the browser knows that it is being told to render the content within the XML file according to a specific and precise set of visual formatting instructions. When the external style sheet is missing, as in Figure 4.6, the XML tags are ignored and only the character data within them is displayed. Although the XML output shown in Figure 4.7 is admittedly not very impressive, it does now contain a linked style sheet. No style information is shown only because we have not yet provided any CSS rules for that style sheet.

In order to have more visually appealing output, the gamer needs to create the "gamestyle.css" file and save it in the same directory as the XML file (or otherwise modify the href attribute to point to an alternate directory). The gamer might begin by adding a formatting command to display text within the title elements in a red, bolded, Helvetica font. That CSS rule looks like this:

```
title
      {
      color: red;
      size: 18px;
      font-family: Helvetica, Arial, Sans-Serif;
      font-weight: bold;
      }
```

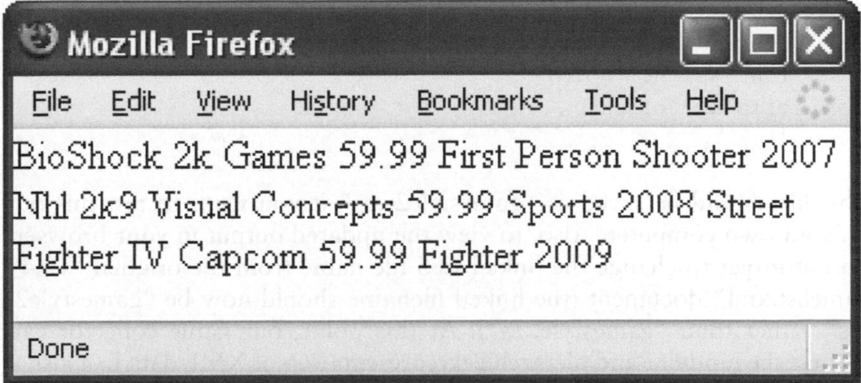

Figure 4.7 XML Content with Missing Style Sheet

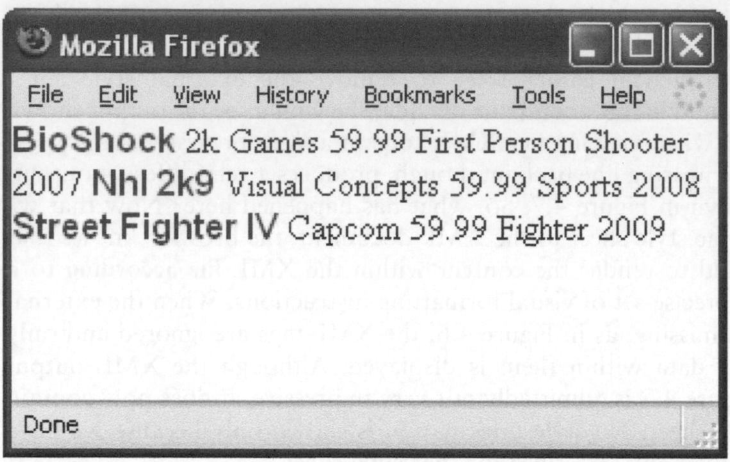

Figure 4.8 XML with CSS

The rule is applied in exactly the same way as it would be in an HTML file. In this case, title is the selector, and color, size, font family, and font weight are the properties. Red, 18px, bold, and the list of fonts are values. As you might expect, the output of this rule will be similar to that shown in Figure 4.7, with the exception of the new rule being applied to the data inside the <title> tags. The output of "gamestyle.css" is shown in Figure 4.8.

In order to make this more readable, the gamer should add block-level formatting instructions to add spacing before and after the XML tags. By adding a CSS rule which specifies the value block for the display property, she will be instructing the browser to add a line break both before and after the content of that element. The new code looks like this:

```
title
        {
        color: red;
        size: 18px;
        font-family: Helvetica, Arial, Sans-Serif;
        font-weight: bold;
        display: block;
        }
```

This file should be saved as "gamestyle2.css" if you want to test this out on your own computer. Also, to view the updated output in your browser, do not forget to change the linked CSS file name from the original "Ch4_gamelist.xml" document (the linked filename should now be "gamestyle2.css" rather than "gamestyle.css"). At this point, our game collector can enforce the modular and hierarchical representation of XML data in a visual fashion, by using line breaks to show the manner in which each catalog item is encapsulated. Her page so far is captured in Figure 4.9.

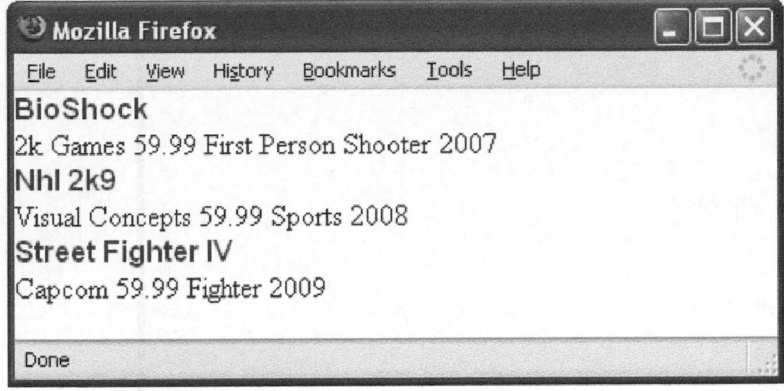

Figure 4.9 Block-level Formatting

Though the display has improved, the information represented in developer, purchase_cost, genre, and release_date is still pasted together on a single line. To finish her style sheet, she can add another CSS rule to format and apply block-level formatting to each of these elements.

```
developer, purchase_cost, genre, release_year
      {
      color: blue;
      size: 14px;
      font-family: Helvetica, Arial, Sans-Serif;
      display: block;
      position: relative;
      left: 10px;
      }
```

Here, in "gamestyle3.css", she is applying the same style to four different selectors by separating them with commas before the opening curly bracket of the CSS property and its value list begins. The new output (Figure 4.10) now reflects the encapsulation of information within each game container as defined in her original XML file. This is accomplished using both block-level formatting for spacing as well as relative positioning to offset each unit of text 10 pixels from the left hand margin.

At this point, the gamer has a structured XML file for archiving her list of games as well as a *separate* mechanism for viewing this list in a browser in a fashion other than the typical XML hierarchy produced by default in a Web browser (as you will see if you refer back to Figure 4.6). When adding new games to her database, she does not need to worry about the details of formatting and layout. These CSS rules will automatically be applied to any new game element within the XML catalog. Similarly, if she were to change the formatting or layout options of her display page, she would not need to worry about changing anything in the XML data file. She would

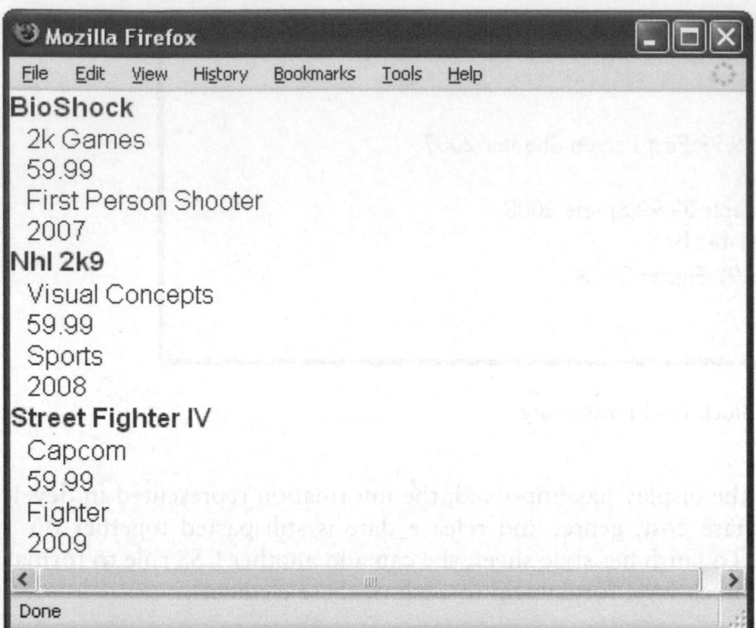

Figure 4.10 Fully-formatted XML Data

only need to modify the linked external style sheet. This convenience is another advantage of separating semantically meaningful content from its visual presentation.

As a final example of using CSS with XML, let us consider the side-by-side comparison of two visual representations of this video game data as produced by more sophisticated style sheets. Here is a new style sheet that can be saved as "gamestyle4.css":

```
title
    {
    background: url(images/icon_1.gif) no-repeat;
    color: #000;
    font-family: Astonished;
    font-size: 55px;
    font-weight: bold;
    padding-left: 15px;
    margin-top: 25px;
    display: block;
    }
developer, purchase_cost, genre, release_year
    {
    background: url(images/bg_1.gif) repeat-x;
    color: #BD1A8D;
```

```
    font-family: Astonished;
    font-size: 24px;
    font-weight: bold;
    border: 1px dashed #000;
    padding: 3px 3px 3px 10px;
    margin: 3px;
    display: block;
    position: relative;
    left: 10px;
    }
```

In this CSS code, we see that additional formatting and layout instructions have been provided in order to generate a more substantial aesthetic for our XML document. Most notably, we are using a new CSS property, named url, to embed small images into the background of each element. In this case, "icon_1.gif" is a small circle that fades from purple to white and "bg1.gif" is an even smaller image containing a gradient that fades from gray to white. This image is tiled across the background of each developer, purchase_cost, genre, and release_year element.

Here is the last style, which should be saved as "gamestyle5.css":

```
title
    {
    background: #CCDEF3 url(images/bg_2.gif) no-repeat;
    color: #000;
    font-family: Georgia;
    font-size: 28px;
    border: 2px dotted #207AD3;
    padding-left: 15px;
    margin-top: 25px;
    display: block;
    }

developer, purchase_cost, genre, release_year
    {
    color: #4F4F4F;
    font-family: arial;
    font-size: 14px;
    font-weight: bold;
    border-bottom: 1px dotted #34A047;
    border-left: 1px dotted #34A047;
    padding: 3px 3px 3px 10px;
    margin: 5px;
    display: block;
    position: relative;
    }
```

Note the changes that are introduced in "gamestyle5.css". We are no longer using the original graphic for the background image, but we are instead using a new image named "bg_2.gif". This is a blue gradient that will be used as a background for the title text of each element. In addition, we have changed the font from "Astonished" for all elements to "Georgia" for the title and "Arial" for the additional elements. Font sizes have been reduced and borders and padding have been slightly adjusted.

The output from this comparison of "gamestyle4.css" (on the left) and "gamestyle5.css" (on the right) is shown in Figure 4.11. Although each Web page is using the same exact XML data with only minor variations in imagery, the two presentations clearly present different rhetorical messages. By carefully varying the stylistic rules contained within these two new CSS files ("gamestyle4.css" and "gamestyle5.css") an author is able to craft different presentation packages for different informational purposes and audiences.

The left panel of the screen capture in Figure 4.11 ("gamestyle4.css") uses a more ornate style and might cater to a younger audience or even to the collector herself (if used internally for her personal collection). A non-standard and slightly unusual font is used, which makes the textual

Figure 4.11 CSS and Visual Rhetoric

information stand out and seem more impactful. Bolder purple colors (which you can see on our website in the color version of Figure 4.11) suggest excitement and energy. Not surprisingly, conversion to grayscale (resulting in the image you see produced in this book) alters the rhetorical subtext yet again, and the overall aesthetic presentation could be described as gloomy or even apocalyptic (the shaky style used by the primary font is mildly suggestive of instability and disorder).

We could read even further into the rhetorical message of the left panel if we so desired; for instance, we might compare the visual style with the rhetorical substance of each game that is represented by the textual XML tags. In Bioshock, which is a first person shooter game that takes place underwater, the visual message is somewhat consistent with the textual content. The shaky font can be associated with water droplets and the overall aesthetic is certainly consistent with the dark and gloomy artistic style found in most video games of this type. NHL 2K9, on the other hand, is a sports game which would not fit so well with this type of design. So, in terms of an overall strategy for formatting and layout, this visual style would be more suitable for some games than others. Obviously, it is not always feasible to create separate CSS rules for every XML element in a document, but it is certainly possible to find a visual strategy that works well across the entire document to display information in a fashion suitable for your audience and for their informational needs.

On the right hand side of this screen capture, however, a more conventional design is used by "gamestyle5.css". This type of aesthetic is more common in corporate settings and would probably be more likely to blend in with existing website content. In the online version shown on our website, the blue color is calming and nonthreatening, but even in the grayscale version printed here, the visual style is shown to be smoother and less chaotic. Most audiences will be familiar with the Arial and Georgia fonts and will interpret the overall aesthetic as more traditional and "corporate"-looking. Clean lines suggest confidence and conformity. Several rhetorical questions can be generated from this side-by-side comparison:

- If each of these websites were selling products, which site would you be more likely to purchase from?
- Which website would your parents be more likely to purchase from? Which would teenagers be more persuaded by?
- What do the fonts suggest about the ethos of the author? Do both sites appear to be constructed by credible and trustworthy authors?
- How does each website influence your emotions? Is one more boring than the other? Do either of the sites make you feel a certain way? If so, how might companies take advantage of these feelings?

As we have shown in this example, even when working with primarily textual documents such as XML, we can generate and respond to different rhetorical situations simply by varying our use of rules and properties in

CSS. Even small images can produce large rhetorical changes in a document. Oftentimes, the selection and application of an appropriate visual style is one of the most complicated and time consuming tasks for an XML author. When possible, it is very helpful for technical document authors and graphic designers to work together closely during this part of the design process to ensure a consistent rhetorical message and an optimal mode of presentation for XML content.

See the Additional Online Resources section of this chapter for links to additional online examples of CSS in action and for a link to a full listing of all CSS properties. We will focus the remainder of this chapter on a style sheet language developed specifically for XML: the XSL.

XSL

Although using CSS with XML will work just fine, it is also useful to learn to use XSL as that technology was developed from the ground up by the W3C specifically to provide an accompanying style sheet language for XML. In the first half of this chapter, we explained that style sheets for HTML are generally straightforward to work with. One needs only to decide upon a series of rules to define the functionality of a CSS and then to decide how to implement that sheet, either directly within their HTML document, or by using an externally linked file.

Unfortunately, the same process in XML is slightly more cumbersome, if only for the number of acronyms and abbreviations one must be familiar with in order to implement XML style sheets: XPath, XPointer, XSL, XSLT, and eXtensible Style Sheet Language Formatting Objects (XSL-FO) are just a few of the most important ones we discuss here. We will briefly define XPath and XPointer in this chapter, but these will be explained in more detail in Chapter 5.

Like CSS, XSL is a style sheet language, but it is specifically designed for XML documents. XSL is made up of three parts:

1. XPath, which is a language for *navigating* in XML documents.
2. XSL-FO, which is a language for *formatting* XML documents. XSL-FO is also referred to simply as XSL.
3. XSLT, which is a language for *transforming* XML documents.

XPath is a language that is used to access and describe certain parts of an XML document. Using what are known as path expressions, XPath allows a document designer to create formulaic instructions for accessing portions of an XML document. XPath is therefore concerned with navigating XML content.

XSL-FO is to XML what CSS is to HTML. XSL-FO, which stands for eXtensible Style Sheet Language Formatting Objects, is now simply known as XSL. XSL-FO and XSLT are often used together to both format and transform XML data from one form to another.

In the remainder of this chapter, we focus on XSLT. XSLT is perhaps the most powerful subset of XSL because it deals with transformations. A transformation can be defined as the movement of XML data from one file format into another. Let us begin our look at XML transformations by considering a new XML document named "Ch4_garden.xml":

```
<?xml version="1.0" encoding="ISO-8859-1"?>
<!– Chapter Four example using XSL –>
<!DOCTYPE garden [
 <!ELEMENT garden (fruit|vegetable)*>
 <!ELEMENT fruit (name,supplier,units_available,price_unit,
 price)>
 <!ELEMENT vegetable (name,supplier,units_available,price_
 unit,price)>
 <!ELEMENT name (#PCDATA)>
 <!ELEMENT supplier (#PCDATA)>
 <!ELEMENT units_available (#PCDATA)>
 <!ELEMENT price_unit (#PCDATA)>
 <!ELEMENT price (#PCDATA)>
]>

<garden>
      <fruit>
            <name>Vine Red Tomatoes</name>
            <supplier>Henderson Farms</supplier>
            <units_available>45</units_available>
            <price_unit>Bushel</price_unit>
            <price>25.00</price>
      </fruit>
      <fruit>
            <name>Watermelon</name>
            <supplier>Rand Melon Supply</supplier>
            <units_available>350</units_available>
            <price_unit>Individual</price_unit>
            <price>1.50</price>
      </fruit>
      <vegetable>
            <name>Green Leaf Lettuce, Black-seeded Simpson
            </name>
            <supplier>A&amp;H Local</supplier>
            <units_available>2000</units_available>
            <price_unit>Individual</price_unit>
            <price>.25</price>
      </vegetable>
```

```
    <vegetable>
        <name>Green Leaf Lettuce, Grand Rapids</name>
        <supplier>A&amp;H Local</supplier>
        <units_available>700</units_available>
        <price_unit>Individual</price_unit>
        <price>.65</price>
    </vegetable>
</garden>
```

This example is both well-formed and valid as it adheres to a DTD within the file (refer back to Chapter 2 if you need a refresher course on DTDs) and is syntactically correct. Although this is a perfectly valid hierarchy of information that might be useful for a grocer looking to assess his inventory, it is much easier for a computer to "read" than a human (see Figure 4.12).

By using an XML style sheet, we can transform this XML content into normal HTML content by using XSL code. XSL documents are similar to XML documents except they end in the ".xsl" file suffix and are crafted using a special set of XML elements and attributes designed for the transformation process. We transform XML documents by creating an XSL document that searches our XML content for patterns and then applies a template to replace XML tags with HTML tags. Here is what the XSL document "Ch4_garden.xsl" looks like in its entirety:

```
<?xml version="1.0" encoding="utf-8"?>
<xsl:stylesheet version="1.0" xmlns:xsl="www.w3.org/1999/XSL/
Transform">
    <xsl:template match="/">
        <html>
            <head>
                <title>Fruits and Vegetables</title>
            </head>
            <body>

                <!– transform fruits into a table –>
                <h1>Fruits</h1>
                <table border="1">
                <tr>
                    <th>Name</th><th>Supplier</th>
                    <th>Units</th><th>Pricing Unit
                    </th>
                    <th>Price</th>
                </tr>
                <xsl:for-each select="garden/fruit">
                    <tr>
            <td><xsl:value-of select="name"/></td>
            <td><xsl:value-of select="supplier"/></td>
```

```
                <td><xsl:value-of select="units_available"/>
                </td>
                <td><xsl:value-of select="price_unit"/></td>
                <td><xsl:value-of select="price"/></td>
                      </tr>
                   </xsl:for-each>
                   </table>

                   <!- next, transform veggies ->
                   <h1>Vegetables</h1>
                   <table border="1">
                   <tr>
                         <th>Name</th><th>Supplier</th>
                         <th>Units</th><th>Pricing Unit
                         </th>
                         <th>Price</th>
                   </tr>
                   <xsl:for-each select="garden/
                   vegetable">
                         <tr>
                <td><xsl:value-of select="name"/></td>
                <td><xsl:value-of select="supplier"/></td>
                <td><xsl:value-of select="units_available"/>
                </td>
                <td><xsl:value-of select="price_unit"/></td>
                <td><xsl:value-of select="price"/></td>
                      </tr>
                   </xsl:for-each>
                   </table>

                </body>
          </html>
      </xsl:template>
</xsl:stylesheet>
```

Now, we will move through this document and explain each part of the code. First, we provide our XML declaration as usual:

```
<?xml version="1.0" encoding="utf-8"?>
<xsl:stylesheet version="1.0" xmlns:xsl="www.w3.org/1999/XSL/
Transform">
```

In this case, we are using UTF-8 encoding, which is an 8-bit Unicode Transformation Format that can encode Unicode characters. Unicode is a set of around 100,000 characters which can display most of the world's pictographic symbols. Using Unicode gives us much flexibility in working

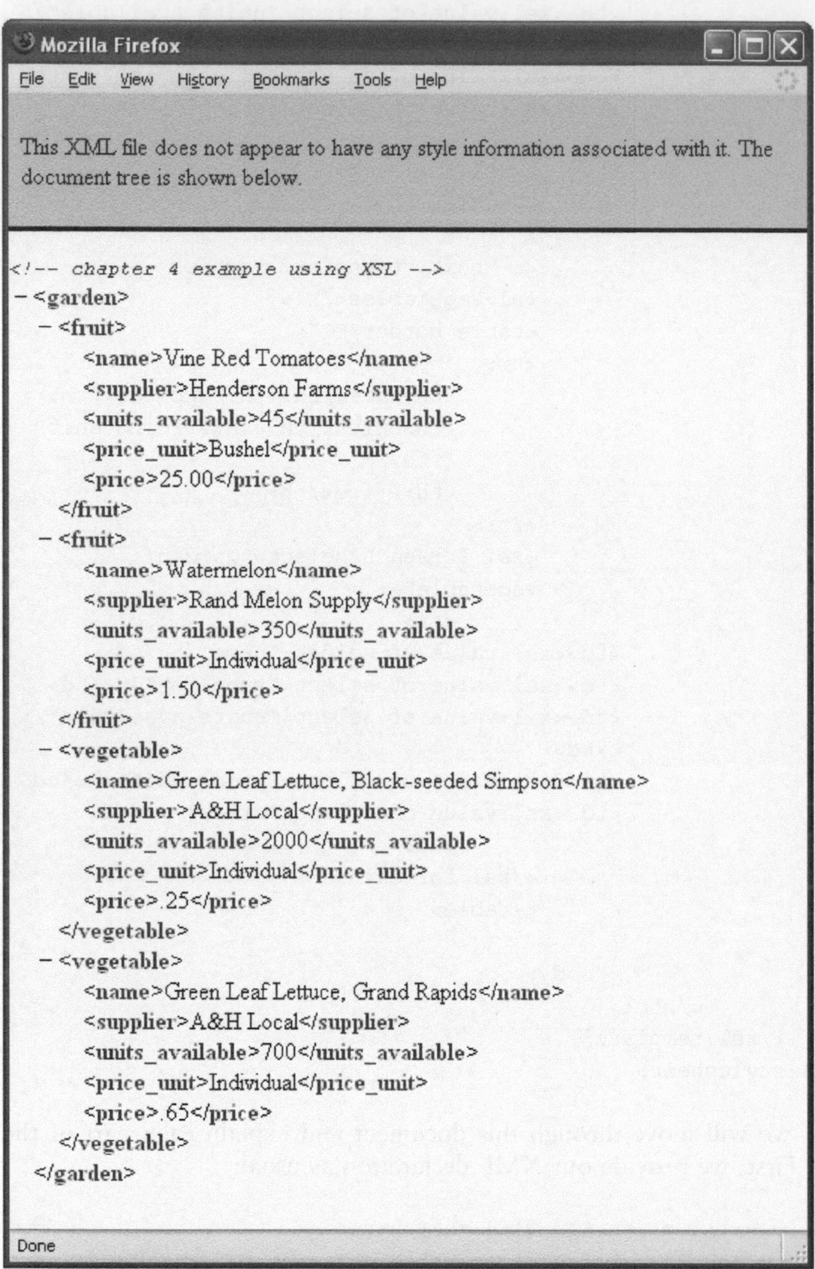

Figure 4.12 XML Garden File

with international languages or other specialized applications that require non-Latin character sets.

In the second line of code, we link to a special namespace using the "xmlns" attribute and the "xsl" prefix. This information tells the parser that we will be linking all xsl prefixed tags to the XSL namespace. Namespaces are described in more detail in the next chapter.

```
<xsl:template match="/">
```

The <xsl: template match="/"> code looks for the root element in the XML document. It will then insert the HTML tags, data, and the comments that begin with <!— and end with —> into the browser or parser.

The template tag and match attribute are used for searching for specific tags within the XML file and then applying a template of HTML or XML code to those particular tags or their data. In this case, we are just searching for the document or root element and then adding in code to create a new HTML document and begin a table that will hold the fruits from the XML file. In HTML, the <tr> tags are used to create new table rows, the <th> tags are used to create table heading columns, and the <td> tags are used to create normal table columns. We include a value of 1 for the table border attribute to visually reflect the separations between columns, rows, and headings. The HTML code we need so far looks like this:

```
<html>
        <head>
                <title>Fruits and Vegetables</title>
        </head>
        <body>

                <!— transform fruits into a table —>
                <h1>Fruits</h1>
                <table border="1">
                <tr>
                        <th>Name</th><th>Supplier</th>
                        <th>Units</th><th>Pricing Unit</th>
                        <th>Price</th>
                </tr>
```

Our next task is to extract the data from the fruit elements and display it in the HTML table we have created. We can use the "for-each" XSL tag to accomplish this along with an XPath expression for the select attribute. The "for-each" tag acts as a looping mechanism to find each data node that is of type garden, then fruit. The front slash is used to show the progression down the XML hierarchy and is used in XPath to search for particular elements within the document tree. The parser will then follow down the XML tree looking for each element of type garden, fruit, and then extracting the data from the children elements using the "value-of" tag and the "select" attribute. The "value-of tag" instructs the parser to remove the values of the name, supplier, units_available, price_unit, and price elements and paste them in between new table columns <td> and </td>. This entire process will be repeated for each garden, fruit element within the XML document. Finally, the fruits table will be closed using the standard </table> HTML tag. Here is the new code with the XSL instructions added:

```
<xsl:for-each select="garden/fruit">
     <tr>
          <td><xsl:value-of select="name"/></td>
          <td><xsl:value-of select="supplier"/></td>
          <td><xsl:value-of select="units_
          available"/></td>
          <td><xsl:value-of select="price_
          unit"/></td>
          <td><xsl:value-of select="price"/></td>
     </tr>
     </xsl:for-each>
</table>
```

This process is then repeated for the vegetable elements. The only detail that has changed here is that the parser will be searching for elements of type garden, vegetable rather than garden, fruit. Here is the code for the vegetables:

```
<!– next, transform veggies –>
<h1>Vegetables</h1>
<table border="1">
<tr>
     <th>Name</th><th>Supplier</th>
     <th>Units</th><th>Pricing Unit</th>
     <th>Price</th>
</tr>

<xsl:for-each select="garden/vegetable">
     <tr>
          <td><xsl:value-of select="name"/></td>
          <td><xsl:value-of select="supplier"/></td>
          <td><xsl:value-of select="units_
          available"/></td>
          <td><xsl:value-of select="price_
          unit"/></td>
          <td><xsl:value-of select="price"/></td>
     </tr>
     </xsl:for-each>
</table>
```

At this point, all of the vegetables and fruits have been extracted from the XML document. The remainder of the code closes the open HTML tags and then the XSL template tag that was used to select the document root at the beginning of this process. Finally, the style sheet tag is closed so that the document is fully well-formed (syntactically correct). Here are the last few lines of XSL:

```
</body>
</html>
</xsl:template>
</xsl:stylesheet>
```

Now, we must find a parser that is capable of applying our transformation to our original data file. This XSL transformation can be applied automatically, by linking the XSL style sheet to the XML file, or manually, by using an XML editor such as XML Blueprint to apply the style sheet and preview the results. To link the style sheet directly to the XML file, we must add the following line to the top of the XML file:

```
<?xml-stylesheet type="text/xsl" href=
"stylesheet_name.xsl"?>
```

The first attribute, type, specifies that the linked document will be textual and will contain XSL content. The next, href, specifies the exact filename of the document to be linked. So, for this example, the beginning of our XML file now looks like this:

```
<?xml version="1.0" encoding="ISO-8859-1"?>
<!— Chapter Four example using XSL —>
<?xml-stylesheet type="text/xsl" href=
"Ch4_garden.xsl"?>
<!DOCTYPE garden [
  <!ELEMENT garden (fruit|vegetable)*>
  <!ELEMENT fruit (name,supplier,units_available,price_unit,
  price)>
  <!ELEMENT vegetable (name,supplier,units_available,price_
  unit,price)>
  <!ELEMENT name (#PCDATA)>
  <!ELEMENT supplier (#PCDATA)>
  <!ELEMENT units_available (#PCDATA)>
  <!ELEMENT price_unit (#PCDATA)>
  <!ELEMENT price (#PCDATA)>
]>
```

When we open this file in Mozilla Firefox®, we now see the familiar HTML output shown in Figure 4.13. We have effectively applied a style

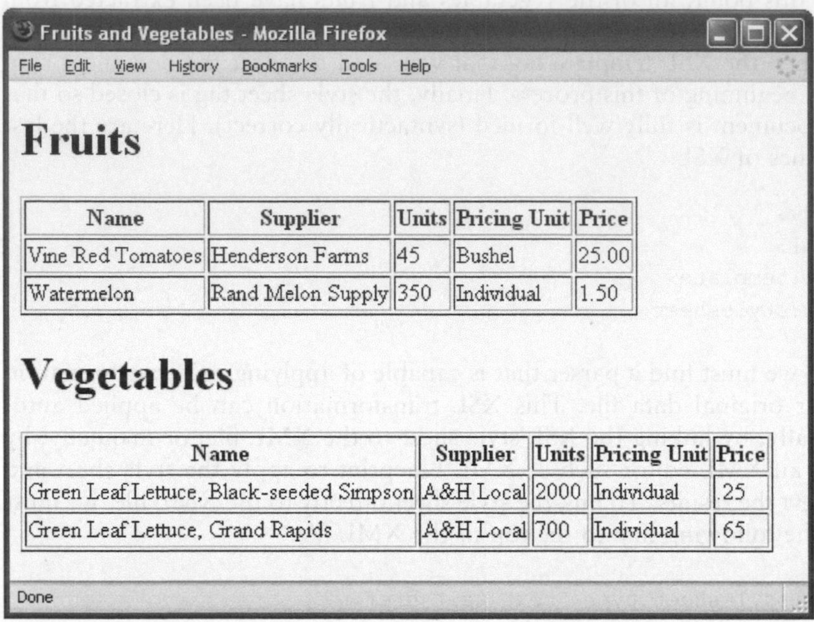

Figure 4.13 XML to HTML Conversion Results

sheet to transform a document from XML to HTML format. This output can then be further customized and beautified using CSS.

XSL transformations are capable of transforming one XML document into another XML document, or for translating XML documents into several other types of formats such as eXtensible Hypertext Markup Language (XHTML), scalable vector graphics (SVG), or Virtual Reality Modeling Language (VRML) —not just HTML. This transformative feature is useful when knowledge management systems must exchange their data across networks; in general, the systems must agree upon a common set of elements and attributes and then one document must be transformed into a format suitable for interpretation by the other system. The ability of XSL transformations to produce any XML-based file format, such as an SVG image file or a VRML file makes this technology extremely powerful and well-suited for advanced knowledge management systems that are implemented using the World Wide Web.

We will return to the topic of XSL in Chapter 5 when we discuss advanced technologies such as XLink, XPath, and XPointer. Since the XSL family contains such a complex and powerful language for transformation and formatting, we need to study some advanced topics such as XML namespaces and XLink before approaching this topic in more detail.

Chapter Summary

Knowledge of visual formatting techniques is very important for those wishing to convey a professional ethos using the World Wide Web as a delivery medium. CSS allow a designer to quickly and efficiently make enterprise-wide changes to what graphic designers call the "look and feel" of a website or portal. Without technologies such as CSS and XSL, making global changes to visual themes would be a more arduous task. Though XSL has a more complicated syntax than CSS, it is better suited for traversing and applying formatting rules to the embedded tags contained within XML data.

CSS and XSL also make websites more efficient: since the code can be stored in external files and linked a single time from each page, the additional time needed to find commands within each file in the site's hierarchy can be trimmed away. Less text in each file means a smaller file size, which in turn means faster loading and transfer times between server and client computers. Overhead is therefore reduced in several different ways.

The front half of this book was primarily focused on theoretical ideas about how information and knowledge can be structured and studied from multiple perspectives and contexts. It also contained an introductory tutorial on XML and DTDs. Starting with this chapter, we began our transition into some of the more advanced technical topics related to XML technologies. These technical topics introduce powerful tools for knowledge managers and professional communicators who wish to work with very large data sources or who hope to build knowledge management tools and practices that span large organizations. With that in mind, we will continue to discuss some of the more advanced topics in XML such as schema, XPath, and namespaces in Chapter 5. Following this last look at XML in isolation, the remainder of the book will focus on fully realized implementations of XML, including three custom-designed XML parsers, and will conclude by examining the ways in which XML may be useful to one's career and studying an interview of five practitioners who use XML in their professional workplaces.

Discussion Questions

1. What additional rhetorical decisions must a writer make when constructing a site for an international audience? How might visual elements need to be tweaked in order to be appropriate for one culture or another? Discuss some specific examples.
2. Write down some of the rhetorical characteristics of different categories of websites (banks, activism websites, environmental agencies, commercial vendors, book stores, etc.) Can you identify common visual characteristics or styles that apply to these categories? In what ways are these characteristics effective or ineffective?

3. Using recent news stories as inspiration, discuss a situation in which the visual rhetoric of a situation led one or more audiences to interpret that situation in a certain way. How might visual elements be changed in order to adjust those first impressions towards another perspective?
4. Consider the problem of negotiating meaning across different organizational units. What types of informational tasks must be performed in order for one data source to work successfully with another? Using ideas from Chapter 1, make a list of the various factors that may influence the way in which certain informational units are created and/or suppressed in the visual translation from one data format to another.

Activities

1. Find a popular website such as CNN, Slashdot, or MSNBC.com and view the source of the home page. See if you can find the location of the linked CSS file. View its source. How is this website handling the separation of descriptive and visual content?
2. Build a simple Web page based on the examples provided in this chapter. Experiment with CSS properties such as *font size*, *font family*, and *color* to design several different external style sheets for your page. Link each sheet to your site using the method explained in this chapter. Watch how quickly and easily you can change the entire visual dimension of your page simply by changing a line of HTML code.
3. Visit the Visual Rhetoric Portal at www.tc.umn.edu/~prope002/visualRhet.htm. Browse through some of these examples and write down how visual elements provide a rhetorical framework for the messages contained in these texts. How might these sites differ with different visual themes and formatting? Are the styles of these sites appropriate for the types of messages they are trying to communicate? Load up several different examples at the *CSS Zen Garden* and discuss these sites using your knowledge of visual rhetoric.
4. Download the "Ch4_garden.xml" file from the website or find the example you saved on your computer earlier in this chapter. Experiment with the "Ch4_garden.xsl" file to produce different transformations of this data. Can you transform the fruits to be listed in an unordered list? Can you make all of the vegetables italicized and bold? Try using your knowledge of HTML, CSS, and XML to produce more sophisticated transformations using the garden example or any of your own XML examples from Chapter 2 or your own projects.

References

Garrett, Jesse James. "Ajax: A New Approach to Web Applications." October 8, 2006. http://adaptivepath.com/publications/essays/archives/000385.php.

Leithead, Alastair. "'Blasphemous' Balls Anger Afghans". 2007. BBC News. Retrieved August 27 2007. http://news.bbc.co.uk/2/hi/south_asia/6964564.stm.

Scott, Linda M., and Patrick Vargas. "Writing with Pictures: Toward a Unifying Theory of Consumer Response to Images." *Journal of Consumer Research* 34.3 (2007): 341–56.

Warnick, Barbara. "Online Ethos: Source Credibility in an 'Authorless' Environment." *American Behavioral Scientist* 48.2 (2004): 256–65.

Whitburn, Merrill D. *Rhetorical Scope and Performance: The Example of Technical Communication*. Stamford: Ablex, 2000.

Additional Online Resources

1. W3C CSS tutorial: www.w3.org/Style/Examples/011/firstcss
2. A list of all CSS properties listed alphabetically: www.blooberry.com/indexdot/css/propindex/all.htm
3. A list of HTML hexadecimal color codes: http://html-color-codes.com
4. CSS Zen Garden: www.csszengarden.com
5. W3C Schools XSLT tutorial: www.w3schools.com/xsl
6. W3C XML Path Language (XPath) Recommendation: www.w3.org/TR/xpath
7. W3C Schools XSL-FO Tutorial: www.w3schools.com/xslfo/xslfo_intro.asp
8. W3C XSL Page: www.w3.org/Style/XSL
9. XML Blueprint Software: www.xmlblueprint.com

5 Advanced Concepts in XML

Namespaces, Schemas, XLink, XPath, XPointer, DITA, and DocBook

Chapter Overview

So far, we have seen the ways in which XML can be used to structure data in rhetorically meaningful ways and then display that data in a visually appealing fashion. In this chapter, we will learn about some of the advanced tools and techniques available to technical communicators and other professionals working with XML. We provide an overview of several advanced topics and show how they are important to the overall landscape of rhetorically thoughtful XML design.

We combine these topics into a single chapter largely because of their interrelationships. For example, one cannot easily work with schemas without understanding namespaces. Similarly, it is impossible to get very far with XLink or XPointer without understanding how XPath expressions are formed. For this reason, we have grouped these topics into a single unit. Our last two topics in this chapter, DITA and the DocBook language, are popular implementation schemes based on XML. DITA and DocBook are used for authoring, organizing, and delivering structured technical information. They operate using DTDs or schema for validation and are useful in single sourcing applications and for building CMS.

Although this chapter is named "Advanced Concepts in XML," none of the material discussed here is particularly complex or confusing in principle. While the basic ideas behind these principles are easy, the syntactical implementation can be somewhat difficult to grasp, at least at first. When mastered, however, these techniques add a significant amount of flexibility to a document creator's repertoire. For this reason, we feel it is important to introduce these topics and to discuss them at least from a broad perspective. Once one is exposed to the basics of namespaces, schemas, and the other "x-languages" (XPath, XPointer, and XLink), it is easy to develop further competency with these techniques based on the particular needs of one's own projects. Similarly, a basic understanding of DITA and/or DocBook provides a document author with a standardized repertoire of element names and a common encapsulating strategy that allows information to be reused and shared with different types of parsers and translation tools. DITA and DocBook also require authors to chunk information into discrete units; this practice helps to develop the types of rhetorical skills necessary for this alternate form of writing and the technical skills necessary for working in structured writing environments.

Namespaces

In Chapter 2, we discussed how namespaces can be used to differentiate between common names stored in different databases. Specifically, we discussed how namespaces could be used to differentiate between XML elements used for the WHO and the HHS. In this chapter, we will further examine namespaces in order to understand how they integrate with more advanced concepts associated with XSL, XLink, XPath, and XPointer.

As we learned in Chapter 2, an XML namespace is an encapsulating unit that defines the scope in which particular elements are defined and used. This encapsulating unit is *virtual* in the sense that it is created by a line of code that tells an XML parser or software program which XML document to use as a point of reference when looking at elements and attributes. Namespaces are not unique to XML. In fact, many programming languages (such as C++) have long used namespaces to differentiate between repeated names in programs that might have different meanings in different algorithmic contexts. Knowledge managers also need to be familiar with this convention as there are numerous situations in which different organizational groups may use their own jargon or terminology to refer to the same concept or idea from a particular perspective. This issue is especially important when document designers need to combine multiple XML documents into a single database or into a unified collection of databases.

The W3C writes about this problem in terms of "recognition" and "collision." When multiple tag collections are using the same element names to describe their data, the software programs that interpret and act upon that XML data may have a hard time determining the correct way to react. Recognition and collision problems both contribute to a sense of rhetorical ambiguity in that they obfuscate meaning and complicate the process of working with combined documents.

A recognition problem is experienced when XML tags are used outside of the original document in which they are defined, perhaps resulting in their misinterpretation or misuse. For example, consider a situation in which the logistics manager of a grocery store chain needs to combine the main grocery database with an engineering reference database. The manager wants to investigate a new method for manufacturing the crates that are used to ship apples from her main warehouse in Kansas to a particular store location anywhere in the United States. The engineering process is somewhat complicated and requires a substantial amount of information from both databases. From the main grocery database, she needs information about typical stocking demands for different store locations, the life cycle of perishable goods, classes, and categories of apples and the typical bushel weights for these categories, and other types of information relevant to the shipping process. From the engineering database, she needs to select information describing the capacity of certain box sizes, the placement and locations of screw holes, the amount of raw materials necessary for the production of the crates, and so forth.

In this example, a recognition problem could be encountered for any of the XML tags defined for the individual items residing in the original grocery store or engineering databases. The new combined database might be unable to properly classify element names such as "apple_type" and values such as "Earligold", "Delicious", and "Fuji", which are meaningful names and values only within the master grocery database. Similarly, the new combined database might not recognize engineering elements like "screw_type" and values such as "wood", "machine", and "sheet metal". Recognition problems like these are common when multiple XML documents are combined into new forms. Without a means of linking a traveling XML document to its original starting point, the original context of meaning from which that document emerged is impossible to recognize.

A collision problem is found when multiple tags with the same name but different semantic meanings are present in the same document. For instance, in the example above, the grocery and engineering databases might each have used the tag <type> to describe both apples and screws. In this case, how would the new document be able to tell one element from another without requiring a detailed examination of the element's associated information? Another example might be the element "produce". This element could refer to a noun (as in a specific type of food) in the grocery database and a verb (as in the act done to manufacture a product) in the engineering database. Without a mechanism for defining these elements within their original contexts, the combined database has no way of easily differentiating one meaning of an element from another.

Collision problems are also problematic when different libraries use different DTDs or schemas to enforce the validity of the data. A namespace is a mechanism for removing this ambiguity and for addressing both the recognition and the collision problems. A namespace therefore serves as the link from traveling XML documents to their original, contextually-rich, and properly named and identified locations. Much like a passport defines a traveler's country of origin (and, by rhetorical association, her likely primary language, her cultural characteristics, and certain societal conventions like the form of currency she uses and certain legal rights she may hold) a namespace provides a link from XML documents back to their carefully defined origins. However, namespaces are much more precise than passports in that they clearly and unambiguously define the meaning of elements as they were originally created by a document designer. A certain amount of guesswork, generalization, and stereotyping is involved when trying to learn about a person from her passport profile, but when we use a namespace, we can be directed to the actual source of the data where we can inspect it. For example, we could determine in the example we just described just what is meant by the term "<type>". Is it an apple, or a screw, or something else entirely? Because of this rhetorical power, namespaces are critically important when any problems of recognition or collision are likely to occur.

The importance of clear and precise language in combined documents is also important considering the interdisciplinary practices of the many different discourse communities that use XML to communicate information. For example, in certain scientific communities, there are words that have different meanings depending on the discourse community in which they are used. Even a general word like "research" may have different contextual meanings to a group of biologists, librarians, psychologists, or sociologists. A biologist may immediately think of laboratory data and microscope slide imagery, a librarian might associate the word with a thorough literature review, a psychologist with empirical data and experimental designs, and a sociologist with field studies and focus groups.

A namespace is nothing more than the XML equivalent of stating "I am talking about this word X from the context of Y," where Y represents a particular group's or organization's point of view. To continue our example, using a namespace would be like stating "I am talking about the word 'research' from a librarian's point of view." We now have a better idea about what this word "research" entails in this context—performing electronic searches, using the card catalog, or perhaps even poring over old volumes looking for a specific passage of text. By switching our namespace from "librarian" to "psychologist," we would then alter the various connotations associated with the word. We would then think about different activities and tasks associated with psychologists, such as working with participants and performing statistical analysis on various types of data sets. These perceptions and generalizations may or may not be accurate, but with XML, we can provide very clear and unambiguous guidelines without falling prey to the same types of biases and stereotypes we as humans are susceptible to.

In the case of an ambiguous reference, the XML namespace definition defines the context in which a particular element or attribute exists. For instance, it is perfectly likely that both an XML document and an HTML document would contain a TITLE element. In order to differentiate between the HTML <title> tag and the XML <title> tag, a namespace can be used to identify which DTD (or schema, as we will learn about later in this chapter) should be consulted in order to enforce the validity of that tag. You may recall from Chapter 2 that a *valid* XML document is one that conforms to the requirements of its DTD or schema; a *well-formed* XML document is one that is written according to the syntactical requirements of XML.

An XML namespace definition is recognizable through the special XML attribute "xmlns". Xmlns is a "reserved" word, meaning that it cannot be used as a user-defined name for an element or attribute and that it has special meaning for XML parsers. The value of this xmlns attribute will be a URI,

Table 5.1 URI Examples

URI Example	*Description*
www.rhetoricalxml.com	a URI for the website of this book
mailto:rudy@mail.ucf.edu	a URI used to launch an e-mail client
tel:+1–888–555–1234	a URI used to hold a telephone number
urn:oasis:names:specification: docbook:dtd:xml:4.1.2	a URI using a URN rather than a URL

which is defined by The Internet Society Network Working Group in the RFC 3986 standards document as a sequence of characters that identifies an online or physical resource. Most Internet users are probably most familiar with the URIs used to identify addresses on the World Wide Web, though URIs can also hold addresses used for FTP or even the clickable e-mail links used to launch a mail client from a website. Examples of URIs are shown in Table 5.1.

As discussed in RFC 3986, a URI can be further decomposed into the well known URL as well as the somewhat lesser known Uniform Resource Name (URN). URLs contain more information than URNs in that the protocol requires that in addition to identifying a resource by name, URLs must also "provide a means of locating the resource by describing its primary access mechanism (e.g., its network 'location')" (The Internet Society online). While these details are important to document experts working on Internet technologies or network protocols, most professionals will do well simply to remember that URI is a more general form of URL, or that a URL is a more specific instance of a URI that includes a mechanism for network location. Many people confuse the two, so it is useful to know the difference.

The W3C Recommendation of August 16, 2006 specifies that XML namespaces are composed of the xmlns attribute, a namespace prefix, and a URI value. An example namespace declaration is found in Chapter 2 and is repeated here to illustrate the basic syntax:

```
xmlns="www.hhs.gov/"
```

It is important to remember another important fact about namespaces: the URI associated with a namespace is simply an identifier and does not contain any information that must be extracted by the XML file or used to verify the validity of the markup. In other words, there is no special file that will be downloaded from the Health and Human Services website and inserted into our XML file. This is merely an expanded name used to further qualify and classify the code into different units and to avoid potential conflicts or misrepresentations. It is also a useful scaffold for a human reader that may be examining the data and looking for a resource from which to gather more information about it.

Returning to our previous example from Chapter 2, we can see how to link a given element to our namespace:

```
<natural_causes xmlns="www.hhs.gov"/>
4.65 million
</natural_causes>
```

In this instance, we are including a prefix to indicate that the <natural_causes> tag is defined within the HHS namespace. If a similar natural_causes element was used by the WHO or even by the default XML file we were working in, this namespace declaration would specify to the XML processor exactly which XML tag we were referring to.

To see a second example illustrating why namespaces are important in XML documents, we can return to our XSL example from Chapter 4. The first two lines of that example are repeated here:

```
<?xml version="1.0" encoding="utf-8"?>
<xsl:stylesheet version="1.0"
xmlns:xsl="www.w3.org/1999/XSL/Transform">
```

Note that the line immediately following the XML declaration uses the xmlns namespace declaration to bind the URI "www.w3.org/1999/XSL/Transform" to the prefix "xsl". This allows the parser to recognize that any elements using the "xsl" prefix should be treated according to the XSL Transform conventions. These conventions can be found by following the included URI, though, at the time of writing, this page only leads to a placeholder for an eventual XSL schema definition (we discuss schemas in the next section). Ideally, the URI would lead to a page with documentation describing the allowable actions a parser can take and the rules it must follow when executing a transformation on an XML database. Although the URI does not provide DTD or schema information about the XSL tags, it does provide the parser with an exact and unambiguous encapsulating unit (or perspective) from which to examine these particular tags. In addition, an individual reviewing the code would have the option to visit this URI in order to learn more about the XSL specification.

Schemas

The XML Schema Language, also known as the XML Schema Definition (XSD), provides a means for describing the structure and required contents for an XML document. Eric van der Vlist describes them as a formalization of the constraints, expressed as a set of rules, which apply to XML documents (1). Although they can be used for various tasks such as describing ontologies (shared vocabularies of terms referring to the same concept), helping to guide support systems for XQuery (another emerging language), binding data to database-driven applications, or helping with guided editing tasks; their most popular use by far is as an additional mechanism for validation (van der Vlist 1–4). We discuss two popular schemas for technical document authors, called DocBook and DITA, at the end of this chapter.

Schemas are similar to DTDs, which were discussed in Chapter 2. Like DTDs, schemas are used to verify the validity of XML data and to ensure that XML information is structured according to a set of precise specifications. Unlike DTDs, however, schemas are unique in that they are written in the same XML syntax used by the data they describe. Because they must be expressed using XML's rules, schemas require a different syntax than DTDs.

Like DTDs, XSDs/schemas are a collection of rules that specify the patterns allowed in XML documents. They contain rules that define the elements and attributes allowed in the document and their data types, the order and number of child elements, and the default values and rules for repetition for each child element. One of the primary advantages of XSDs over DTDs is that schemas provide better support for validating data types such as strings of characters, numbers, or even specialized patterns (such as what one might produce for a date or postal code). In addition, since XSDs are specified in native XML code, they are more intuitive for a document designer already working with XML. This is in contrast to the makeup of DTDs, which use their own special language and rules. With schemas, we can use the same rules as we use with XML to produce well-formed (syntactically correct) documents.

Simple and Complex Types

XML schemas differentiate between complex types, which are elements that contain other elements, and simple types, which are elements that do not contain other elements. For example, consider the following segment of XML code from Chapter 2:

```
<memo>
      <author>
            <given_name>Condoleezza</given_name>
            <family_name>Rice</family_name>
      </author>
      <addressee>
            <given_name>Colin</given_name>
            <family_name>Powell</family_name>
      </addressee>
      <subject>Speech</subject>
```

```
    <date>September 22, 2001</date>
    <line>Please read the speech the President will give
    tonight.</line>
    <line>It might surprise you.</line>
</memo>
```

In this example, the memo element is a complex type because it contains other tags such as <author>, <addressee>, <subject>, <date>, and <line>. Similarly, addressee is a complex type because it contains the tags <given_name> and <family_name>. The tags <given_name> and <family_name>, however, are simple types because they do not contain any additional tags.

We can use an XSD to replace the DTD we created in Chapter 2. Here is that original code, which used a DTD and contains an XML file that was created using that DTD template:

```
<?xml version="1.0" standalone="yes"?>
<!DOCTYPE health_status_mortality [
<!ELEMENT health_status_mortality (cause_specific_per_
one_hundred_thousand_population, age-standardized_
mortality_rate_per_one_hundred_thousand_population)>
<!ELEMENT cause_specific_per_one_hundred_thousand_
population ( HIV-AIDS, TB_HIV_negative_people, TB_HIV_
positive_people ) >
<!ELEMENT HIV-AIDS  (#PCDATA)>
<!ELEMENT TB_HIV_negative_people (#PCDATA)>
<!ELEMENT TB_HIV_positive_people (#PCDATA)>
<!ELEMENT age-standardized_mortality_rate_per_one_
hundred_thousand_population (non-communicable_disease,
cardiovascular_disease, cancer, injuries)>
<!ELEMENT non-communicable_disease (#PCDATA)>
<!ELEMENT cardiovascular_disease (#PCDATA)>
<!ELEMENT cancer (#PCDATA)>
<!ELEMENT injuries (#PCDATA)>
]>
<health_status_mortality>
<cause_specific_per_one_hundred_thousand_population>
<HIV-AIDS>&lt;10</HIV-AIDS>
<TB_HIV_negative_people>&lt;1</TB_HIV_negative_people>
<TB_HIV_positive_people>&lt;1</TB_HIV_positive_people>
</cause specific per one hundred thousand population>
<age-standardized_mortality_rate_per_one_hundred_
thousand_population>
<non-communicable_disease>460</non-communicable_disease>
<cardiovascular_disease>188</cardiovascular_disease>
     <cancer>134</cancer>
     <injuries>47</injuries>
```

```
</age-standardized_mortality_rate_per_one_hundred_
thousand_population>
</health_status_mortality>
```

By using the alternate syntax required by schemas, we can rewrite this example using more specific constraints. To begin, we include our standard XML declaration and then bind the "xs" prefix to the namespace found at "www.w3.org/2001/XMLSchema" (the W3C's recommendation for XML Schema).

```
<?xml version="1.0" encoding="UTF-8"?>
<xs:schema xmlns:xs="www.w3.org/2001/XMLSchema">
.../...
</xs:schema>
```

After the schema namespace has been defined in the example above, we must specify a collection of complex and simple types that describe the allowable content and ordering rules for elements contained within this XML document. Our root element in this example is health_status_mortality, so our schema needs to first address this element. Since this is a complex type containing additional elements, the <xs:complexType> tag is used immediately following the <xs:element name="health_status_mortality"> tag in the schema document.

Sequencing

Following this same example, we next use the <xs:sequence> tag to specify the sequential sub-elements (or children) associated with that parent element. These are listed in the order in which they must appear in the XML document. In this case, the children elements are named cause_specific_per_one_hundred_thousand_population and age-standardized_mortality_rate_per_one_hundred_thousand_population, each of which are also complex types, because they too contain sub-elements.

The full version of our schema document now appears as follows:

```
<?xml version="1.0" encoding="UTF-8"?>
<xs:schema xmlns:xs="www.w3.org/2001/XMLSchema">
<xs:element name="health_status_mortality">
     <xs:complexType>
     <xs:sequence>
     <xs:element name="cause_specific_per_one_hundred_
     thousand_population">
          <xs:complexType>
          <xs:sequence>
     <xs:element name="HIV-AIDS" type="xs:string">
     </xs:element>
```

```
        <xs:element name="TB_HIV_negative_people" type=
        "xs:string"></xs:element>
        <xs:element name="TB_HIV_positive_people" type=
        "xs:string"></xs:element>
        </xs:sequence>
        </xs:complexType>
        </xs:element>
        <xs:element name="age-standardized_mortality_rate_per_
        one_hundred_thousand_population">
        <xs:complexType>
        <xs:sequence>
        <xs:element name="non-communicable_disease" type=
        "xs:integer"></xs:element>
        <xs:element name="cardiovascular_disease" type=
        "xs:integer"></xs:element>
        <xs:element name="cancer" type="xs:integer"></xs:element>
        <xs:element name="injuries" type="xs:integer">
        </xs:element>
        </xs:sequence>
        </xs:complexType>
        </xs:element>
        </xs:sequence>
        </xs:complexType>
</xs:element>
</xs:schema>
```

XML schema documents are saved using the .xsd file extension. If you want to try editing this document on your own computer, you can save it as "ch5_health.xsd".

Enforcing Data Types

In the previous example, we added specific instructions limiting the types of data that can be accepted by our elements. For example, the TB_HIV_negative_people and the TB_HIV_positive_people elements are designed to accept string data, which will accept sequences of characters, numbers, or combinations thereof. Examples of strings include the text "this is a string," and the text "strings can have numbers like 23 as well as special sequences like < or >." It is important to use strings for these elements' data types because these particular elements may need to contain less than or greater than symbols, as shown in our sample XML file. The <non-communicable_disease>, <cardiovascular_disease>, <cancer>, and <injuries> tags, on the other hand, restrict their data to integer numbers (negative or positive whole numbers) only. Examples of integers include the numbers 14, –5, 0, and 137. The number 14.23 is *not* an integer.

Finally, we need to provide a directive in our XML file to instruct the parser to use our schema document when validating our XML. The XML file with this directive ("ch5_health.xml") is shown here:

```
<?xml version="1.0" encoding="UTF-8" ?>
<health_status_mortality xmlns:xsi="www.w3.org/2001/
XMLSchema-instance" xsi:noNamespaceSchemaLocation="health_
status_mortality.xsd">
      <cause_specific_per_one_hundred_thousand_population>
      <HIV-AIDS>&lt;10</HIV-AIDS>
      <TB_HIV_negative_people>&lt;1</TB_HIV_negative_people>
      <TB_HIV_positive_people>&lt;1</TB_HIV_positive_people>
      </cause_specific_per_one_hundred_thousand_population>
      <age-standardized_mortality_rate_per_one_hundred_
      thousand_population>
            <non-communicable_disease>460</non-communicable_
            disease>
            <cardiovascular_disease>188
            </cardiovascular_disease>
            <cancer>134</cancer>
            <injuries>47</injuries>
      </age-standardized_mortality_rate_per_one_hundred_
      thousand_population>
</health_status_mortality>
```

Validation using Schema

After an XML file has been linked to a schema, we can use any XML parser with schema support to validate our XML document, just as we would do with a DTD. The output from one such software application is shown below in Figure 5.1. Note that the Output Window now displays a message indicating that our XML file has been found valid according to our "health.xsd" schema.

When validation fails, the parser should return an error showing what part of the document is in violation of the schema. If we were to change the value of the data contained within the <cardiovascular_disease> tag, for instance, to "<188" (which equates to "<188" after the HTML entity is translated) the validation should fail, because the less than symbol is not an integer. Indeed, running that new input against the schema returns the following error:

```
XML Document is not valid against XML Schema
Error parsing '<188' as integer datatype. The element:
'cardiovascular_disease' has an invalid value according to
its data type.
```

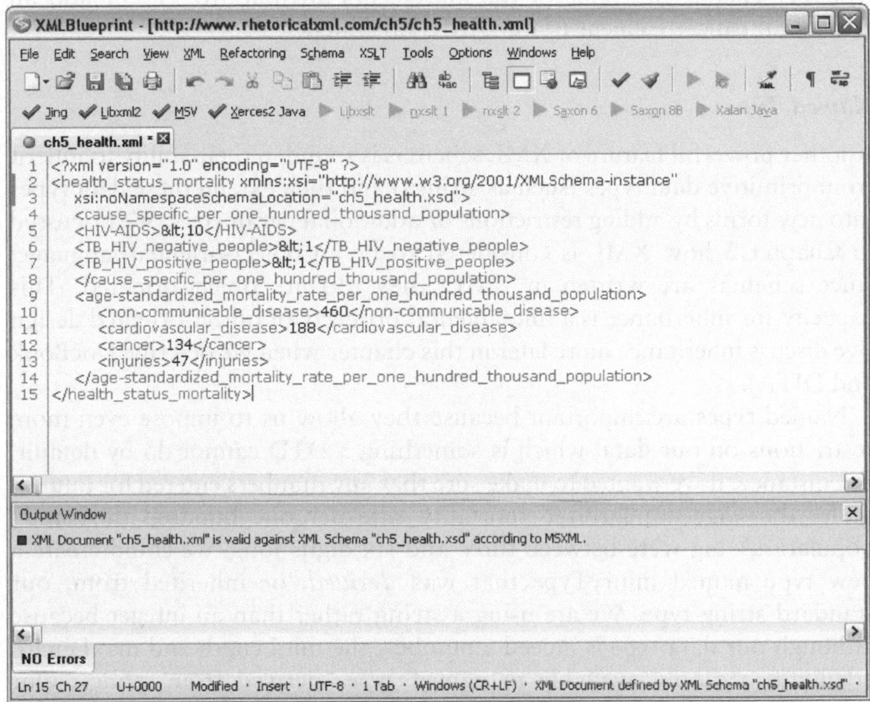

Figure 5.1 Schema Validation Results

Cardinality

Cardinality, or the number of times in which an element may occur, is enforced using the minOccurs and maxOccurs attributes. Recall from Chapter 2 that DTDs enforce cardinality using special characters such as the plus symbol "+" and the asterisks "*". To specify the appropriate cardinalities with schema, we must provide numeric values for the minOccurs and maxOccurs attributes. To show that an element can occur as many times as we wish, we use the special value "unbounded." Here is some schema code that we could use to ensure that the number of cancer elements specified in an XML document should be at least two, with no upper limit:

```
<xs:element name="cancer" type="xs:integer" minOccurs="2"
maxOccurs="unbounded"></xs:element>
```

Unless we returned to our original XML document and added an additional <cancer> tag, the example we presented earlier would no longer be a valid document. Since we are now asking for at least two cancer elements and we only provide one in our earlier example, we would need to either change the value of the minOccurs attribute to "1" or add an additional cancer element to our XML document.

Named Types

Another powerful feature of XML schemas is found in their ability to inherit from primitive data types (such as strings) and then extend these data types into new forms by adding restrictions or additional parameters. We discussed in Chapter 3 how XML is considered to be an object-oriented language; since schemas are written in XML, they too are object-oriented. This capacity for inheritance is a fundamental property of object-oriented design (we discuss inheritance more later in this chapter when we describe DocBook and DITA).

Named types are important because they allow us to impose even more restrictions on our data, which is something a DTD cannot do by default. For instance, if we wanted to make sure that any numbers entered for injuries under the <age-standardized_mortality_rate_per_one_hundred_thousand_population> tag were between three and six digits long, we could create a new type named injuryType that was *derived*, or inherited from, our standard string type. We are using a string rather than an integer because although our data type is indeed a number, the minLength and maxLength attributes are only available to named types derived from strings. Our schema definition for this type looks like this:

```
<xs:simpleType name="injuryType">
        <xs:restriction base="xs:string">
                <xs:minLength value="3"></xs:minLength>
                <xs:maxLength value="6"></xs:maxLength>
        </xs:restriction>
</xs:simpleType>
```

We can then link our element defined in the XML schema to this new named type using the code that follows. It is important to notice that we do not use the namespace prefix "ns" with the named types we create. This is because these items are not associated with the W3C's schema data types, but with our own that we define in our own local (document level) namespace. We do not need to use a different namespace prefix because the local namespace is assumed if no explicit prefix is provided. This named type must also be created outside the <xs: element name="health_status_mortality"> element. It should be in between the prefix definition and the health_status_mortality element.

Here is the code to link the named type with our element:

```
<xs:element name="injuries" type="injuryType">
</xs:element>
```

With this new named data type, we provide additional feedback to any XML document creator that tells them when the data they are entering is not correct according to the schema. For example, using the injuryType named data type we define above, our schema expects to see a string that is between three and six characters in length. If an author were to accidentally type "10" as opposed to "100," they would be presented with an error message similar to this (the specific error message may vary according to the parser one is using): "minLength constraint failed. The element: 'injuries' has an invalid value according to its data type." Likewise, if they typed a value that was too long, such as "1000000," they would see an error message similar to this one: "maxLength constraint failed. The element: 'injuries' has an invalid value according to its data type."

Using custom patterns, we can build our schema to enforce an even greater degree of control upon our documents. For example, if we wanted to create an ID attribute that was ten digits long and accepted only digits ranging from zero to five, we could add a new named type called "idPattern" that looks like this:

```
<xs:simpleType name="idPattern">
    <xs:restriction base="xs:string">
        <xs:pattern value="[0-5]{10}"></xs:pattern>
    </xs:restriction>
</xs:simpleType>
```

The numbers provided in the brackets, [0–5], specify the allowable range that these numbers can have in the XML document. A user can type in the values 0–5. The number in the curly braces, {10}, specifies that a total of ten digits must be provided. So, using the pattern restriction imposed above, the value "0123450123" would be accepted as valid data because it contains only the digits zero through five and is a total of ten digits long. The value "0123456789" would be rejected because it contains the values "6," "7," "8," and "9," all of which are outside the allowable range of characters. Similarly, the value "01234" would be rejected because it is too short.

As we can see from these examples, named types are useful for customizing the default data types used in the W3C recommendation for schemas. They allow us to further customize the length of string data and to enforce custom patterns for element values. As knowledge managers, such capabilities are useful for helping data providers to understand precisely what should be entered into XML documents as well as for troubleshooting any errant data that may be entered by mistake.

To review, schemas function exactly like DTDs in principle, but they are advantageous in that they are written using the same well-formed syntax of the XML documents they are verifying. In other words, they are required to use the same syntax as general XML documents. They also provide additional mechanisms for specifying more precise data types. Another useful side benefit of using a native XML format is that XSL transformations can now be applied to schema, making it easy to transform one schema into another or even to display the schema using HTML elements such as tables or lists.

XPath

It is common for some large XML documents to contain hundreds or even thousands of elements and attributes. Even with a potent XML editor that has the capability to display the relationships between elements in a graphical hierarchy, it can be a challenge to locate certain elements or attributes that are nested deep within the database of elements. The XPath provides a means for locating information at any point within an XML document.

Recall the garden example from the previous chapter. In a wild and overgrown garden, it can be difficult to locate the precise areas in which fruits and vegetables are planted. Using this garden example as a metaphor, XPath can be likened to a planting grid overlay that shows exactly where every item was originally planted and provides a map showing that particular plot in relation to other items in the garden. A better analogy is perhaps the classical problem of finding hidden pirate treasure. A treasure hunt is simply not possible without some general directions leading one to likely search locations. In other words, X marks the spot, but without a good treasure map (and the ability to read it) there is not a very good chance of finding that buried gold. Similarly, with XPath, you have to understand the syntax in order to find the (potentially valuable) buried information within an XML document.

XPath refers to the elements and attributes within an XML file as nodes. The node concept is likely familiar to HTML designers; nodes of information are often arrived at through the process of "chunking" in which similar items are grouped together on the same page, providing for a rough encapsulation of semantic content. We discussed this in detail in Chapter 3 when we considered different types of classification systems such as Aristotelian (binary), psychological, and linguistic strategies. In the XPath data model (XDM), there are seven different kinds of nodes that can be used to classify data. These seven types are called document, element, attribute, comment, namespace, text, and processing nodes.

Document nodes are the highest level, or "root," elements that encapsulate an XML document. As you would expect, element nodes contain information about elements, and attribute nodes contain information about attributes. Similarly, namespace nodes encapsulate namespace information and comment nodes encapsulate comments. Text nodes contain XML character content. Finally, processing nodes contain special XML processing instructions.

In addition to nodes, XPath uses components known as axis specifiers to specify the direction in which an XML document should be traversed.

Types of axis specifiers include ancestor, ancestor-or-self, attribute, child, descendant, descendant-or-self, following, following-sibling, namespace, parent, preceding, preceding-sibling, or self. These specifiers are included in XPath expressions to specify the routes to be taken when traversing an XML tree.

An XPath expression is composed of three parts:

1. An axis specifier, which provides the direction to travel within the document hierarchy.
2. A node test, which specifies the types of nodes (comments, text, processing information, or any/all of these) that should be considered during the trip.
3. A predicate expression, which uses square bracket notation to further filter the sequence, if necessary.

In order to create an expression, one must combine the "steps" required to move from one node to another and separate them using the "/" character.

To understand how XPath works, it is important to remember the tree analogy from Chapter 2. Here we learned that an XML file can be represented as a family tree with parent elements, children elements, and siblings (nodes that are on the same level of the hierarchy). The following code is reprinted from the "person.xml" example in Chapter 2. As reflected in Figure 5.2, the spouse, daughter, and two son elements are all considered siblings since they reside on the same level of the hierarchical tree. In the XML code, these elements are all defined "beneath" the immediate_family element. This representation of XML elements in a hierarchical format is also known as the Document Object Model, or DOM.

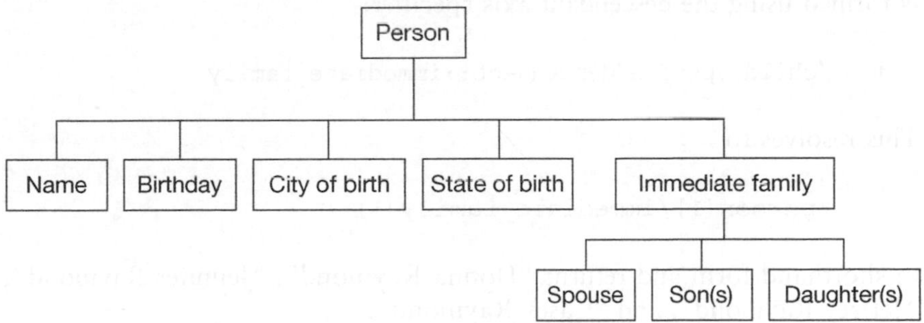

Figure 5.2 Tree Representation of "person.xml"

```
<person>
      <name>John Raymond</name>
      <birthday>July 27, 1970</birthday>
      <city_of_birth>Des Moines</city_of_birth>
      <state_of_birth>Iowa</state_of_birth>
```

```
    <immediate_family>
            <spouse>Donna Raymond</spouse>
            <daughter>Jennifer Raymond</daughter>
            <son>Jeffrey Raymond</son>
            <son>Casey Raymond</son>
    </immediate_family>
</person>
```

This family tree analogy is important to help understand the traversal procedure when working with axis specifiers in XPath. For example, if we wanted to create an XPath expression to find the data contained in the <name> tag, we would use the child axis specifier twice. The expression would look like this:

```
/child::person/child::name
```

We could also use the shorthand version of this XPath query and write this new expression as follows:

```
/person/name/
```

Or, we could use the fully qualified version in this fashion:

```
/person[1]/name[1]
```

In each case, the XPath expression leads us to the element containing the text "John Raymond". A more complicated task is to find all of the descendants of the immediate family element. In this case, our XPath query is formed using the descendant axis specifier:

```
/child::person/descendant::immediate_family
```

This resolves to

```
/person[1]/immediate_family[1]
```

in shorthand form and returns "Donna Raymond", "Jennifer Raymond", "Jeffrey Raymond", and "Casey Raymond".

If we wanted only to access the node in which the value "Casey Raymond" is stored, we could use this XPath syntax:

```
/child::person/child::immediate_family/child::
son[last()]
```

In this example, we are using a special predicate expression named "last()" that will select the last node in the document hierarchy with this element name. If we know the exact number of elements named "son", we could also use indexed notation to provide an XPath expression that is easier to remember:

```
/person[1]/immediate_family[1]/son[2]
```

Many parsers have XPath support built in, since this syntax makes it so convenient to locate nodes buried within the document hierarchy. For large XML libraries with complex, nested structures of information, XPath expressions are indispensable for document designers. As with namespaces, XPath expressions are also valuable when multiple XML documents are combined together for a specific informational purpose. When an author needs to search a document that he or she did not originally create, XPath simplifies the process of accessing buried nodes by using a specific syntax that will work equally well for all well-formed XML documents.

Linking in XML

The XML Linking Language, or XLink, is a set of standards that defines rules for linking XML elements. In HTML, links provide the *hyper* part of hypertext, allowing a document's reader to navigate between and beyond documents and websites in a nonlinear—and usually nonhierarchical—fashion. Navigation *between* documents would be found in documents that link to one another; we might describe navigation *beyond* documents as navigation integrated with multimedia or video components or navigation linked to virtual worlds like *Second Life* (Linden Research online) or *There* (Makena Technologies online). This ability to link to unforeseen document media types as well as to an enormous amount of existing textual documents is what makes hypertext so powerful.

Links are what drive the connectivity of the Internet's documents on a technical level. Similarly, they also contribute to the rhetorical dimensions of online document sharing. Jay David Bolter suggests that because of their ability to define relationships and meaningful connections among various electronic resources, the links themselves "constitute the rhetoric of the hypertext" (29). The decentralized nature of hypertext and the grassroots community involvement that often surrounds new ideas on the Web is directly attributable to the nature and functionality of hyperlinks. This was another fundamental idea Tim Berners-Lee expressed and that we observed at the genesis of the hypertext markup language. Aside from the constraints imposed by one's employer, there is no central authority that approves or disapproves one's linking to a particular resource or making connections between one organizational unit and another in cyberspace. Depending on the country in which one lives, access to these outside resources may be a problem, however. In other words, the links can be made, but not necessarily followed.

Though they are in contrast to the rigid taxonomical structure of predefined classification systems, many emergent types of knowledge management can benefit from the freedom of hypertext linking practices. For example, social bookmarking websites like *del.icio.us* provide a means for likeminded individuals to discover new resources and make new connections to existing sources based upon the browsing and linking habits of other individuals on the World Wide Web. These resources can then be presented to other users using metadata and common keywords that identify and classify themes within the linked resources. In this case, knowledge is encapsulated using the interests of compatible individuals rather than some existing framework for information classification. Since much language and terminology on the Internet is organic and evolves over time, social bookmarking sites are useful for keeping track of resources that refer to a specific subject or topic using contemporary language, or for finding resources that may be particularly popular during a given moment in time. Though additional keywords may be necessary to update a resource with its most recent descriptors, it makes more sense to open this keyword generation mechanism to the general public rather than relying on a single organization or individual to keep the database updated.

Functionally speaking, links are remarkably simple considering their ability to connect source materials from any networked location. They provide connections between resources that exist on computers residing anywhere in the world. The difficulty of moving objects from one location to another in physical space is irrelevant to a hypertext link. The same physical hardware and networking standards that connect computers through the Internet enable a seamless and oftentimes transparent browsing experience from one resource to another, unless a resource has been removed or otherwise secured from the general public.

Linking in HTML

To understand the concepts behind XLink, it is important to first understand how links work in traditional HTML documents. There are different types of links used in HTML. First, there is the traditional clickable link that is formed using an <a> element. The format of this link looks like this:

```
<a href="www.rhetoricalxml.com">This is a link to the text's
website</a>
```

If you are not familiar with hand coding links in HTML, you might try typing this example into a text document and saving it as "links.html". Open the document in your Internet browser to see how the link is rendered from HTML text to a clickable hyperlink. In this case, the reader of a hypertext document needs to explicitly click the fragment of text that says "This is a link to the text's website" in order to "follow" the originating link to its targeted destination. Here, clicking on the link leads the reader to the companion website for this book.

Relative and Absolute Linking

Links in HTML can be relative or absolute. Relative links are links that are defined relative to a particular location on the server computer. For example, a link defined

```
<a href="xmlstuff/xmlnotes.html">Link to notes<a>
```

is relative. If one were to create this link and store it in a file named "relative.html", one's browser will attempt to find a directory named "xmlstuff" and a file named "xmlnotes.html" within whatever current directory "relative.html" resides in.

Absolute links, on the other hand, specify an exact path. To make this example absolute, we could rewrite it as follows:

```
<a href="www.rhetoricalxml.com/ch5/xmlstuff/xmlnotes.html">
Link to notes<a>
```

This example can be saved as "absolute.html". With absolute linking, we are clear about exactly where a resource is located and we specify a precise server name (or directory path).

In general, relative links are useful for developing HTML files when they need to be moved from a local computer to a server computer. This makes them fairly portable. One must be careful with the direction of slashes, as they may need to change direction depending on the type of server computer the files reside on (Windows and Linux servers use different notation for path expressions, for example). Absolute links are useful when certain files are deeply nested in multiple locations but they all refer to the same navigational content, or when a clear and unambiguous pathname is necessary for security reasons. They are also useful when the developer does not have control over the installation location of the HTML files. For example, if you tried these examples yourself, you should notice that the "relative.html" file does not contain a working link (unless you painstakingly recreated the "xmlstuff" directory and the "xmlnotes.html" file or downloaded this directory to your local hard disk drive) while the "absolute.html" file does work (assuming our Web server is online and functioning properly). This is because your relative link was searching for the files on your local computer while the absolute link was searching for the files on our website's server.

Linking Images and Multimedia

Another type of link used in HTML is found when resources are loaded into the document from outside files such as JPEG, GIF, PNG, WAV, MIDI, SWF, or MP3 files. These types of links are usually loaded automatically and do not require the reader to click on anything in order to connect to the new resource. The exception to this is found when a browser plugin is missing or outdated; when this happens, the reader may need to explicitly click a link to download the updated software.

Here is an example of this second type of link:

```
<img src="achre_seal.gif">
```

If you would like to test this second example, create a new file named "links2.html" and type in the line of code above into that file. Save it on your computer along with the image file (or change the image name to reflect an image already on your local computer). You can download the test image named "achre_seal.gif" from our website (the file is listed with the Chapter 5 content). Save this image file in the same directory and see what happens when you load the file into your browser. You should see the official seal for the ACHRE. If you see a broken image icon, you may not have the image file stored in the same directory as your HTML file. In this case, you should include the path name in your filename or copy the image into the same directory as your HTML file.

XLink

XLink, the XML Linking Language, provides additional options for working with hyperlinks in XML documents. In addition to the simple links supported by HTML, XLink supports extended links for tying multiple resources together. XLink became a W3C Recommendation in June of 2001.

A basic XML file containing a set of XLinks is shown below. Note the use of the "xmlns" directive to link the "xlink" prefix to the W3C's official XLink website.

```
<?xml version="1.0" encoding="utf-8"?>
<xml_resources xmlns:xlink="www.w3.org/1999/xlink">
  <homepage xlink:type="simple" xlink:href="http://xml.
  silmaril.ie/">The XML FAQ Page</homepage>
  <homepage xlink:type="simple" xlink:href="www.rhetoricalxml.
  com">Rhetoricalxml.com</homepage>
  <homepage xlink:type="simple" xlink:href="http://html-
  color-codes.com/">Hexadecimal Color Chart</homepage>
</xml_resources>
```

You can see that the "xlink" prefix in this example is specifying two different attributes: "type" and "href". The "href" attribute functions exactly like the "href" attribute in HTML; it contains a normal URL string which is an Internet Web address. The type attribute, though, is something unique to XLink. Assigning this attribute a value of "simple" means that the link is a traditional HTML link—clicking the link will lead directly to the resource specified in the href attribute. We could also add an xlink: show="new" attribute that would instruct the parser to open these resources in a new window, much like the target="new" attribute of HTML (if you have not used this particular attribute in HTML before you should try adding it to your "links1.html" file and observing what happens).

Unfortunately, due to some political problems between XML developers and XML enthusiasts and problems with backwards compatibility, XLink has not made as much progress as other associated XML technologies like XSD and XPath. One of the major problems is that it uses a new syntax for links, making the traditional HTML linking syntax (e.g., <a href="test.html">) mostly incompatible. XLink is, however, implemented in some widely used electronic communication languages, such as the eXtensible Business Reporting Language (XBRL) and some versions of the DocBook schema. For this reason, it is wise to be familiar with the syntax and to understand this recommendation of the W3C. XPointer, which we discuss in the next section, is also somewhat unstable in terms of implementation and adoption, but it is worthwhile to study how it works as it too has implications for the technical manipulation of combined documents.

XPointer

The XML Pointer Language, or XPointer, is a notation system for XML that is even more specific than XLink. It allows one to access nodes that may be buried deep within XML databases, using an addressing system with precise syntax. Like HTML links, these addresses can be specified in absolute or relative terms. In an absolute addressing scheme, the pointer will return a result independent of the location of other terms (such as the current location of the pointer in the XML document or DOM). In *Learning XML*, Erik T. Ray discusses the four types of absolute location terms used with XPointer: `id()`, `root()`, `origin()`, and `html()` (84). Each of these terms can be used to point to commonly accessed parts of an XML document (see Table 5.2).

In a relative addressing scheme, the location of a referenced element is dependent upon the current location within the XML hierarchy. XPointer contains four different specifications:

1. A framework which specifies the basic functionality of the language.
2. An addressing methodology for positional elements.
3. A scheme for namespaces.
4. A scheme for XPath-based addresses.

Table 5.2 Absolute Terms

Absolute Term	*Description*
Id	Points to an element within the XML document with a given id attribute.
Root	Points to the root node of the XML document (the top of the document).
Origin	Points to the element from which a pointer is originating.
Html	Used with HTML documents to locate named anchor or link elements.

The framework for XPointer builds upon XPath and enables direct access to fragments within an XML document. Although it is currently lacking in mainstream browser support, XPointer has some interesting implications for document design that bear discussing.

The basic idea behind XPointer is to enable extended URLs in XML links that allow a parser to directly access information that may be buried deep within the XML document. The concept here is not new; in HTML, the same sort of feature can be enabled by using the anchor tag and name attribute. For example, consider the following simple HTML document:

```
<html>
<head><title>Anchored Tag Examples</title></head>
<body>
      <h1>Anchored Tag Examples</h1>
      <p>This is a simple demonstration of using anchored tags
      to directly access different portions of an HTML document
      in a browser.  By default, a link to this page will open
      up this document as you might expect, with the heading at
      the top of the page and then the paragraph text (the
      content you are reading now) following after that.  Using
      the anchored links shown below, however, will position
      the page so that the tagged information using the name
      attribute appears at the top of the browser window.  Give
      it a try!</p>
      <a name="content1">Content Node 1</a>
      <p>If this were a real document, some information about
      the content 1 node would be displayed here.</p>
      <a name="content2">Content Node 2</a>
      <p>If this were a real document, some information about
      the content 2 node would be displayed here.</p>
      <br />
      <ul>
      <li><a href="anchored_example.html">Click here to access
      the HTML document as usual</a>
```

```
    <li><a href="anchored_example.html#content1">Click here
    to access Content Node 1</a>
    <li><a href="anchored_example.html#content2">Click here
    to access Content Node 2</a>
    </ul>
</body>
</html>
```

Save this file as "anchored_example.html" and open it in your browser. Click on the various links and you will see how the anchor tag and name attributes function in normal HTML documents. Since there is not very much text displayed on this page, you may need to experiment with your browser window's size in order to see the true effects. Screenshots taken after the various links have been clicked are shown in Figures 5.3–5.5. Note the way in which direct access to the inner portions of the HTML document is enabled through the use of named anchors within the document.

This simple example showcases the ability of an HTML document to link to embedded information that is buried within a page's nest of tags and attributes. XPointer is the equivalent of a named anchor, but for XML documents. The basic structure of an XPointer is even the same: a URL, followed by the hash symbol (#), followed by a named identifier. The XPointer is actually more powerful, though, in that it can link to *any* element within an XML document, rather than just to the <A> element, which is

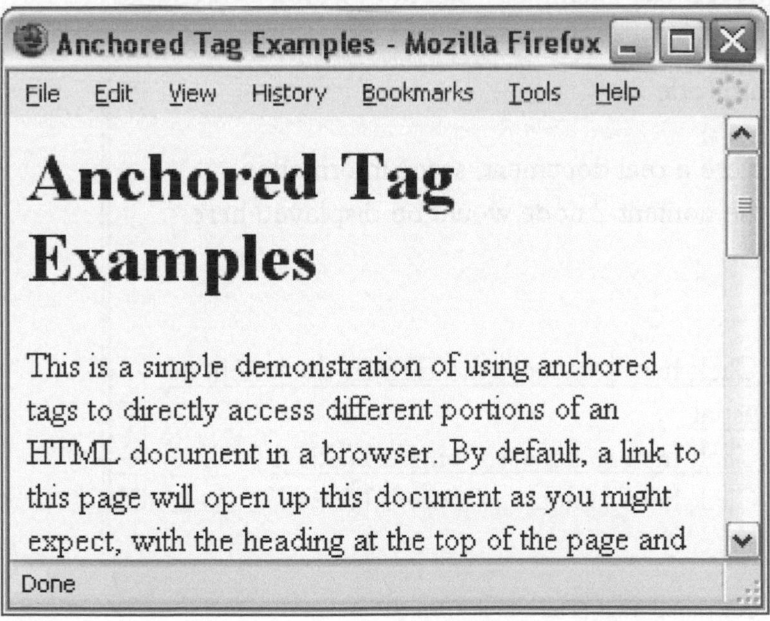

Figure 5.3 Anchored Tags (First Link Clicked)

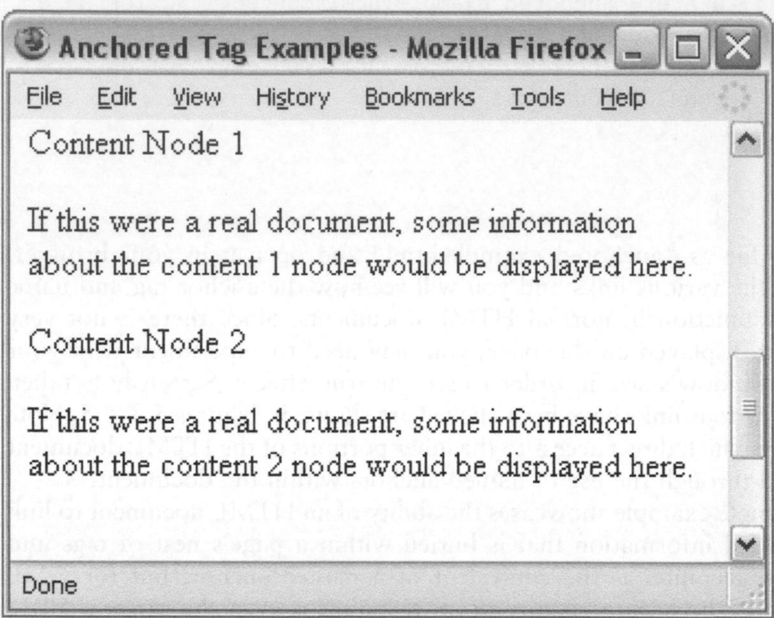

Figure 5.4 Anchored Tags (Second Link Clicked)

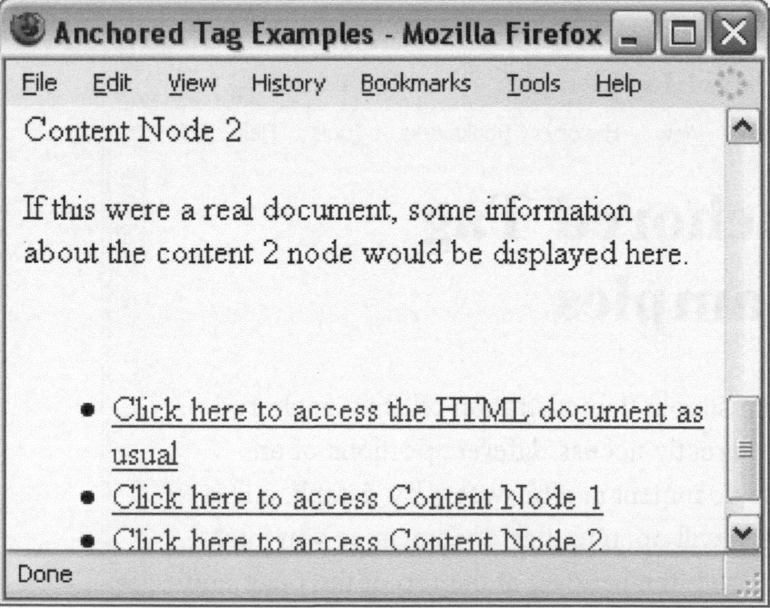

Figure 5.5 Anchored Tags (Third Link Clicked)

what HTML is limited to. XPointer can also be used to point to sets of elements or even to ranges of text between nodes in an XML document. Several different examples of XPointer in use are shown in Table 5.3. These XPointers are being used to access portions of the XML document "garden. xml", which is slightly revised from Chapter 4 and reprinted here with minor modifications. Specifically, the name element from the Chapter 4 version has been turned into an id attribute in this version. The id attribute is then used as a point of reference for our XPointers.

```
<?xml version="1.0" encoding="ISO-8859-1"?>
<!— Chapter Five example —>
<garden>
     <fruit id="Vine Ripe Tomatoes">
          <supplier>Henderson Farms</supplier>
          <units_available>45</units_available>
          <price_unit>Bushel</price_unit>
          <price>25.00</price>
     </fruit>
     <fruit id="Watermelon">
          <supplier>Rand Melon Supply</supplier>
          <units_available>350</units_available>
          <price_unit>Individual</price_unit>
          <price>1.50</price>
     </fruit>
     <vegetable id="Green Leaf Lettuce, Black-seeded Simpson">
          <supplier>A&amp;H Local</supplier>
          <units_available>2000</units_available>
          <price_unit>Individual</price_unit>
          <price>.25</price>
     </vegetable>
     <vegetable id="Green Leaf Lettuce, Grand Rapids">
          <supplier>A&amp;H Local</supplier>
          <units_available>700</units_available>
          <price_unit>Individual</price_unit>
          <price>.65</price>
     </vegetable>
</garden>
```

As is evident from Table 5.3, XPointer can use XPath expressions for locating particular elements within an XML hierarchy. The two languages work well together and can be used to maximize the efficiency with which particular data nodes can be extracted from an XML database.

Since browser support for the XML Pointer Language is still shaky, it is difficult to experiment much with XPointer without using a custom XML parser or a specialized tool designed for this purpose. XPointer was also

Table 5.3 XPointer Examples

URL Structure	*Description*
garden.xml#xpointer(id("Watermelon"))	An XPointer that points to the fruit with an id of "Watermelon" in the XML document.
garden.xml#xpointer(/garden/1)	An XPointer that points to the first fruit element in the XML document.
garden.xml#xpointer(/1/2/1)	An XPointer that points to the supplier element contained within the second fruit element.
garden.xml#xpointer(//vegetable)	An XPointer that points to all elements named vegetable in the XML document.
garden.xml#xpointer(id("Watermelon"). descendant(1)	An XPointer that locates the first descendant element of the element with an id of "Watermelon." This would point to the supplier element in this example.

briefly hindered in its evolution from a working draft to a W3C Recommendation in 2003. A 2001 article by Leigh Dodds suggests that a Sun Microsystems patent is partly to blame for the slow implementation of XPointer (online). This legal issue is an excellent example of the complex rhetorical space constantly being negotiated between open source Web technologies and large technical corporations. Though XML thrives on notions of interoperability and communication between disparate organizations, the complex legal and political contexts of information can make these endeavors much more complicated than they originally appear to be.

In the final section of this chapter we focus on two advanced validation technologies that are used in structured writing and single sourcing environments. These two technologies are DocBook and DITA. Both DocBook and DITA are used in writing environments to help XML authors maintain consistency, standardization, and interoperability. Each has both DTD and schema methods of validation.

DocBook and DITA

Overview

There are numerous frameworks, DTDs, schema, and architectures designed to help with the standardization and distribution of XML-encoded content. In this section, we briefly discuss two products—DocBook and DITA—that may be of particular interest to readers working with large technical documents. It should be noted that these are only two popular products in this area; there are numerous other software products and technology suites that can be used for adding structure, standardization, and portability to XML documents. We also do not cover these two products in great detail as there are many technical tutorials on the World Wide Web that do a great job of explaining their intricate details, inner workings, and syntax (we provide links to some of these websites at the end of this chapter). We do, however, want to introduce these two tools as some of the advanced toolsets available to document authors.

Though their comparison can spur debate among different camps and groups of XML enthusiasts, DITA and DocBook are in fact very similar. Both offer sets of standards for dealing with XML documents, both are used to encode technical information using XML, and both have schemas and DTDs available to validate documents for properly formed and ordered content. Both have style sheets enabling DocBook or DITA-compliant XML files to be transformed into human readable file formats such as XHTML or PDF. Perhaps most importantly, both technologies have contributed to the progress of structured writing as a viable communication strategy in software documentation and technical writing.

DocBook and DITA are especially useful as tools for using in single-sourcing applications, which we learned about in Chapter 3. Recall the motto for single sourcing: "write once, publish many times." In order to enable this type of single sourcing system, one needs to plan for a specific type of authoring that includes a focus on chunking, modularity, and structured writing. Unfortunately, this shift from a traditional "craftsman model" of writing to a modular type of writing introduces some significant challenges. For example, as Norman Walsh and Leonard Muellner note, structural authoring introduces an entirely new process for authoring information, a process very different from the traditional approach of using a word processor to create a single, linear document (8). They also discuss the added expense of authoring tools and, since structured writing separates semantics from appearance, the additional work and expertise required to create effective style sheets (8). Traditional roles and professional contexts are also influenced by single sourcing writing procedures (see Michael Albers' *Single Sourcing and the Technical Communication Career Path* for a detailed discussion of the professional implications of this shift for technical communicators).

Both DITA and DocBook are quite popular and do a fine job of addressing some of the rhetorical problems that are introduced when any document author is free to create his or her own set of tags. One such problem is the creation of multiple tags that refer to the same concept or unit of information as referenced in different documents—we discussed these types of issues at length in Chapter 3. We can think of this issue as a tradeoff between rhetorical power and interoperability between systems. While XML is a standardized language, the languages *generated* by XML are not. As Priestley, Hargis, and Carpenter note, "when you create a new markup language (using XML to define its markup and rules), you shut yourself off from interchange with the rest of the world; when you adopt a standard markup language, you lose the benefits promised by content-specific markup" (354). In this sense, you are both empowered (in terms of your ability to choose any semantic identifiers you feel are appropriate for your data) and limited (in terms of a loss of ability for wider outside recognition) by the extensible nature of XML.

One of the primary benefits of using DocBook or DITA is that you can take advantage of standardized tags at a more general level while still having some amount of flexibility and customization available at a more specific level. Often this is accomplished through inheritance relationships. An inheritance relationship is when a basic information type (such as a node designed to store general definitions for a technical vocabulary) is specialized to take on a particular informational need. Using this strategy, general descriptors that are commonly used by document creators (such as author name, title, subtitle, and so on) can be standardized, while these specialized nodes can be invoked to handle custom needs.

For example, a general definition node defined using a schema or DTD could be designed to hold elements such as an identifier (the name of the term being defined), a list of keywords (a list of descriptive terms used as additional metadata), and a definition (the extended definition in paragraph format). For the sake of this example, let us call this module "vocab_term". This structure might work well for eighty percent of all cases in a technical document. For the other twenty percent, however, there might need to be an additional element to hold cross referencing information and a placeholder for an image link. These definitions would need to include additional text describing where similar concepts were discussed in a document as well as a visual representation of the concept being described. Rather than creating an entirely new structure for these cases, the new module, which we will call "vocab_term_crossref_image", can inherit the base features of the more general "vocab_term" module and then specialize or customize this base module with different features. The "vocab_term" module is therefore fairly interchangeable, while the "vocab_term_crossref_image" module is less interchangeable, but quite specialized.

Like namespaces, inheritance is a feature that has a long history with advanced programming languages and it is a key characteristic of modular document design (sometimes called object-oriented document design). Features such as inheritance allow a document designer to still create extensible documents, but also to use standards for those document components. This helps when these deliverables need to be recognized and interpreted by other parsers outside the namespace of one's immediate project.

An in-depth analysis of DITA and DocBook is beyond the scope of this book (though we do suggest several useful online resources for any interested readers at the end of this chapter). Since these tools are often used by technical communicators and information architects working with XML, however, it is useful to understand at least some of the basic concepts behind them. We will first consider the central strategies behind the DITA architecture.

DITA

The DITA is an XML-based architecture named in part after the famous natural scientist Charles Darwin. Darwin used many classification techniques over his long career as a scientist and author of biological and geological texts. DITA, developed by IBM in early 2001 and originally introduced as a series of technical articles, is used for writing and managing information using a predefined set of information structures that are broken down into topic types named tasks, concepts, and references. Each of these topic types is derived from a more general module, named "topic." Document authors create new types in DITA through a process called specialization (Priestley, Hargis, and Carpenter 358). Specialization is DITA's answer to the process of inheritance as it allows one to inherit base elements and then specify new elements according to particular informational needs. This also helps to explain the name for this technology since inheritance and specialization were key ideas guiding the research of Charles Darwin. DITA 1.0 was formally approved as an OASIS standard in 2005 and DITA 1.1 followed in 2007.

DITA DTDs and schemas carefully enforce the organization of topics. The DITA topic type uses only three required elements: an id attribute, a title, and a body. The id element contains an identifier for the topic module and the title contains a title for the module. The body tag contains any number of elements and will likely contain many HTML elements that are used for formatting and displaying text. For example, the body element might contain numerous HTML tags such as <b> (to bold text), <p> (to create a new paragraph), <table> (to begin a table), and <ul> (to begin an unordered list).

An important feature of DITA is the standardization of XML element names. Although DITA uses a specific vocabulary for the authoring process, the topic modules (as well as elements derived from the topic modules) are all represented using standard XML. So, a general topic module using the required elements will always look something like this:

```
<?xml version="1.0" encoding="utf-8"?>
<topic id="1">
      <title>A Compelling Title</title>
      <body>A <i>body</i> element with some <b>tags</b> and
      HTML markup code.</body>
</topic>
```

In addition to the three required elements of id, title, and body, the topic can also optionally contain additional elements such as shortdesc (which can be used to provide a short description), prolog (which can hold additional metadata), and related-links (which can hold a collection of links that are similar in nature to a particular module). Elements within DITA topics are further specialized into three different types: concepts, tasks, and references. In each of these data structures, the root element (formerly named topic) is renamed to concept, task, or reference (see Figures 5.6–5.8).

In Figure 5.6, we see that the concept type is very similar to the topic type. The primary difference is that the body element in topic is renamed to conbody in the concept type. Concept elements answer "what is this?" types of questions. These element types are used for introducing background information about a subject. A concept module looks like this:

```
<?xml version="1.0" encoding="utf-8"?>
<!DOCTYPE concept SYSTEM "concept.dtd">
<concept id="toaster_overview">
<title>Introduction to the SuperChar 3000</title>
<shortdesc>The SuperChar 3000 is a deluxe toaster manufactured
to the highest quality standards.  It is capable of
simultaneously toasting 32 slices of bread using four heat
settings.</shortdesc>
<conbody>
      <p>SuperChar 3000 Models:</p>
      <ul>
            <li>SuperChar 3000a (Base Model)</li>
            <li>SuperChar 3000b (High-Speed Model)</li>
            <li>SuperChar 3000c (Collector's Edition)</li>
      </ul>
</conbody>
</concept>
```

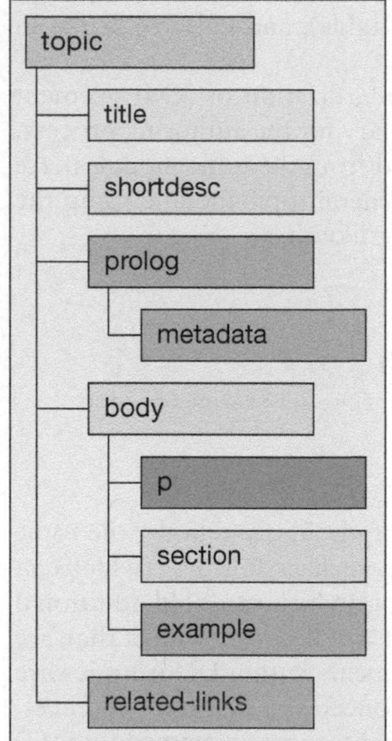

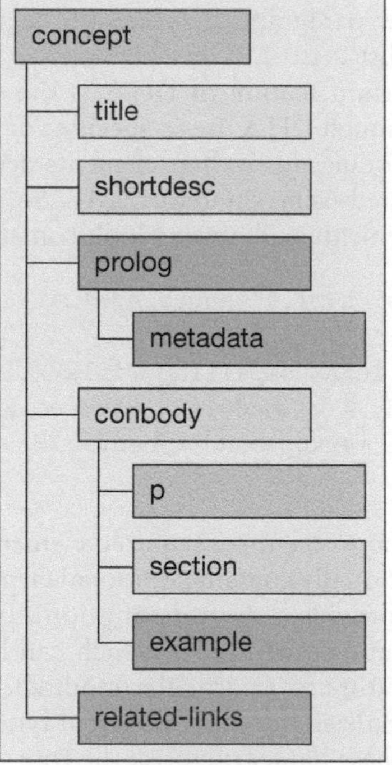

Figure 5.6 Concept (Specialized from Topic)

In this example, we are linking the "concept.dtd" file (DITA's concept DTD) in the second line of code. We use this DTD to validate our document and make sure it is acceptable using DITA's list of rules and allowable element names. DITA also includes XSD schema files, so we could also choose to validate our XML file using a schema by linking the schema file using the techniques described earlier in this chapter. As we covered schema in detail here in Chapter 5, we use DTDs with these XML examples as a review in linking DTD to XML files.

Figure 5.7 shows how the task type is specialized from the topic type. The task type is more specialized than the concept type because its purpose is more action-oriented than descriptive. Task types are used to store

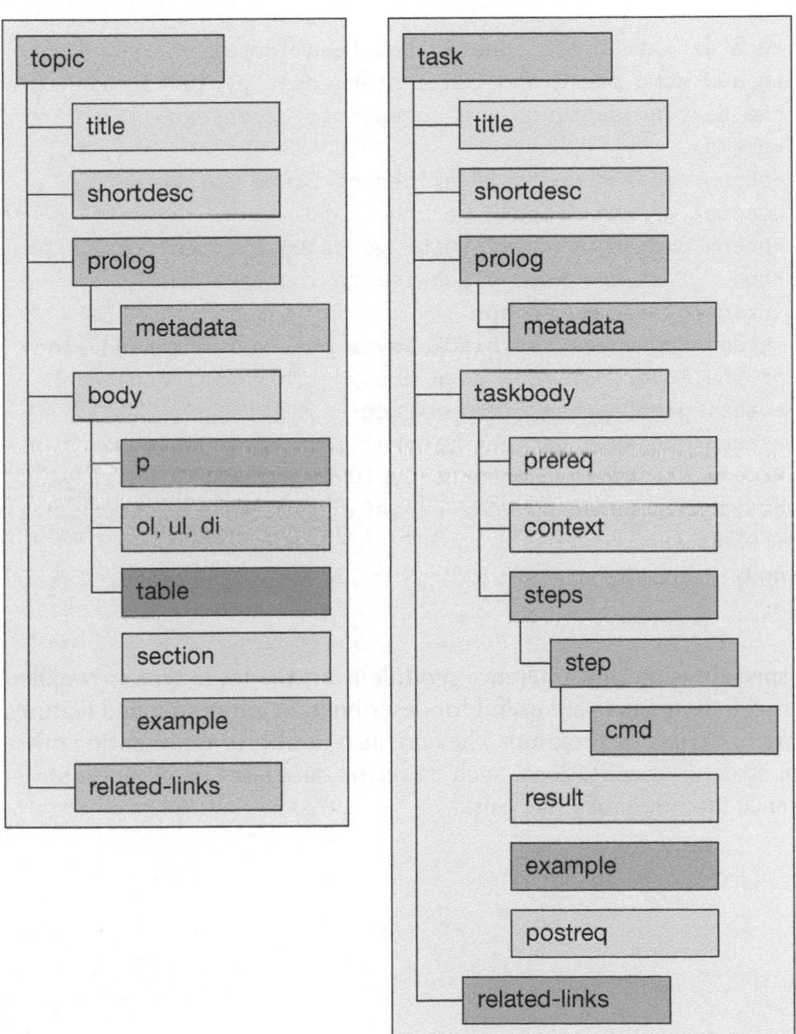

Figure 5.7 Task (Specialized from Topic)

information related to procedures and contain entities useful for performing actions related to a specific action. They answer "how do I do this?" types of questions. For example, Figure 5.7 shows that the taskbody entity contains a subentity named steps; each steps entity then holds information related to substeps and commands. A context element is used to provide background information about the task. A sample task module looks like this:

```
<?xml version="1.0" encoding="utf-8"?>
<!DOCTYPE task SYSTEM "task.dtd">
<task id="toast_test">
<title>Making Toast with the SuperChar 3000: Testing Your
Toaster</title>
<taskbody>
      <context>After the toaster has been installed, you need
      to add some bread and run a test toast cycle to make sure
      the heating settings are acceptable.</context>
      <steps>
      <step><cmd>Add up to 16 slices of bread to the
      toaster.</cmd></step>
      <step><cmd>Adjust heat settings using the red rotary
      knob.  Turn the knob clockwise to increase heat
      intensity.</cmd></step>
      <step><cmd>Press the black lever on the right hand side
      of the SuperChar 3000 down firmly.  This will engage the
      toasting mechanism.</cmd></step>
      <step><cmd>Wait patiently until your toast is done.
      Repeat the heat adjustment cycle as necessary until your
      bread is toasted to your satisfaction.</cmd></step>
      </steps>
</taskbody>
</task>
```

The specialization of a reference module from the topic type is revealed in Figure 5.8. References are useful for describing the properties and features of a system, device, or program. They are also useful for representing other types of structured collections, such as recipe databases or bibliographies. A reference module looks like this:

```
<?xml version="1.0" encoding="utf-8"?>
<!DOCTYPE reference SYSTEM "reference.dtd">
<reference id="toaster_features">
<title>SuperChar 3000 Specifications</title>
<refbody>
<refsyn>Italicized features are only available in the deluxe
models (versions b and c).</refsyn>
<properties>
      <property>
            <proptype>Slice Capacity</proptype>
            <propvalue>16</propvalue>
            <propdesc>Number of bread slices that can be
            toasted at once.</propdesc>
      </property>
      <property>
            <proptype>Special Features</proptype>
            <propvalue>High-capacity bread load, precision
            engineered parts, three year warranty, <i>super-
            fast toasting cycle</i></propvalue>
            <propdesc>Special features used in this product
            line.</propdesc>
      </property>
</properties>
</refbody>
</reference>
```

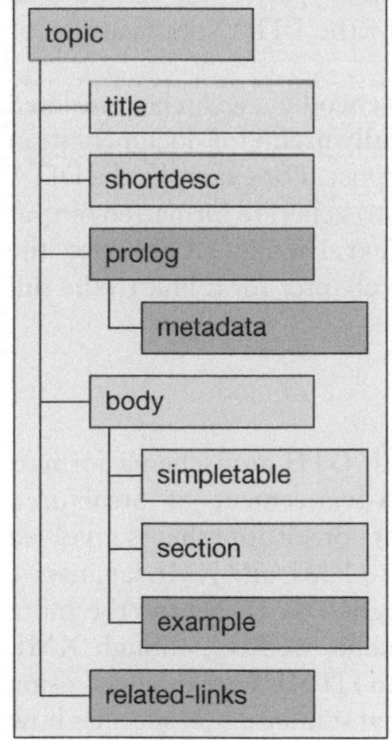

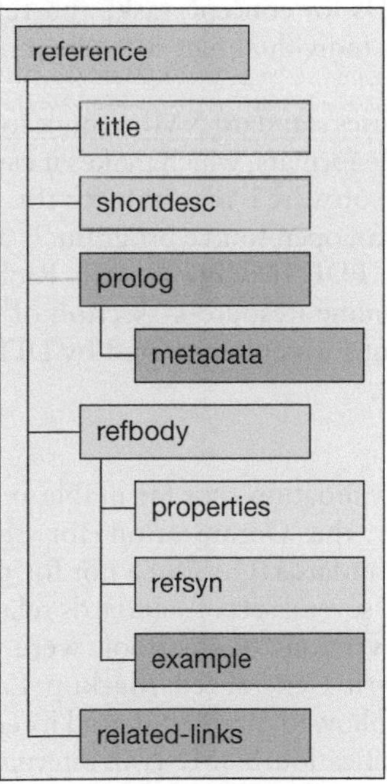

Figure 5.8 Reference (Specialized from Topic)

The reference type is similar to the concept type in its renaming of the body element to a more precise variant (in this case, refbody). It extends the topic by adding in new elements to handle properties, property values, and descriptions. It also includes a special element, named "refsyn," that enables a document author to include special notes regarding syntax that pertain to the information contained within the reference element.

DITA Maps

Once a collection of DITA topics has been authored, it is arranged using a DITA map. DITA maps contain nested lists of topicrefs, which are links to DITA topics. A sample DITA map looks something like this:

```
<map id="toaster" title="toaster instructions">
      <topicref href=" toaster_overview.xml" type="concept" />
      <topicref href="toaster_features.xml type="reference" />
      <topicref href="toast_test.xml" type="task" />
</map>
```

DITA maps can then be used to select different modules for different informational needs. Unique DITA maps can also be created to group together different topics for specific types of formatted output. The full process of compiling and linking a DITA project is somewhat complicated (it involves, among other things, installing the Java Development Kit). You can, however, download the test files from our website along with the individual DTDs for concept, task, and reference types. These can be used to validate the individual files for adherence to the DITA specification for each type.

Since DITA uses standard XML syntax for its encoding, it can be translated to numerous file formats, which makes it especially useful for documentation tasks. Custom software is available for this purpose. For example, the DITA Open Toolkit, an open source program, is able to generate formatted output types such as PDF, DocBook, and Rich Text Format (RTF). See the "Additional Online Resources" section of this chapter for a link to the full listing of output formats supported by DITA.

DocBook

DocBook is a validation tool (available in both DTD and schema format) maintained by the Organization for the Advancement of Structured Information Standards (OASIS), a not for profit consortium that is involved with advancing several sets of standards related to XML and XML languages. The very first versions of DocBook were designed as DTDs for the more general Standard Generalized Markup Language (SGML), though XML versions soon followed. Both XML and its cousin HTML were designed using the SGML specification. SGML is an international standard that specifies how markup languages should be defined (Walsh and Muellner 3).

DocBook is used to validate the form and structure of XML content according to a set of rules defined by the schema's authors (in this case, the OASIS DocBook Technical Committee). DocBook uses elements and organizational strategies derived from printed text, so it is a popular tool for authoring books or documentation projects with complex content (such as software language manuals or computer hardware reference books). Though *this* book was not written using the DocBook standard, it certainly could have been. The claim has been made that DocBook is probably the most widely used XML documentation format in use today (Miller and Clarke 179).

The full specification of DocBook can be overwhelming for beginners. It contains hundreds of elements and specific rules about how these elements should be combined and ordered within documents. There is, however, a simplified version of DocBook that is more accessible to novice users. This version, known as the Simplified DocBook Document Type, contains a smaller number of elements and was originally designed to have the same number of tags and the same expressive power as HTML (OASIS online). At the time of this writing, though, the Simplified DocBook Document Type was available only in DTD format.

While it is certainly possible to learn about DocBook by accessing the current DTD or schema directly and reading through the elements, attributes, entities, and rules (and we encourage you to do so), it can still be confusing for beginners, even in the simplified version. It is much easier to follow a brief example that uses only a few of the available elements in the DocBook specification. Here is a very simple example of some content marked up using valid Simplified DocBook XML code:

```
<?xml version="1.0" encoding="utf-8"?>
<!DOCTYPE article PUBLIC "-//OASIS//DTD Simplified DocBook
XML V1.1//EN" "www.oasis-open.org/docbook/xml/simple/1.1/
sdocbook.dtd">
<article>
      <title>SuperChar 3000 User Guide</title>
      <subtitle>A Toaster for the New Millennium</subtitle>
      <articleinfo>
            <author>
                  <firstname>John</firstname>
                  <surname>Doe</surname>
            </author>
            <authorblurb>
                  <para>
                        John Doe is a toast lover and technical
                        communicator from Orlando, Florida.
                  </para>
            </authorblurb>
      </articleinfo>
```

```
    <section>
        <title>Introduction</title>
        <para>The SuperChar 3000 is a deluxe toaster
        manufactured to the highest quality standards.
        It is capable of simultaneously toasting 32 slices
        of bread using four heat settings.</para>
    </section>
</article>
```

If you would like to try editing the file and adding additional tags, you can download this file ("ch5_docbook.xml") from our website. While this is admittedly a very primitive application of DocBook, it does allow us to see why the DocBook DTD is so useful. First, note that all of our information is contained within an element named article. Within this element, we can create a structured article using the types of information normally associated with smaller documents (author name, blurb, sections, section headings, paragraphs, and so forth). Although we can provide a sense of order to our document by manually placing sections after one another in the XML file, we are also creating modular containers for certain sections that can be reused as parts of other documents. For example, the introductory paragraph describing the toaster's features in the user manual might be repurposed as an advertisement for the user manual of another home appliance from the SuperChar product line.

We now also have a means for ensuring consistency and dealing with potential ontological collisions with our element names. For instance, note that the XML tag used to hold the authors last name is typed in as <surname>. If we were to try using a <lastname> tag for this element, our XML file would be marked as invalid, since the DocBook DTD is expecting to see that information encapsulated using the <surname> tag. This consistency is especially important in collaborative structured writing environments where multiple authors may be contributing to different sections of our document.

Like DITA, DocBook has numerous XSLT style sheets already created for it, so translating DocBook documents into other formats such as XHTML or PDF is common practice. More information about DocBook and its uses can be found in Norman Walsh and Leonard Muellner's *DocBook: The Definitive Guide*. There is much more to be learned about this robust validation toolset and we have only scratched the surface in this overview.

Chapter Summary

In this chapter, we considered some of the more advanced topics and technologies associated with XML. Although the syntactical requirements of these technologies are often challenging, there are real world equivalents to these procedures that we all use on a daily basis, and that enable us to interact with one another more efficiently and collegially. For example, we all communicate with one another using common vocabularies, social behaviors, and protocols, and these protocols may change from one discourse community to another. By using such tools as namespaces, schema, XLink, XPath, and XPointer, we are giving computers the same tools for recognition, seeking, searching, and verification that we ourselves use to evaluate the credibility and accessibility of our information sources. By being aware of existing and robust schema and DTD architectures, such as DocBook and DITA, we learn to use interchangeable tags and coding strategies that will allow multiple, different databases to be recognized and used for structured writing tasks. These predesigned DTDs and schemas are valuable for authors working with large combined documents that frequently undergo revision. They are also useful in single sourcing applications.

As we began to see in this chapter, there are new issues introduced when we combine, share, and repurpose XML documents for the sake of information sharing and knowledge dissemination. Fortunately, the World Wide Web has given us an efficient and stable mechanism for exchanging our XML documents with one another. Unfortunately, since our computers do not yet understand human language, we must describe our online information with semantic language that is meaningful for computer programs. This is the idea behind the Semantic Web as expressed by Tim Berners-Lee that we discussed earlier in Chapter 3. However, human agency is still very much involved in this process—the act of selecting, grouping, and managing XML tags during information authoring is still fundamentally a rhetorical act, even if the final product will be processed mostly by computer software programs.

It is this processing stage that we will focus on in Chapter 6. To explore the concepts behind XML processing, we will build some basic XML applications using a custom parser written in an open source Internet scripting language. Though this process is more cumbersome than using a prepackaged XML parser, a standard Web browser, or existing validating architectures, such as DocBook or DITA, it ultimately gives the designer even more control and flexibility when using XML for a specific purpose. Mastering advanced technologies associated with XML, such as the topics presented in this chapter, is an important first step for forward-thinking technical communicators and information architects looking to design websites or technical documents for Web 2.0 and the Semantic Web.

Discussion Questions

1. Identify several terms that have different meanings in different discourse communities. In what ways are problems of recognition and collision handled? What mechanisms exist in the real world for dealing with these inconsistencies? Are these mechanisms at all similar to the way that namespaces function in XML?

2. Jay David Bolter suggests that links contribute largely to the rhetoric of hypertext. What other types of hypertext elements influence the rhetorical context of information, and in what ways? For example, how do bulleted lists force a designer to begin thinking about compartmentalization and modularization of content? How do the bold and italics tags influence the rhetorical style of one's message? Refer to Chapter 4 if necessary and consider the rhetorical implications of separating form from descriptive content.
3. In our discussion of XPointer, we explained that a legal patent from Sun Microsystems held up the formal W3C Recommendation for XPointer. What other types of legal issues have hindered developments on the World Wide Web, and how have these issues influenced the credibility (ethos) of the involved stakeholders? How have these issues been resolved? If necessary, do some research on your own and take some notes to be used for discussion.

Activities

1. Revisit your university student and personnel database from Chapter 2. Replace the DTD you designed in that example with an XSD (schema). Prepare a short document discussing the advantages and disadvantages of each method (a simple bulleted list for each category would be a good start).
2. Create your own version of the "anchored_example.html" file from this chapter and create several types of links (internal and external links as well as absolute and relative links). Now, consider the limitations of traditional HTML links in relation to the level of control in an XML document. How would a working XLink implementation help to solve these problems?
3. Using the same student and personnel database mentioned in Activity One, write XPath expressions to select the following nodes:
   a. All students referenced in the XML file
   b. All faculty members referenced in the XML file
   c. The e-mail addresses of all students
   d. The e-mail addresses of all faculty members
   e. The name of the last student listed in the XML file
   f. The name of the last faculty member listed in the XML file
4. Find some real world examples of DITA or DocBook being used in technical documentation by performing a keyword search with your favorite search engine. After finding some information on these products, answer the following questions:
   a. What challenges do these architectures pose to the creation and distribution of technical content?
   b. How would using one of these tools differ from using one's own custom library of XML elements and attributes?

   c. What are the advantages and disadvantages of DITA as compared to DocBook? Are the products competitors? Is one better suited than the other for addressing particular types of rhetorical situations or serving particular types of audiences?

5. In this chapter we discussed DITA and DocBook, two sources of standards for XML documents. Research some additional sources of XML-based standards, such as TEI (Text Encoding Initiative), OeB (Open eBook), and EAD (Encoded Archival Description). Make note of the major features of these standards and list some of the applications of these standards for specific purposes. Write a short essay that:
   a. explores the intersections and relationships between rhetoric, specific discourse communities (librarians, archivists, scientists, etc.), and the standardization of XML content, or,
   b. provides an overview of one of these technologies and discusses its use for a particular type of project. Why are standards necessary for this type of project?

References

Albers, Michael J. "Single Sourcing and the Technical Communication Career Path." Technical Communication 50.3 (2003): 335–43.

Bolter, Jay David. *Writing Space: Computers, Hypertext, and the Remediation of Print*. 2nd ed. Mahway, New Jersey: Lawrence Erlbaum Associates, 2001.

Del.icio.us. July 23, 2008. <http://del.icio.us/>.

Dodds, Leigh. "XPointer and the Patent." 2001. O'Reilly XML.com. November 6, 2007. <www.xml.com/pub/a/2001/01/17/xpointer.html>.

Linden Research, Inc. "Second Life: Official Site of the Free 3D Online Virtual World." July 23, 2008. <http://secondlife.com/>.

Makena Technologies, Inc. "There: The Online Virtual World that is Your Everyday Hangout." July 23, 2008. <www.there.com/>.

Miller, Dick R., and Kevin S. Clarke. *Putting XML to Work in the Library: Tools for Improving Access and Management*. Chicago: American Library Association, 2004.

OASIS. "The Simplified DocBook Document Type." 2004. The OASIS Technical Committee. July 22, 2008. <www.oasis-open.org/docbook/specs/cs-docbook-simple-1.1.html>.

Priestly, Michael, Gretchen Hargis, and Susan Carpenter. "DITA: An XML-Based Technical Documentation Authoring and Publishing Architecture." *Technical Communication* 48.3 (2001): 352–67.

Ray, Erik T. *Learning Xml*. Sebastopol, CA: O'Reilly, 2001.

The Internet Society. "Uniform Resource Identifier (Uri): Generic Syntax." 2005. The Internet Society Network Working Group. October 28, 2007. <http://tools.ietf.org/html/rfc3986>.

van der Vlist, Eric. *Xml Schema*. Sebastopol, CA: O'Reilly, 2002.

W3.org. "Namespaces in Xml 1.0 (Second Edition)." 2006. W3C. October 31, 2007. <www.w3.org/TR/REC-xml-names/>.

W3Schools.com. "Introduction to XML Schemas." November 4, 2007. <www.w3schools.com/schema/schema_intro.asp>.

Walsh, Norman, and Leonard Muellner. *Docbook: The Definitive Guide*. Sebastopol, CA: O'Reilly & Associates, 1999.

Additional Online Resources

1. DITA Downloads: www.ibm.com/developerworks/web/library/x-dita6/x-dita_downloads.html
2. DITA Online Community: http://dita.xml.org
3. DITA Open Toolkit: http://dita-ot.sourceforge.net
4. DITA Output Formats: http://dita.xml.org/output-formats
5. DITA Resources for Beginners: http://dita.xml.org/wiki/resources-for-beginners
6. DITA Users Group: www.ditausers.org
7. DocBook: The Definitive Guide: www.docbook.org
8. OASIS: www.oasis-open.org
9. Simplified DocBook DTD: www.docbook.org/xml/simple/1.1
10. The DocBook Project: http://docbook.sourceforge.net
11. Validator for XML Schema: www.w3.org/2001/03/webdata/xsv
12 W3C XML Path Language (XPath) 2.0: www.w3.org/TR/xpath20
13. W3C XML Schema Primer: www.w3.org/TR/xmlschema-0
14. W3C XML Schema: www.w3.org/XML/Schema.html
15. W3C XPath Data Model (XDM): www.w3.org/TR/xpath-datamodel/#dt-node
16. W3C XPointer Framework: www.w3.org/TR/2002/PR-xptr-framework-20021113
17. W3C XPointer Working Draft: www.w3.org/TR/xptr
18. XML schema tutorial: www.w3schools.com/schema/default.asp

6 Focused Implementations

Using PHP to Design Custom Parsers for XML Projects

Chapter Overview

At this point in the book, we have learned about theoretical models for thinking about data representation and information exchange as well as about applied technologies for encoding and representing modular units of data. We will now learn to synthesize these ideas in order to solve problems in an applied environment. This chapter discusses XML parsers and then introduces three examples of XML parsers that can be used to process and act upon particular XML elements and attributes as they are scanned from a file. The first and simplest example uses existing XML files—structured as Really Simple Syndication (RSS) newsfeeds—as source content for a basic news display page that is used to update website visitors about news or events in a streaming fashion. This means that the content is pushed out to the users automatically—they can then choose whether or not to activate the link and receive more detailed content.

When we discuss our second parser example, we outline a process for creating a CMS for keeping track of digital assets. Here we outline a process for building such a system using only XML and plain text files as our data sources. One of the benefits of such a system is that its data store is portable and does not require any special software in order to be accessible.

The final parser example involves building a single sourcing system for a software documentation project. As we construct this parser, we revisit the ideas discussed in Chapter 3 to observe specific ways in which XML can be used as a supporting technology for single sourcing systems. This shows how content can be repackaged and repurposed to serve different rhetorical and informational needs for end users of varying skill levels. As we discussed in more detail in Chapter 3, this is an example of what Ann Rockley calls a level three, or "Dynamic Customized Content" single sourcing system.

This chapter is the lengthiest in the book because it involves applying many of the topics we have discussed to specific, real world types of problems. In order to create a useful XML parser, we must understand the general problem space as well as the rhetorical implications of our informational decisions. In addition, to *truly* understand the informational context of a metadata system, we must work on both sides of the equation:

as information designers, producers, or information architects; and as hypothetical consumers or end users of the system. In these roles we must rely upon our technical acumen, our usability knowledge, and our understanding of rhetoric.

The technical challenges of working with custom-designed parsers can be considerable. Like the aesthetic dimension we discussed in Chapter 4, the programmatic demands of building custom parsers often require a different type of thinking than we are used to. For instance, since machines are now an audience we must serve, we need to figure out how to *write* the XML data rather than just how to read it. We cannot assume a human will be writing all of the XML code for the parsers we design. In addition to building software that is able to read and act upon metadata, we generally must also include capabilities for writing metadata and for translating user-generated content into XML elements and attributes that our parser can understand.

Many organizations and authors are resistant to editing raw XML files directly using a text editor. For one thing, it is reassuring for an author to have a graphical user interface between herself and the XML document when she is adding content or metadata. Familiar graphical interfaces are always less intimidating than raw XML files for non-expert users. For another, computers do not make mistakes, as long as they are programmed correctly. So, it is much less likely that a computer would incorrectly code an XML file than a human user who has to deal with phone calls, unexpected office visits, multitasking, other projects, and any number of other unforeseen distractions. While there still may be an opportunity to introduce error during content addition, at least syntactical errors related to the correct structuring of XML tags can be minimized.

In each of the three projects we discuss in this chapter, we include both an XML writer and an XML reader, or parser. Due to the length of the code samples for the latter two projects, we discuss only portions of the code within this chapter and include the unabridged code listings in Appendices C and D. The full code for our first example, though, is shorter and somewhat less complicated. It is therefore included within this chapter in its entirety.

While we believe that studying these coded examples will provide much insight into the computational side of the interactive cycle between human users and XML databases, we also want to point out that the coded examples in this chapter will be challenging for readers without a background in computer programming. In particular, the examples we provide for the CMS (Project 2) and the single sourcing demonstration (Project 3) are admittedly more complicated than the shorter XML examples we discussed in previous chapters—even those advanced topics in Chapter 5. Rather than reducing the complexity of these examples, we chose to include them, and also to provide significant commentary in areas which might be difficult for some readers to understand. The unabridged code listings in Appendices C and D are also heavily commented (marked up with additional explanatory text) throughout. We hope that these projects will serve as rough prototypes or starting points for individuals hoping to complete applied projects, but without a good idea of how to get started with using XML content for solving real world types of problems. Our intent is for this chapter to serve as a cookbook of sorts for the theorist–practitioner or symbolic-analyst who wishes to take advantage of XML techniques in order to advance and improve knowledge management techniques in his or her own organization.

XML Parsers

An XML parser is a software program or a component of a software program that processes and acts upon one or more XML documents. It specifies precise procedures to be followed when particular XML elements or attributes are encountered. For example, in a grocer's database, a parser might be designed that would iterate, or make continued cycles through, the XML document's hierarchy of inventory elements. The parser could be programmed to print out the data content of every element with an onsale="true" attribute value and an instock="true" attribute value to a special Web page listing the daily specials. This software program acts as the critical link between an XML repository and any users who are looking to extract information from that collection of metadata. For customers looking for grocery items on sale, the XML document itself would not provide that information in an accessible and meaningful way. Instead, a document designer would rely upon a rule-based software program to iteratively pass through the XML content and extract any items that were coded as being both on sale and in stock. This is what the parser does. The parser can then apply CSS formatting or XSL transformations in order to tailor that information for particular audiences.

Minimal XML parsers are built into Internet browsers, like Microsoft Internet Explorer®, but more comprehensive browsers are also available. A parser may be commercially distributed, such as the parsers designed by IBM or Oracle for the Java programming language, or it can be open source and freely available, such as the popular Expat XML parser for the C programming language. It can also be validating, with support to check XML data against a defined DTD or schema, or non-validating, without this support. It is possible to develop customized parsers based on the needs of one's own projects or informational needs; in this chapter, this is the approach we take. We choose to discuss examples of custom-defined parsers in order to demonstrate the flexibility and customization possible when one builds their own non-validating or validating XML parser.

The parser is vitally important for an information designer to understand because this alone determines what the end users, readers, or audiences actually *see* based on their interface decisions. Details like element names and schema validation are largely irrelevant to most information seekers. What they are looking for is a clear and direct path to the information they need. They would like the information revealed in the interface they are using at the time, not hidden in an unfamiliar file format with (seemingly) complicated XML notation. As Michael Albers notes in his introduction to the edited collection *Content and Complexity*, the best interface is the one that disappears, leaving the information clearly defined without distractions from the interface (6). Thus, a parser has a vital, but difficult, duty: to make the structure and syntax of the XML document disappear by interpreting this content and structuring it in a visually and rhetorically appealing fashion for the audience that hopes to extract information from it.

Recall from Chapter 4 that XML has evolved as a syntactically standardized language for *computers* to exchange as a means of internetwork communication. When a human tries to interpret this data, she may easily become lost in the clusters of code from DTDs, schemas, elements, attributes, and style sheet instructions that reside in a typical XML file. It is the parser's job to translate this complex language into something a human being can more readily interpret and understand. This may involve transforming the XML file into an HTML file using XSLT, moving XML content from its hierarchical form into a more digestible tabular representation, or even using sophisticated visual display techniques to graphically represent the data stored within a document.

The role of the XML parser is often downplayed or even overlooked entirely in XML literature. This is true for a variety of reasons. For one thing, as noted previously, XML parsers can come in a variety of flavors and packages. In addition to being validating or non-validating; custom-designed, open source, or commercial; parsers can range from complex generalized graphical user interface applications for Windows, Macintosh, or Linux to more application-specific and portable implementations that run entirely on the World Wide Web in a language such as Perl, Java, or Hypertext Preprocessor (PHP). For another, designing and building a usable parser is more complex than writing an XML file. This is because parsers are implemented using programming or scripting languages that contain entirely new sets of syntactic rules that one must learn how to use. A third reason is that parsers are sometimes considered to be rather dull and boring. Since XML is all about data, the parser may be viewed as a mere tool or vessel to hold or transport this data and it can be difficult to understand why the parser is so rhetorically significant.

Some of these things are true. Standard parsers *are* rather boring, and there does not seem to be much rhetorical substance to a standard software utility that simply reprints XML content from a file to a user's browser window. With a *customized* parser, however, software is used to extend and shape the messages embedded in XML files in new and exciting directions. By specifying how, when, and under what circumstances data can be extracted from elements and attributes, the parser is analogous to the ancient Greek rhetor. The classical rhetorician could deliver a speech that was either stunningly powerful or woefully ineffective. Only with appropriate decisions in regard to arrangement, style, and delivery would the audience be receptive to the speech. The parser similarly adds an element of agency to XML. It must also follow the rules of arrangement, style, and delivery to effectively present information to an audience. In this sense, it can be seen as something more than a mere vessel, since it has this ability to transform and manipulate data on the fly. Parsers specify the expressive and rhetorical potential of XML documents. More robust and flexible parsers have more rhetorical potential.

Because of the importance parsers play in the discursive flow of information between machines and humans, we believe it is important to understand both the technical underpinnings of XML parsers and the rhetorical ideas that can be used to build customized parsers for one's own research or professional projects. With this importance in mind, we spend a good deal of time in this chapter discussing both theoretical preproduction strategies and the applied programmatic skills necessary to build rhetorically capable XML parsers.

Rhetorical Approach

As this book is fundamentally about the rhetoric of XML, we approach each of the three parser projects in this chapter using two phases. In the first phase, we discuss some of the rhetorical considerations that are important when making decisions about how to best serve our anticipated audiences using metadata. We employ three different rhetorical heuristics in these examples, from an ad hoc rhetorical analysis of user needs in the first example to specific rhetorical strategies taken from information design research used in the second example. We then use a distillation of our ideas from Chapter 3 in the third example.

We include these three different strategies in order to demonstrate the unique pairings that emerge when different applied technologies are paired with different rhetorical perspectives. These pairings generate distinct perspectives on the information ecologies and stakeholders that are involved in a communicative act. The second phase of each project is then concerned with a detailed technical discussion of how to use available technologies to design a parser capable of serving these informational needs. This combined approach embodies the spirit of the theorist–practitioner and symbolic-analyst that we believe is so important for technical professionals working with XML tools and technologies.

These rhetorical considerations are especially important when designing the parsers, even primitive parsers like the ones we discuss in this chapter. Why is this? As we discussed previously, because XML is often a format that is communicated from computer to computer rather than from computer to human, it is the parser that takes on the heavier burden of presenting meaningful information to a human user. The parser must reshape the information that has been culled from an XML document or merged from multiple metadata sources.

Though it is overly reductive and simplifies the complex social nature of information, the traditional communicative paradigm is a useful construct for understanding the role of the parser in a rhetorical act. The model described here uses a "contractive" view of technology wherein an information "receiver" is seen in a relatively passive role and information itself is chunked into discrete and unambiguous units (see Chapter 1 for a discussion of contractive versus expansive communication technologies). When communicating using XML, an information "sender" is responsible for thinking carefully and logically about how to structure data in a fashion that facilitates the extraction of useful information from a data source. The message itself then resides as potential within the XML document that this sender creates. It is the parser's responsibility to enable the extraction of *information* from the collection of *data* that exists in the document. This further facilitates the formulation of new knowledge through the combination process (explicit to explicit knowledge transfer) as defined by Nonaka and Takeuchi and discussed previously in Chapter 1.

On the other end of the transaction, a "receiver" will attempt to locate nodes of data that are contextually relevant and useful for a specific type of task. The parser is therefore the final instrument that can be used to shape a message before it is interpreted by a receiving agent. When artifacts from the XML document are presented to the user along with their selected information (artifacts such as snippets of code describing elements or their relationships), this "extra" information serves as communicative "noise" that reduces the quality or fidelity of the message itself. In other words, the XML rules and relationships are irrelevant to most information seekers and only serve to confuse or frustrate them in their quest for information. The parser must transform raw XML content into a message interpretable by an individual.

Technological Approach

Because of its affordability, availability, relative ease of use, support for XML, and popularity, we chose to use the server-side Internet embedded scripting language known as PHP for the three project exercises in this chapter. Server-side means that the language requires a special software program on the server in order to run correctly. The server will "preprocess" the script files and return normal HTML code to users' browsers. We will introduce PHP later in this chapter and explain why it is a good choice when learning how to construct custom XML parsers for your projects. We also provide a simple overview of PHP and its associated XML capabilities. It should be noted, however, that different languages have different strengths and weaknesses as parsers. For example, the Java programming language would be a better choice for working with an object-oriented application that required XML data integration and validation. Similarly, if a Web-based application was designed that needed to minimize strain on the server computer, a client-side alternative such as JavaScript might be a better solution. Client-side means that the code runs directly in the users' browsers and that it does not require any special server software. The ability of XML to integrate with such a large variety of programming and scripting language further speaks to its importance as an agent of interoperability and structured communication.

Although PHP is very forgiving to beginning programmers, we recognize that not all readers will have a background in programming concepts such as looping, repetition, and working with complex data structures (although some of these concepts were discussed in Chapters 4 and 5 with advanced topics such as XSL transformations). For this reason, we have designed the examples in this chapter to be as modular as possible. In many cases, you can copy and paste code from these examples and modify a few *variables* (which we will discuss shortly in this very chapter) in order to implement working, but customized and personalized, versions of these parsers for your own projects. The parsers can then be reconfigured to work with your own XML data sources. You can also see a listing of the full source code for Projects 2 and 3 at the end of this book in Appendices C or D, or you can download the source code for each project from this text's accompanying website.

Getting Started

In earlier chapters, the examples we presented could be typed up in any editor capable of saving in the plain text format—Notepad, Wordpad, BBedit, XML Notepad, Dreamweaver, Stylus Studio, XMLBlueprint, or XMLSpy, to name a few popular editors. In this chapter, however, you will need to use a specific software suite in order to follow along with the examples. This is because we are moving from a more abstract domain of data representation into the application-specific domain of data parsing.

The first specific software we will be using in this chapter is called PHP. We will discuss PHP in more detail soon, but, for now, just know that PHP is a software program that runs on a server computer. When the server computer's Web server (sometimes known as an http daemon) encounters files of a particular type (generally files that end in the ".php" suffix), it sends these files to PHP for processing. PHP then filters out the programming instructions, executes them, and returns normal HTML code to any client computers that are requesting pages stored on the server. This is a very important thing to note, because it allows PHP files to be served to any computer capable of processing HTML code, even older browsers without the latest plug-ins and security updates. This ability to return standard HTML files to client computers, as generated from complex programming scripts on a server computer, is a major benefit of server-side languages like PHP.

The next software program we will be relying upon is Apache. Apache is the world's most popular Web server. Apache supports PHP integration, but PHP will also work with several other configurations of operating systems and services, including Windows Server distributions running Internet Information Services. For our purposes, we will use Apache as our Web server and PHP as our scripting language. Rather than setting up a high-capacity rack mounted server and opening it to multiple computers on the Web, we will in effect be installing a virtual "web server" on a local computer such as a desktop PC or laptop. This allows us to test and develop the parser without immediately releasing what may be incomplete or untested parsers to the entire Web community. The installation program we will be using to set up our development computer is called XAMPP, which is an integrated installation package developed by a group of individuals calling themselves the Apache Friends Network.

XAMPP is an Apache distribution. A distribution is a custom installation suite that can be installed on one's computer. XAMPP comes preconfigured with support for PHP, MySQL, and Perl. At the time of this writing, the distribution could be downloaded from www.apachefriends.org/en/xampp.html. We will update the link on our associated website if this should change, but you should also always be able to find the latest version simply by searching for the keyword "xampp" using your favorite search engine.

If you would like to follow along and build the examples in this chapter on your own, you should download and install XAMPP. While we will not cover the installation in detail here, we do provide a link on our website to detailed installation instructions for various operating systems. After you install the software, you should start the XAMPP control panel. Figure 6.1 is a screenshot of the Windows version of the XAMPP control panel. There

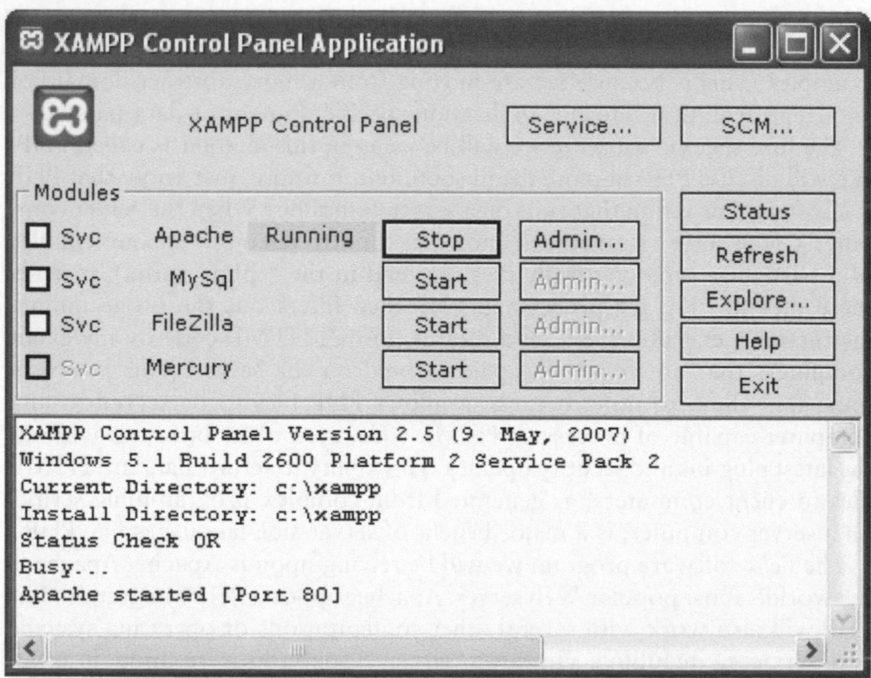

Figure 6.1 XAMPP Control Panel

are versions available for other operating systems, but they will look somewhat different. Macintosh users may already have Apache and PHP installed and may need only to activate the Apache service.

If you used the default settings in Windows during installation, the XAMPP control panel is likely in a program group named "Apache Friends." Start the XAMPP control panel and make sure the Apache service is running. If it is not running, click the "Start" button to initiate the service. If you are using a firewall, you may need to click "Allow" or "Unblock" in your firewall software to enable the service when it starts for the first time. You will not need to use any of the services other than Apache to build the XML parsers in this chapter.

There are a few important things to note about Apache Web server when installing it for the first time. First, when you save HTML or XML or PHP files that you would like to be accessible by the service, you need to save them in a special directory that Apache has access to. By default, this directory is "c:\xampp\htdocs" on a PC computer. You can create folders within this directory and then navigate to files saved within that folder using your Internet browser. In Figure 6.2, we have created a folder named "xmlparsers" that will be used to house our three projects for this chapter.

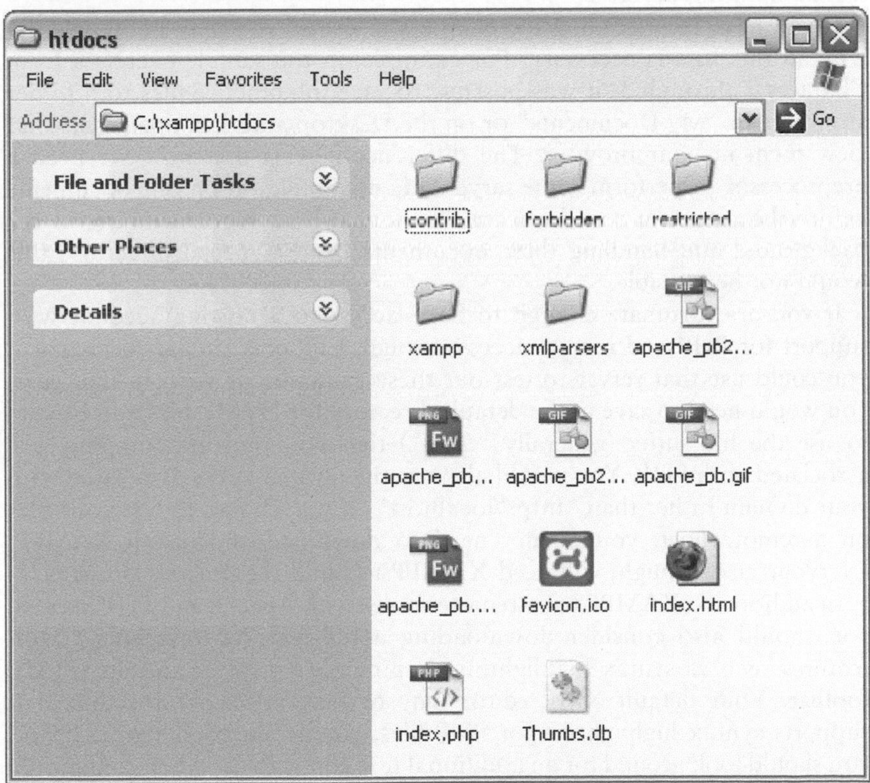

Figure 6.2 Htdocs Directory

You should create the same folder on your own hard drive ("c:\xampp\htdocs\xmlparsers") if you wish to follow along with the three parser design exercises.

The second important note about Apache concerns the way in which hypertext, XML, or PHP files are opened from a browser. To view a saved file, you will need to open the URL "http://localhost/folderpath/filename", where folderpath is the folder structure you have created in the htdocs folder, and filename is the PHP, HTML, or XML file you have saved within that folder. Mechanically speaking, the "http://localhost" portion of the URL tells Apache to look inside the default htdocs folder (in our case, "c:\xampp\htdocs") when processing and displaying documents.

Finally, note that the direction of slashes changes when we are talking about online documents versus documents on our local hard drives. In general, directory paths on Windows machines are indicated using backslashes, while Internet (and Unix/Linux) directories use the forward slash to show a file's location within a directory hierarchy. It is certainly confusing at first, but easy enough to fix through trial and error experimentation. If you forget the rules and a file does not seem to be loading up in your browser, try changing the direction of the slashes as a first step in the troubleshooting process.

If you have been reading the chapters in order, you may be wondering why all this fuss is necessary. For example, in the sample exercises from Chapters 2 through 5, it was possible to save the practice files to a folder anywhere in "My Documents" or on the Desktop and then open them and view them in your browser. The difference here is that now we need a preprocessor to perform some server-side manipulation on our documents before they are sent to a browser. Without a Web server running in the background and handling these documents from a special directory, this would not be possible.

If you are fortunate enough to have access to a hosting account with support for PHP and remote access through FTP or a similar mechanism, you could use that server to test out these examples as well. In that case, you would need to save in the default directory for HTML files and be sure to use the file suffix (generally ".php") that your hosting company has associated with PHP. You would also use the normal URLs associated with your domain rather than "http://localhost". If you choose to test your files on a remote host, you do not need to download and install XAMPP. Everyone else, though, will need XAMPP to build the practice projects.

In addition to XAMPP (or a remote server with Apache and PHP access), you should also consider downloading a full-featured text editor with features such as syntax highlighting, line number display, and an HTML toolbar. Your default XML editor may or may not work for this; if it supports syntax highlighting for PHP files, you are in good shape. If not, you should look around for an additional text editor to use when configuring your parser. You can find a list of text editors for PHP on the rhetoricalxml.com website. The examples for this chapter were produced using EditPlus for

Windows, which is a comprehensive text editor with syntax highlighting, FTP, and S-FTP (secure FTP) support. It also includes a seamless Web browser that runs from within the text editor and an HTML toolbar. EditPlus supports syntax highlighting for other file formats, including XML, which makes it a good choice for editing our source XML files.

In order to verify that your setup is working correctly, perform the following test. First, open your text editor and type the following text into a new document:

```
<html>
<head>
      <title>Test PHP Document</title>
</head>
<body>
      <?php echo "My first XML parser is coming soon!"; ?>
</body>
</html>
```

Be careful with the syntax. Blocks of PHP begin with the <?php sequence and end with the ?> sequence. Note that we can nest PHP tags within HTML tags, as we did here within the body tags. We could also add an <h1> tag before the beginning of the PHP statement and a closing tag </h1> at the end of the statement in order to display the text in a large heading format. You need to be sure to include the semicolon after the end of the second quotation mark. This symbol is used in PHP to indicate the end of certain types of statements (excluding special types of control structures, like loops).

Save your file with the name "test.php". Since your file is now using the .PHP file suffix, your text editor should recognize the file contents and display color coding for your document, which makes it easier to spot errors and inconsistencies in your file. In this case, since we already created a subfolder named "xmlparsers" within our htdocs document root directory, we are going to save the file in "c:\xampp\htdocs\xmlparsers\test.php". Next, open up your preferred Web browser and navigate to "http://localhost" by typing that URL into the location bar of your browser.

If XAMPP is installed correctly, the URL will redirect to "http://localhost/xampp" and you should see a screen similar to that shown in Figure 6.3. This screen provides access to demonstration applications and provides useful links to additional documentation and security information. This version of XAMPP also provides information about the current version of PHP that is running on the server. Here, we are using PHP version 5.2.6 and XAMPP version 1.6.7.

Since the test document is saved within a subdirectory of the "htdocs" directory, we must provide additional information to the browser in order to navigate to our test document. The full directory path of our file is "c:\xampp\htdocs\xmlparsers\test.php" and the full URL is "http://

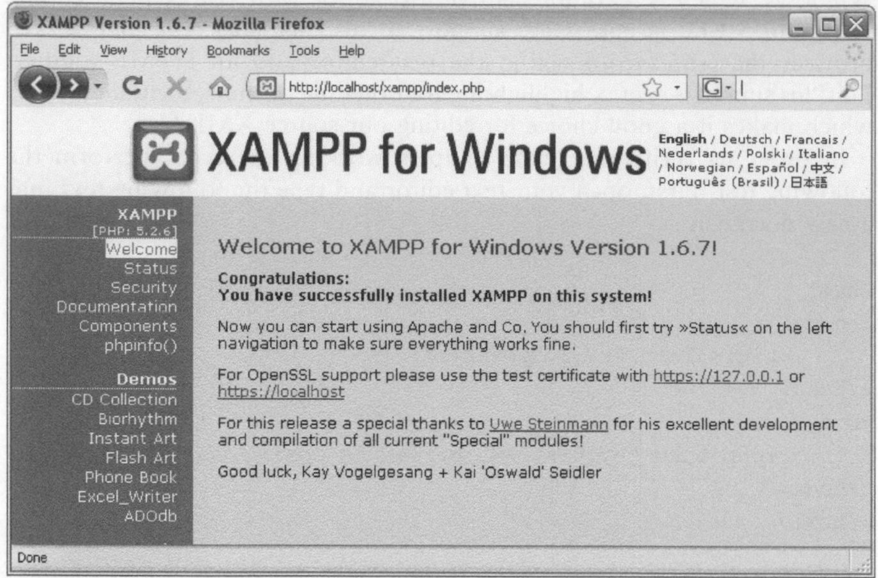

Figure 6.3 Localhost in Browser

localhost/xmlparsers/test.php". It is important to understand that the "http://localhost" portion of the URL is equivalent to the "c:\xampp\htdocs" portion of the local directory. This means that any directory structures you create within the htdocs directory on your local computer must be mirrored in the URL when you are testing out your documents.

Loading "http://localhost/xmlparsers/test.php" into your browser should produce the same output as that shown in Figure 6.4.

If you are not able to produce this screen on your own, these are some likely problems:

- XAMPP is not installed successfully. Check the XAMPP documentation for troubleshooting tips and detailed installation instructions.
- The XAMPP service is not started in the XAMPP control panel. Make sure the service is running by ensuring that the text "running" appears next to the Apache checkbox in the XAMPP control panel. Refer back to Figure 6.1 if necessary.
- You are not saving your files in the correct directory or you are not typing the URL correctly into your browser. Check each of these things carefully. If you do not remember the document directory you used during installation, you can try using your computer's search feature and searching for a folder named "htdocs". Then, you can save your files directly into this directory and they will resolve to "http://local host/folderpath/filename" as we discussed earlier in this chapter.

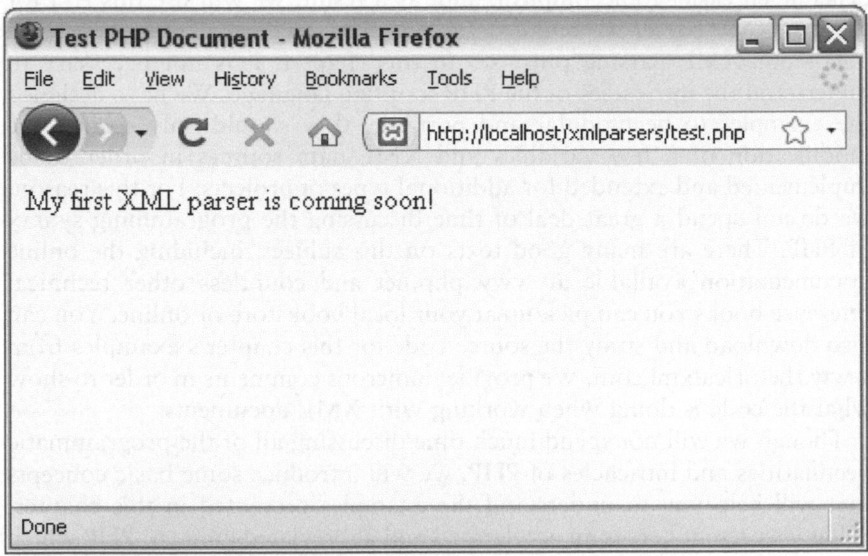

Figure 6.4 Test PHP Document

If you *are* able to produce this screen on your own, you have managed to install XAMPP successfully and you have learned how to save files into the correct directory and load them into your browser for testing. Congratulations. In the next section, we will briefly introduce PHP and SimpleXML and then we will design our first XML parser using a combination of these two resources with the aid of a preliminary rhetorical analysis.

PHP

PHP is an Internet-embedded scripting language. PHP evolved from a language known as PHP-FI, which was written by Rasmuf Lerdorf in 1995 (Php.net online). PHP-FI originally stood for Personal Home Page Tools / Forms Interpreter. You can read more about the history of PHP from the link provided in the Additional Online Resources section of this chapter, but, for now, just know that PHP has been a popular scripting language for building dynamic websites on the World Wide Web for many years.

For several reasons, PHP is a good language to use when designing an XML parser. First, it is extremely popular and widely available in many different distributions (it will run on many different types of servers). Second, it is open source and freely available. Third, and most important for our purposes, PHP has a very intuitive means for working with XML hierarchies. As of PHP version 5, the SimpleXML application programming interface (API) was introduced. SimpleXML makes formerly tedious XML tasks much easier to accomplish, and, as a result, we will use this API for the first XML parser we build.

For our XML parsing purposes in this chapter, it is not necessary to understand the intricacies of the PHP scripting language. We have designed the examples to be modular and portable; they should only require the modification of a few variables and XML data sources in order to be implemented and extended for additional types of projects. For this reason, we do not spend a great deal of time discussing the programming syntax of PHP. There are many good texts on this subject, including the online documentation available at www.php.net and countless other technical reference books you can pick up at your local bookstore or online. You can also download and study the source code for this chapter's examples from www.rhetoricalxml.com. We provide numerous comments in order to show what the code is doing when working with XML documents.

Though we will not spend much time discussing all of the programmatic peculiarities and intricacies of PHP, we will introduce some basic concepts that will help you to understand the examples presented in this chapter. What we present here is an accelerated and abbreviated guide to PHP which contains only the basic techniques you need to know when building an XML parser. The first concept we discuss is the variable.

Variables

A variable is a type of data structure that can take on multiple values (a variable might hold the value 12,100, "Tim Berners-Lee," or TRUE). In PHP, the use of variables is simplified from other languages in that the type of the variable, which may be an integer whole number, a string, a character, or a Boolean TRUE/FALSE value, is not explicitly assigned when it is declared. You can recognize a variable in PHP because it will be a unique name preceded by a dollar sign ($). Here are some examples of variables with values as declared in PHP.

```
<?php
$numberOfElements = 10;
$xmlVersion = 1.0;
$numberOfAttributes =100;
$xmlFileName = "test.xml";
$xmlErrorChecking = TRUE;
$xmlAttributeCharacter = 'Z';
?>
```

Syntax

When writing PHP scripts, one must use a special syntax, or a collection of standard rules for defining instructions in a way that the PHP software can understand. Many of these rules are evident even in this brief variable declaration code example. One syntactical rule common to PHP and other server side scripting languages is the use of delimiters to indicate where PHP code begins and ends. The dollar sign preceding variable names is an example of a delimiter with a special meaning. When a dollar sign is encountered in a document, PHP expects the string of characters following that symbol to be a variable name.

The use of delimiters is important because PHP code and HTML or XHTML code often share the same document files. Sometimes, CSS and JavaScript code is also included in the same file. With up to four (or more) languages residing in a single document, there needs to be some mechanism for differentiating between them so that the Apache software can properly process the document files. This is why delimiters are so important. The PHP code is offset from HTML code using the open and close delimiters. Specifically, the five characters <?php begin a PHP sequence and the two characters ?> end a PHP sequence. When Apache encounters the special PHP delimiters within a document, it sends the document to the PHP software for preprocessing. PHP will then send back a standard document file that any browser can process.

Another syntactical issue to be aware of is the use of semicolons. Semicolons are used in PHP to terminate certain types of instructions. For instance, a semicolon is used to "end" each variable declaration and assignment. They are also used at the end of common processing tasks like providing output to the screen. Consider this code, which uses delimiters and semicolons:

```
<?php
$personsName="Tim";
print("<b>Hello. </b>");
print("<i>How are you, $personsName?</i>");
?>
```

Save this file as "hello.php" in your "xmlparsers" directory. Load it in your browser by navigating to "http://localhost/xmlparsers/hello.php". As you can see, these three instructions create a variable and then print two lines of HTML code. This HTML code is then used to print a greeting to your browser using bold type and the sentence "How are you, Tim?" in italicized type. The name "Tim" is stored in a variable named $personsName, with the dollar sign acting as a delimiter. Try changing the value of this variable to your own first name and refreshing your browser.

Next, notice the semicolon delimiters. The semicolon at the end of each statement alerts PHP to look for the next instruction after the previous statement had finished processing. One must be very careful with semicolons as missing semicolons are the source of much frustration for beginning programmers. Try removing the semicolon after the first print instruction. You will see that PHP returns an error because it has encountered a new instruction without realizing that the first instruction was complete. This is known as a parse error.

CamelCase

If you look closely at the variable names from our previous PHP code examples, you may recognize that there is a pattern dictating how new words in the name sequences sit next to one another. Specifically, the first word of the sequence is in lower case, and each new word in the sequence then has its first letter capitalized to visually separate the words without having to use any spaces. Programmers do this because spaces are not allowed in PHP variable names, but multiword variable names are more descriptive and easier to understand. So, a variable that someone wants to name "cost of living in Seattle" becomes $costOfLivingInSeattle. This practice of writing is known as CamelCase, or camel case, and is used by many script authors to visually simplify the process of recognizing when multiple words are used in variable names. Some programmers differentiate even further between lowerCamel Case, as is used in these examples, and UpperCamelCase, in which the first letter of the first word is also capitalized.

Another variable naming strategy is to use the underscore symbol to separate words, as in $number_of_elements rather than $numberOfElements. Both practices are acceptable and this is largely a stylistic choice on the part of the document designer or programmer.

Data Typing

Since PHP uses what is called dynamic typing, the types of data that are assigned to specific variables are determined by the symbolic composition of the data themselves. For instance, 10 is recognized as an integer number, so the $numberOfElements variable "knows" there are certain types of arithmetic operations that can be performed upon this variable. Similarly, $xmlFileName is assigned a value using quotation marks, so the PHP interpreter will know that this particular variable is a string, or sequence of characters, that can call upon PHP's string processing library. This library can then determine useful information like the length of the string, or whether or not a particular character in the string is capitalized. In our previous variable example, we also have a few additional data types. $xmlVersion holds a floating point number (a number with a decimal component), $xmlAttributeCharacter holds a single character (observe that single quotes are used to encase the character rather than the double quotes used with strings), and $xmlErrorChecking holds a Boolean value, which can hold the values true or false only.

Arrays and Loops

A more complex data structure, which does require a special keyword when it is created, is the array. An array is an ordered collection of data stored in a single data structure. In some languages, arrays must be composed entirely of like data types (all numbers or all strings, for instance). In PHP, you can mix data of different types and store them all in the same array.

When building XML parsers, it is useful to have a good grasp of arrays because multiple XML elements of the same name are often loaded into a single array to make iteration and access to the data contained with the elements that much easier to accomplish.

Arrays in PHP are recognizable through the use of the "array" keyword. In order to create an array with the member variables 1, 12, 256, and 300, we would write a line of code like this:

```
$myArray = array(1,12,256,300);
```

Just like the other variable types we discussed so far, arrays too must end in semicolons after they are defined. To access or print out the values stored within the array, you use the name of the array followed by square brackets and a numerical index indicating the position of the member element in the array. PHP begins its indexing at the number zero, so the following notation is used to access the elements in this example:

```
$myArray[0] contains the number 1
$myArray[1] contains the number 12
$myArray[2] contains the number 256
$myArray[3] contains the number 300
```

You can also loop through an array automatically using a "for loop" or a special type of loop known as a "foreach loop." The foreach loop allows a designer to access or modify array elements without needing to know the exact size of an array and without having to declare the index positions of each and every element inside the array. A simple foreach loop to print the contents of $myArray would look like this:

```
<?php
foreach ($myArray as $memberElement)
    {
    echo $memberElement."<br />";
    }
?>
```

This code generates the following output:

```
1
12
256
300
```

Functions, Arguments, and Variable Scope

A function is another important concept to understand in PHP. A function is an ordered and encapsulated collection of statements that can be accessed using a common function name from within the body of a PHP script. Functions range from simple, with no return value and no ability to receive information from the body of the script, to highly complex, with multiple input arguments and multiple methods for affecting data outside the encapsulated confines of the function's definition. Arguments are special variables that a function uses to interface with other functions or with other parts of a program. They are used to pass data back and forth. When a function executes, values are passed from the main body of the program to the arguments and then the function can use these values for their own calculations. Functions in PHP are declared using the reserved keyword "function." An example of a function in PHP is shown here:

```
<?php
function buildXmlElementTag($elementName, $pcData)
      {
      $string = "<".$elementName.">";
      $string .= $pcData;
      $string .= "</".$elementName."><br />\n";
      echo $string;
      }
?>
```

Though this example is fairly straightforward, there are a few additional syntactical items that will be unfamiliar to beginners. First, the echo statement is an output statement used by PHP to print strings and variable values. It is similar to the print command we used earlier, but it does not require the parentheses. Output statements are those statements which generate some type of information that is printed to the user's browser, to a text or binary file, or to a special console or terminal window in the operating system.

Second, the period (.) in PHP is used as the concatenation operator, which glues data together. The concatenation operator is used to combine strings, characters, numbers, or variables. In this example, we use the period to glue the different pieces of the $string variable together in order to assemble the XML tag as a single unit. Finally, the \n that is pasted at the very end of this variable is a special character sequence used in many programming languages to indicate a new line. By including this \n sequence, we ensure that each element tag generated by the function after a previous tag will begin on a new line. This makes it easier to observe our HTML output using the View-Source functionality of a browser. The <br/> is an XHTML compliant break tag that will also make the tags appear on new lines in HTML view. In other words, the \n creates a new line in source view, and the <br/> creates a new line in HTML view. Note the extra space between the r character and the forward slash character in the break tag. This notation is necessary for the document to validate according to the XHTML parser, and is a convention used for nonpaired tags in XML (tags without ending tags). The $elementName and $pcData arguments are used to copy values from *outside* the function that can be used *inside* the function as parts of the string that is glued together. In this case, these arguments hold the element name and the element data value. Putting all of this together, we have a very primitive element creation function named buildXmlElement Tag that automates the XML code writing process for new elements. This function allows a document designer to focus on more interesting problems like element naming and categorization rather than the repetitive task of writing the same XML element name over and over again.

The backslash in general is another delimiter that is useful to know about. The backslash is used as an escape character to indicate to the PHP interpreter that a special character follows which should be treated differently from that character as it usually exists. We will see some additional uses of the escape character in our second function example, where it is used to differentiate between the quotation marks used for the attribute value in the element tag and the quotation marks used to indicate to PHP when a string has ended. Many other programming languages make similar use of the backslash to indicate escape characters or escape sequences.

Our buildXmlElementTag function enables a designer to use PHP to build a simple XML element with no attributes. In order to "call" the function, or ask it to execute, the designer will use the name of the function followed by parentheses and two argument values to use as assigned values for variables. Arguments are values that the function expects to receive and that it will use as values to copy into the variables declared in the argument list. In this case, the first argument value will be stored in the variable named $elementName and the second argument value will be stored in the variable named $pcData. Both of these variables are called local variables because the values contained within these named variables are only accessible from within the function's namespace. The function's namespace is the body of code in between the beginning and ending curly braces.

Functions are another area of confusion for beginning programmers, and for good reason. It is difficult to understand the relationship between outside variables, or variables that are used in the main body of a program, and inside variables, or variables that are used within the scope of a function body. Variable scope, or the areas of a script in which variables can be "seen" and accessed, is an important concept to understand when thinking about functions and variables. With outside variables, the scope is described as "global." This means that functions cannot access the values of these variables, but the main body of the program can. With variables defined inside functions, the scope is described as "local." This means that only the function itself can see and manipulate variables, but the code outside the function cannot access the variables' values. So, how does a function communicate with the main body of the program? Arguments handle the mapping between global and local variables. Consider the first line of our function definition:

```
function buildXmlElementTag($elementName, $pcData)
```

In this function, $elementName and $pcData are arguments. Since $elementName and $pcData are local variables, the global variables from the main body of the program are "passed" to the function and then converted to local variables that can be used from within the function's namespace. Let us assume we had two global variables named $myElement and $myData that were defined earlier in the body of a PHP program. In order to send these values to our function, we would include them in the function call, like this:

```
buildXmlElementTag($myElement, $myData)
```

Arguments that are passed as parameters handle the conversion between outside and inside variables, but it is not always clear how this happens. Often, this process is better understood through applied examples. In this case, if we wanted to create a tag for an XML element named AUTHOR which contains the data "Benjamin Franklin", we would call the function in the following fashion using literal values:

```
<?php
buildXmlElementTag("author","Benjamin Franklin");
?>
```

We could also call the function using variables, as we described earlier. This new code looks like this:

```
<?php
$elementName = "author";
$pcData = "Benjamin Franklin";
buildXmlElementTag($elementName,$pcData);
?>
```

In either case, the HTML output produced by this function is identical. Here is the output produced by either of these function calls:

```
<author>Benjamin Franklin</author>
```

Because our function uses generic arguments to construct this string, we can call it as many times as we like, with as many different strings as we like. For example, if we wanted to list several additional signers of the Declaration of Independence, we could add the following function calls to our document:

```
<?php
buildXmlElementTag("author","Benjamin Franklin");
```

```
buildXmlElementTag("author2","John Adams");
buildXmlElementTag("author3","Thomas Jefferson");
?>
```

A side-by-side screen capture showing the output of these function calls is shown in Figure 6.5. The left hand image shows how this content would appear in Mozilla Firefox®, while the right hand image shows the content from the View Source perspective. Since we have not included the XML version directive at the beginning of the file, the content is rendered in plain text (the browser does not recognize the generic author tags we are using) rather than in the collapsible and expandable XML tree document hierarchy we have seen in previous chapters. Since we now have the ability to directly create XML tags on the fly using this function, we could potentially use this procedure as an interface for one of our XML parsers.

Functions and Default Arguments

While our previous function has its uses as a simple tag generator, it would be even more powerful if it also had the ability to create attributes for our XML elements. Fortunately, it is not difficult to add this functionality in PHP. A new function, which supports attributes in addition to elements, is shown here:

```
function buildXmlElementTagWithAttribute($elementName, $pcData,
$attributeName="", $attributeData="")
      {
      $string = "<".$elementName;
      if ($attributeName != "")
            {
            $string .= " ".$attributeName."=
            \"".$attributeData."\"";
            }
      $string .= ">";
      $string .= $pcData;
      $string .= "</".$elementName."><br />\n";
      echo $string;
      }
```

This function is slightly more complex, but it is also more powerful. By using what are known as default arguments in our $attributeName and $attributeData arguments, we have now ensured that this function can be called with two, three, or four arguments. If the third argument ($attribute Name) or the fourth argument ($attributeData) is missing, the function will simply substitute the default values (in this case, the empty string " ") for these arguments. This means that if we call the new function with our original values containing elements and data with no arguments, the function will still work.

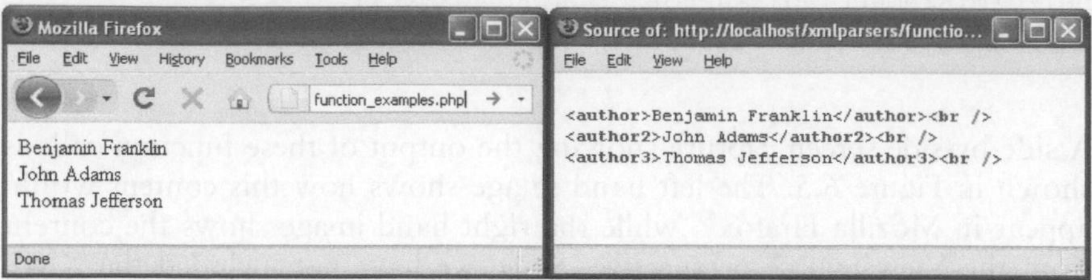

Figure 6.5 Browser View (left) and Source View (right)

Logical Control Using If Statements

In addition to the default arguments, this new function also introduces a new concept: logical control. Using a control structure known as an "if statement," the PHP function can now operate conditionally depending on the value of variables within the program. If statements evaluate the expression in parentheses and then execute the following code fragment depending on whether or not that expression evaluates to true. So, for example, an if statement might check to see if a variable's value was greater than ten. If so, it would execute the instructions surrounded by curly brackets immediately after the if statement. If not, it would ignore these instructions. Else statements can be used with if statements to provide alternate instructions in case the if condition is not true. This statement would take the following form in PHP:

```
if ($variable > 10)
        {
        echo "The value is bigger than ten!";
        }
else
        {
        echo "The value is not bigger than ten!";
        }
```

Returning to our modified function, we see that the following if statement is used:

```
if ($attributeName != "")
                {
                $string .= " ".$attributeName."=
\"".$attributeData."\"";
                }
```

In this case, the variable being evaluated is called $attributeName. The != part of this statement is a special expression in PHP which means "not equal to." So, this conditional portion of the code is ensuring that the

$attributeName value is not equal to the empty string. This means that the concatenation operation defined within the curly brackets will only be performed if the $attributeName variable contains some value (is not empty). This means that the function will include the attribute and its associated value only if data for these items has been provided to the function by way of the function call. The escape sequence \" (a backslash followed by a quotation mark) is also used within the body of the if statement to differentiate between quotation marks for the attribute value and the quotation marks PHP usually uses to discern the end of a string.

We will see many additional uses of the if and else statements later in this chapter when we build our XML parser projects.

Calling Functions with Default Attributes

Here are some new function calls we can use with our newly created buildXmlElementTagWithAttribute() function. These function calls will list some of the other signers of the Declaration of Independence along with additional information about these individuals:

```
<?php
buildXmlElementTagWithAttribute("author","Benjamin Franklin",
"ageAtSigning",70);
buildXmlElementTagWithAttribute("author2","John Adams",
"futurePresidentNumber",2);
buildXmlElementTagWithAttribute("author3","Thomas Jefferson",
"futurePresidentNumber",3);
buildXmlElementTagWithAttribute("author4","Stephen Hopkins");
?>
```

The new output in source view for this function is revealed in Figure 6.6.

PHP and XML

While if statements, loops, variables, and functions are all very useful, they are not the reason we chose to use PHP for our XML parser design exercises. PHP has very good built-in support for XML, and it is especially important for parser designers to know about PHP's XML processing capabilities. Specifically, PHP supports XML using the following technologies:

1. SimpleXML: SimpleXML is a PHP extension for parsing and adding to XML documents using XML-to-object mappings. XML documents are turned into object-oriented data structures that can be manipulated in an algorithmic fashion. This process is discussed in more detail in the next section of this chapter. SimpleXML is usually the easiest extension to use when working with XML in PHP. We will use SimpleXML in Project 1.

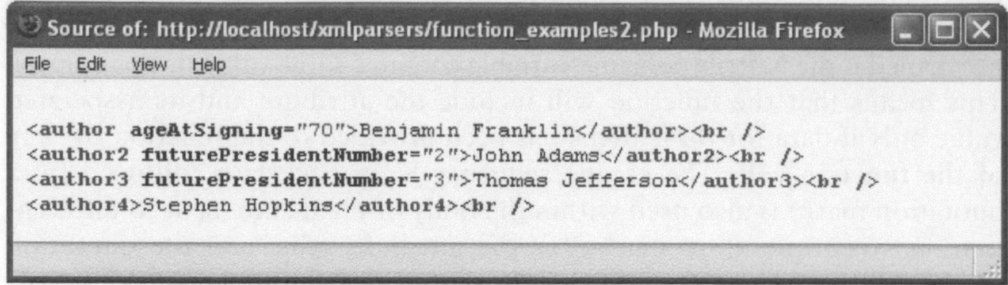

Figure 6.6 Source View of Function with Attribute Support

2. Simple API for XML (SAX): SAX is a lightweight and relatively easy to use API that parses an XML file from top to bottom. SAX is slightly more complicated than SimpleXML, but adds some additional functionality. SAX is used in Project 2.
3. DOM API: Like SAX, the DOM is also an API, but DOM is more sophisticated than SAX in that it builds a full document tree in memory based on an XML file. The DOM is very powerful and supports validation, but it is also difficult to use and can be resource intensive for large XML documents. DOM is our API of choice in Project 3.

Because of these robust XML-handling capabilities and its relatively slight learning curve, PHP is a good language to consider when designing your own customized XML parsers for very specific informational tasks. We will use PHP for the examples in this chapter, but more experienced programmers wishing to design XML parsers may also want to consider fully object-oriented languages such as Java or C++ in addition to PHP.

SimpleXML and Object-Oriented XML

SimpleXML is a popular extension to PHP which provides a toolkit for mapping XML documents into objects that can be directly manipulated by the PHP scripting language. Objects are data structures that have support for both variables and functions, both of which can be encapsulated into a common namespace usable by any instances of those objects. Objects are particularly handy for us as document designers because, as we mentioned in Chapter 3, it has been predicted that we will be moving from document-centered to object-oriented ways of thinking about information and information design (Williams 321). Key concepts which affect both the technological and rhetorical domains of object-oriented information design include encapsulation and information reuse.

Object instances are created from object definitions, or classes, which specify the exact functions and variables available to an object's namespace once it has been defined. It is helpful to think of classes as blueprints and objects as buildings. Classes are recognizable through the use of the class keyword in PHP, while objects are recognizable through the use of the new keyword, which instructs the PHP interpreter to allocate memory for a new object with enough room to hold the variables and functions defined in the class definition. To complicate things further, variables defined inside objects are known as properties, and functions defined inside objects are known as methods. Once an object has been declared, its properties and methods can be accessed using the special notation -> followed by the name of the property or method. For instance, if a property named squareFootage resided within an object named $houseInstance, it would be accessible using this notation:

```
$houseInstance->squareFootage
```

To demonstrate how the translation from an XML document to a SimpleXML object works, we will present a basic example. Consider the following XML file:

```
<?xml version="1.0" encoding="utf-8"?>
<contact_manager>
      <employee>
            <first_name>Pierre-Francois</first_name>
            <last_name>LuCerne</last_name>
            <extension>1001</extension>
            <e-mail>pfl@rhetoricalxml.com</e-mail>
            <birthday>Dec 12, 1949</birthday>
      </employee>
      <employee>
            <first_name>Julia</first_name>
            <last_name>Rodriguez</last_name>
            <extension>1726</extension>
            <e-mail>jr23@domain.com</e-mail>
            <birthday>September 3, 1964</birthday>
      </employee>
      <employee>
            <first_name>Joe</first_name>
            <last_name>Smith</last_name>
            <extension>2488</extension>
            <e-mail>joe_smith@rhetoricalxml.com</e-mail>
            <birthday>July 23, 1978</birthday>
      </employee>
</contact_manager>
```

Once PHP has loaded this XML document using the SimpleXML extension, the above XML file will be translated into an object (or, in this case, an object containing an array of objects) that PHP can manipulate.

The XML document is then accessible using the standard object property notation we discussed previously. Using this example file, the XML is translated into the following data structure:

```
SimpleXMLElement Object [employee] => Array
    (
        [0] => SimpleXMLElement Object
            (
                [first_name] => Pierre-Francois
                [last_name] => LuCerne
                [extension] => 1001
                [e-mail] => pfl@rhetoricalxml.com
                [birthday] => Dec 12, 1949
            )
        [1] => SimpleXMLElement Object
            (
                [first_name] => Julia
                [last_name] => Rodriguez
                [extension] => 1726
                [e-mail] => jr23@domain.com
                [birthday] => September 3, 1964
            )

        [2] => SimpleXMLElement Object
            (
                [first_name] => Joe
                [last_name] => Smith
                [extension] => 2488
                [e-mail] => joe_smith@rhetoricalxml.com
                [birthday] => July 23, 1978
            )

    )
```

Though this looks rather complicated and intimidating, it is actually very logical in terms of access to elements and their data. We need only to follow the paths down the XML hierarchy in order to access particular elements and their data. Assuming our XML data was loaded into an object named $xmlObject, here are some PHP commands that can be used to directly print XML data from this example:

```
<?php
// print out Pierre-Francois
echo $xmlObject->employee[0]->first_name;
// print out 1726
echo $xmlObject->employee[1]->extension;
```

```
// change 1726 to 2556
$xmlObject->employee[1]->extension = "2556";
// print out 2556
echo $xmlObject->employee[1]->extension;
?>
```

We will show how to load an XML document into PHP using SimpleXML when we discuss the first project in the next section. For now, you just need to understand that the SimpleXML extension translates an XML document tree into an object structure with the tree's elements embedded as objects or object properties. This is extremely handy when designing parsers because it gives us a predictable and unambiguous syntax for accessing XML elements, attributes, and data values when working in a programmatic environment.

SimpleXML is enabled by default in PHP version 5 and above and its functions can be called directly from any PHP script with this toolkit enabled. We discuss some specific examples of SimpleXML in use later in this chapter, but, for now, here is a general overview of some of what it can do:

- Translate XML documents from a tree-based hierarchy into an object that can be handled by PHP's standard object-oriented handling procedures;
- Load XML documents from separate files or directly from string variables;
- Convert DOM documents into SimpleXML objects.

Our review of Apache, PHP, and SimpleXML is now complete. Each of these technologies is quite complex; we only touched briefly upon some major features that readers should be aware of as we move through our parser design examples. In the remainder of this chapter, we will discuss these technologies as potential solutions for different types of informational needs. We move from a relatively easy problem in the first project to more complex information design problems in the second and third projects. Regardless of the level of technical difficulty, each project requires a similar amount of rhetorical consideration in both the preproduction (planning) phase as well as in the postproduction (revision and fine tuning) stage. For the most part, we focus on the preproduction and design phases, but we do suggest some ideas for postproduction activities and discussion prompts in the questions at the end of the chapter.

Project 1: RSS Parser

Even when designing a simple parser such as one to display newsfeeds on a website, it is important to think about the rhetorical considerations of the communicative act. While it can be useful to begin this rhetorical inquiry from a particular perspective, perhaps by considering the classical rhetorical canons or using a rhetorician's theoretical model as a starting point, it can also be advantageous to simply take a step back and consider the informational context based on one's prior knowledge. This may include one's immediate knowledge of the data source and of the users who will need to be accessing this data or adding to it. We describe this approach as an ad hoc rhetorical analysis because it does not necessarily have wider applications outside the immediate scope of a particular project.

The first and foremost question asked before selecting a parser, or deciding to build your own custom parser, is obvious: who is the intended audience for this information? A general list of rhetorical considerations is outlined in the Ad Hoc Rhetorical Analysis of XML (RAX) form included in Appendix B. The questions from this form are repeated here for the sake of this sample exercise:

- Who is my primary audience?
- Who are the secondary and tertiary audiences?
- What is the purpose of this metadata system?
- What are the informational needs of my audiences?
- How should the information be arranged and presented for this audience?
- What kinds of backgrounds will my audiences possess? Will they have a high literacy level? What about their level of technical literacy? What style will work well for these needs?
- What vocabularies will my audiences use to identify relevant elements and data nodes within the XML hierarchy?

There are many more rhetorical questions that can be asked here. These will depend entirely on the context of your project. For example, if you are building a parsing system to manage newsfeeds for a hospital, you will need to ask questions that examine the tradeoffs between privacy and accessibility and make document design decisions that are both ethically and legally justifiable. We suggest using this form as a starting point and adding to or adjusting it according to the needs of your own project and audience.

For this project, we are going to create personas to help us visualize an imaginary information context. Personas are fictitious characters that we can create in order to help us visualize the demographic characteristics and informational needs of a typical user. In this exercise, we will create personas for both the designer and for his audience. Our designer persona, Mr. Joe Smith, is a document designer who wishes to build a news communication system for a local newspaper's website. His contact is Ms. Shirley Brown, senior managing editor, who will serve as our primary audience persona. Using the RAX form, Joe produces the following answers to the survey:

- Who is my primary audience? My initial primary audience member will be Ms. Brown, who will review the system I build for appropriateness and sustainability. Once approved, the primary audience will be those townspeople who visit the newspaper's website and wish to be updated about breaking news.

- Who are the secondary and tertiary audiences? Secondary and tertiary audiences include visitors from other locales, administrators in the newspaper office, and perhaps friends and family members I cajole into helping me troubleshoot my system.
- What is the purpose of this metadata system? The purpose of this system is to provide timely news titles and descriptions of breaking news stories that can then direct visitors to the full text versions of those articles.
- What are the informational needs of my audiences? The audiences will need to know the title, date, description, and URL of the full length articles. In addition, the titles and descriptions should be meaningful without requiring too much cognitive investment (reading) from visitors who might be casually browsing the newsfeed. Newspaper writers and editors may need access to directly modify the XML newsfeed document without having to resort to manually editing the XML file.
- How should the information be arranged and presented for this audience? The information parsed from the XML file should be displayed in a fashion that visually separates one news article from the next and uses standard HTML conventions to indicate hypertext links and descriptive titles for these links. The article descriptions should be listed hierarchically under each of their parent item elements within the news article directory.
- What kinds of backgrounds will my audiences possess? Will they have a high literacy level? What about their level of technical literacy? What style will work well for these needs? I don't yet have information on the typical demographic profiles of the newspaper's readership, but this is a great question to ask Ms. Brown when I meet with her next week.
- What vocabularies will my audiences use to identify relevant elements and data nodes within the XML hierarchy? Audiences will come from varied backgrounds, but will likely expect standard journalistic style. This will, in part, address the stylistic concerns of the previous question.

Based on Joe's responses, we can extract the following four design parameters from this rhetorical analysis exercise:

1. The XML file we use for the newsfeed should contain, at the very least, elements to hold:
   a. information about the news article title,
   b. a longer description,
   c. a date of publication, and
   d. a link to the full article on the newspaper website.
2. The data extracted from the XML file should be displayed in a visually appealing and hierarchical fashion and should include semantically meaningful title tags linked to the full news articles. Beneath each title,

the system should include the description element with a brief description of what the news article is about.
3. The project needs to contain functionality for both retrieving XML data (for the audience of news story readers) as well as inserting new XML data (for the audience of news staffers who will be adding to the news repository).
4. A journalistic style, or a style that is as neutral and objective as possible, is necessary for the news data contained within the elements. This particular rhetorical consideration, however, is more relevant to the writers of the stories than to the designer of the metadata system, so it will not pose many design challenges. We do not need to adopt a strict journalistic style for our test XML file, since that is only a temporary data source that we will use to develop the application and ensure that the tagging and structure format is compatible.

In terms of technical implementation, the first thing we will do for this project is build an XML file that can be used for testing. The XML file for this particular purpose does not need to be very complex. It only needs to contain elements to hold those informational needs Joe identified in his rhetorical analysis. Based on these needs, a sample XML file with some fictional news story data might appear as so:

```
<?xml version="1.0" encoding="utf-8"?>
<news_articles>
    <article id="1">
    <title>Bear Loose on Freeway</title>
    <description>A 750-pound grizzly bear was found wandering
    around aimlessly on Interstate 4 this afternoon.
    Officials are attempting to goad the bear into an Animal
    Services&apos; truck but have not had any luck with this
    so far.</description>
    <url> www.rhetoricalxml.com/rssparser/news.php?id=1</url>
    </article>
    <article id="2">
    <title>Stock Market Plunges</title>
    <description>The stock market has taken another hit due
    to rising prices in agriculture (corn and wheat) and in
    the oil industry.  Nasdaq has dropped 600 points and the
    S&amp;P500 dropped 300 points.</description>
    <url> www.rhetoricalxml.com/rssparser/news.php?id=2</url>
    </article>
    <article id="3">
    <title>Toddler Saves Pet Turtle from 10-foot
    Gator</title>
    <description>A three year old boy saved his pet turtle,
    named Harold, from a 400-pound alligator on Tuesday.
```

```
    The turtle had wandered down from the boy&apos;s home
    into a nearby swamp region.</description>
    <url> www.rhetoricalxml.com/rssparser/news.php?id=3</url>
    </article>
</news_articles>
```

Though this is a perfectly acceptable XML file, it does not yet meet the schema requirements of a valid RSS document. RSS is a particular type of XML file that uses a collection of specialized tags to present information in a standardized format. Since they use standard element and attribute names, RSS feeds can be read by a variety of RSS news readers or even by browsers such as Mozilla Firefox®. Because of this, RSS is a logical format for Joe to use in his newsfeed parser application. According to RSS, the toplevel element should be named "channel" rather than "news_articles." In addition, each article should be named "item" rather than "article," and we can provide additional information using a channel description and title tag. The URL element needs to be renamed to GUID, which stands for "globally unique identifier." We can remove the ID attribute since the GUID will provide a unique reference for each item. Finally, we need to encapsulate all of the data in an additional tag named RSS with a version attribute value of 2.0. This new and revised XML file appears here in proper RSS format:

```
<?xml version="1.0" encoding="utf-8"?>
<rss version="2.0">
<channel>
<title>Joe&apos;s Test Newsfeeds File</title>
<description>Local news from your community.</description>
    <item>
        <title>Bear Loose on Freeway</title>
        <description>A 750-pound grizzly bear was found
        wandering around aimlessly on Interstate 4 this
        afternoon.  Officials are attempting to goad the
        bear into an Animal Services&apos; truck but have
        not had any luck with this so far.</description>
    <guid> www.rhetoricalxml.com/rssparser/news.
    php?id=1</guid>
    </item>
    <item>
        <title>Stock Market Plunges</title>
        <description>The stock market has taken another hit
        due to rising prices in agriculture (corn and
        wheat) and in the oil industry.  Nasdaq has dropped
        600 points and the S&amp;P500 dropped 300 points.
        </description>
        <guid> www.rhetoricalxml.com/rssparser/
        news.php?id=2</guid>
```

```
    </item>
    <item>
        <title>Toddler Saves Pet Turtle from 10-foot Gator
        </title>
        <description>A three year old boy saved his pet
        turtle, named Harold, from a 400-pound alligator
        on Tuesday.  The turtle had wandered down from the
        boy&apos;s home into a nearby swamp region.
        </description>
    <guid> www.rhetoricalxml.com/rssparser/news.
    php?id=3</guid>
    </item>
</channel>
</rss>
```

If you would like to follow along with this example on your own, you should first create a folder named "rssfeed" in your "xmlparsers" folder. Then, you need to save this XML file as "rss_feed_sample.xml" within your "c:\xampp\htdocs\xmlparser\rssfeed" folder (or the appropriate folder containing the htdocs directory on your own XAMPP installation). You can also download the Project 1 files from the website and extract them to this directory.

Next, we need to build the parser in PHP. Using the SimpleXML functionality we discussed previously, we know that the first thing we need to do is to load the XML file into an object that PHP can manipulate. Since SimpleXML is such an easy API to manipulate, this process only involves a few steps.

Here are the contents of the PHP file (save this as "displayNews Previews.php") that we will be using as a parser:

```
<?php
/* This variable sets the name for the news parser */
$newsWidgetName = "News Widget 1.0";

/* This variable holds the name of the XML source file */
$xmlFileName = "rss_feed_sample.xml";

/* load the XML structure into an object (data structure) */
$xmlHierarchy = simplexml_load_file($xmlFileName);
?>

<html>
<head>
    <title><?php echo $newsWidgetName; ?></title>
    <link rel="alternate" type="application/rss+xml"
    title="News Feed RSS" href="<?php echo $xmlFileName; ?>">
</head>
```

```
<body>
    <h1><?php echo $newsWidgetName; ?></h1>
    <p>Here are some recent news articles from the XML data
    file named <?php echo $xmlFileName; ?>.  Click on any
    article title to view the full news article associated
    with that news feed.</p>
    <ul>
    <?php
    foreach ($xmlHierarchy->channel->item as $article)
            {
            echo "<li>";
            echo "<a href=\"".$article->guid."\">";
            echo $article->title."</a>";
            echo "<ul>";
            echo "<li>".$article->description."</li>";
            echo "</ul>";
            echo "</li>";
            }
    ?>
    </ul>
</body>
</html>
```

We will not discuss each line of PHP code presented here, but we will touch upon a few new concepts that were not covered in the PHP tutorial earlier in this chapter. First, note the comments in the file, which begin with a /* character sequence and end with the */ sequence. The comments here describe the purposes of the first three variables used in this script. Comments are used heavily both in this first project and in the content management and single sourcing projects later in the chapter.

In PHP, there are three styles of comments that can be used. First is the /* beginning sequence and */ ending sequence that we have already seen. These are good to use when comments span multiple lines, or when you want to temporarily deactivate large chunks of code so that you can troubleshoot or debug areas of your program. PHP also supports single line comments using either double forward slashes: // or the hash mark: #. Comments are important in PHP just as they are in HTML as they provide a space for designers to make notes or annotations explaining why code was created a certain way. Each of the comments below would be valid according to PHP:

```
/* This is a comment that takes up more than one line.  Note
that we must close it when we are done. */

// This is also a valid single line comment in PHP.

# This is a valid single line comment, too.
```

The next important line in the Project 1 script begins with the text link rel="alternate". This line of code links the XML newsfeed into the document so that newsreaders can recognize the document and bookmark the feed accordingly. Figure 6.7 shows the RSS parsing application's output so far as displayed in Mozilla Firefox®. Firefox® is a useful browser for this type of application as it enables a person to bookmark RSS feeds with "live" streaming XML information.

Finally, look at the line that begins with the variable $xmlHierarchy and contains the function simplexml_load_file(). This function is used by the SimpleXML extension to load the XML document into an object. This line of code executes the XML document to PHP object conversion as we discussed in the previous section of this chapter. This function expects a single argument: the name of the XML file to be loaded. We set this filename using the $xmlFileName variable, so we are "passing" the name of this variable as the function's argument.

```
$xmlHierarchy = simplexml_load_file($xmlFileName);
```

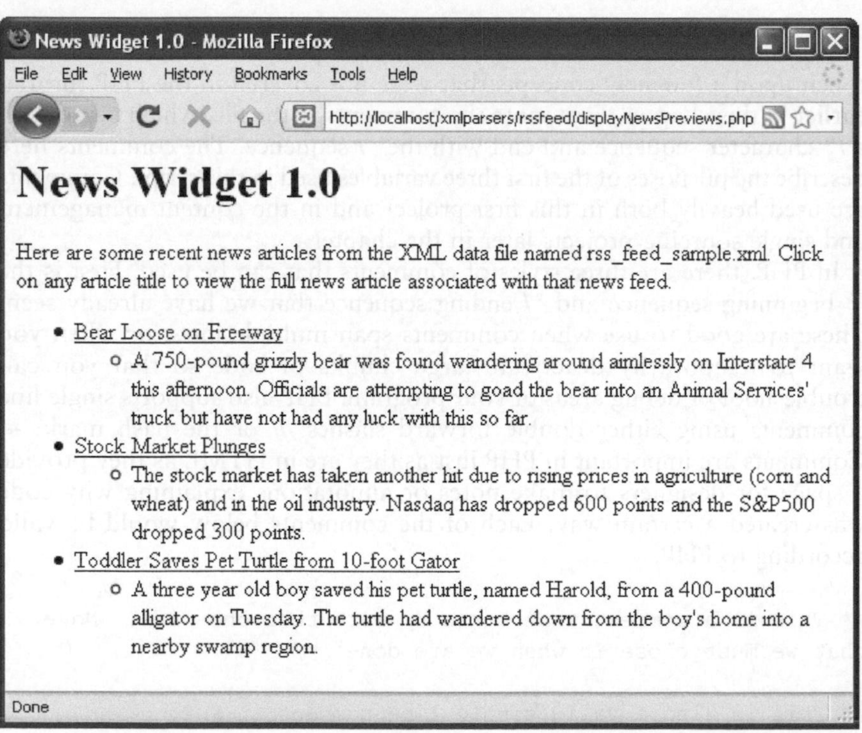

Figure 6.7 Newsfeeds in Mozilla Firefox®

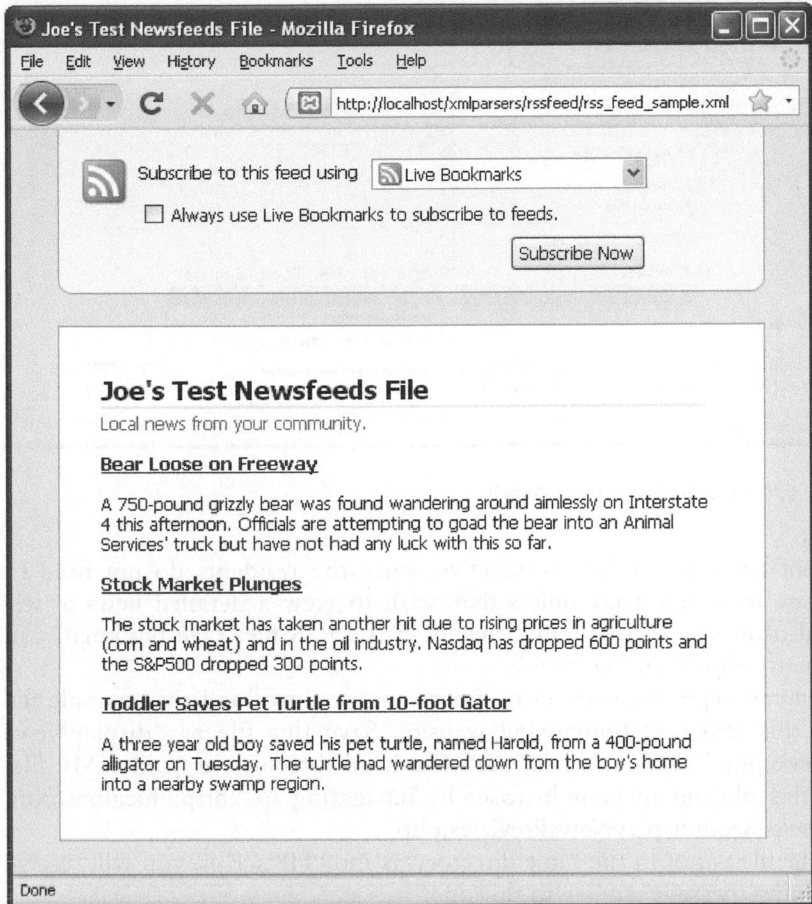

Figure 6.8 Adding a Live Bookmark

From this point on, our XML information will be navigable and accessible as an object-oriented data structure. To make things easy on us, SimpleXML will automatically convert the names of our elements into object properties so that we can easily remember how to navigate to different parts of our XML hierarchy in an object-oriented fashion. As you can see from the previous code listing, we are doing just this when we refer to the element <item> inside the element <channel> by accessing the object property in this fashion: $xmlHierarchy->channel->item.

Live bookmarks are added in Mozilla Firefox® by clicking the small orange icon to the right of the URL and to the left of the outlined star in the address bar. This will bring up the screen shown in Figure 6.8. Once the live bookmark has been added, residents in Joe's community will be able to access streaming news information directly from their browsers. This is

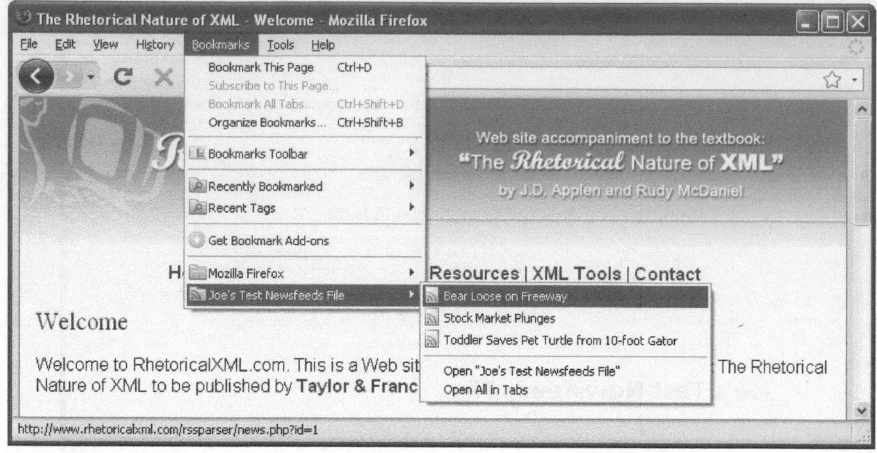

Figure 6.9 Live Bookmarks in Action

efficient from a usability perspective, since the residents do not need to navigate to a new page unless they wish to view a detailed news article linked from one of the item elements in the RSS file. Live bookmarks in action are shown in Figure 6.9.

You can type this code in on your own or download the example file from this text's accompanying website. Save this file as "displayNewsPreviews.php" and store it in the same local directory as your XML file. Test this file out in your browser by navigating to "http://localhost/xmlparsers/rssfeed/displayNewsPreviews.php".

If the file is not in the same directory as the PHP script, you will receive a warning message similar to this one:

```
Warning: simplexml_load_file() [function.simplexml-load-file]:
I/O warning : failed to load external entity "rss_feed_
sample.xml" in C:\xampp\htdocs\xmlparsers\rssfeed\
displayNewsPreviews.php on line 9.
```

This means you need to move the XML file named "rss_feed_sample.xml" into the same directory as your PHP file. Otherwise, you should see the same browser screen that was shown earlier in the News Widget 1.0 screen capture (Figure 6.7).

At this point, we have a working XML parser built to show the output of an XML file structured as a RSS newsfeed. We can act upon specific input, which may be elements or attributes, in order to create a sophisticated handling system for our document. To do this, we are using the following algorithm, or precise set of instructions, for the XML handling portion of our script:

```
While (new elements of type item are available)
      Begin an unordered list
      Look for child elements named title, description, and
      guid.  When these elements are found, the data inside the
      guid element and title element is used to construct the
      hyperlink in the form of <a href="[guid]">[title
      data]</a> where [guid] and [title data] are placeholders
      for the data extracted from these tags.
                  Begin an embedded unordered list
                  Print out the data from the description tag
                  as an indented list item below each title
                  item and link
                  End embedded unordered list
      End unordered list
End While
```

Revisiting our original design specifications and goals, we see that only half of Joe's news system has been implemented. He still needs a mechanism to add news articles directly into the XML document. This feature will eventually be used by news staffers to add new articles. Fortunately, SimpleXML provides support for adding child elements and attributes through its addChild() and addAttribute() functions. Even with these helpful functions, the task of writing to the XML file is slightly more complicated than the task of reading from the XML file. This is primarily because we now must build in support for file input and output as well as work more intimately with the SimpleXML object structure.

The first component that is necessary to add XML writing capability is an HTML form, which will be composed of text input fields and a mechanism for sending data from those fields to a processing page. We can see from our XML file that the elements associated with the news articles include ITEM, TITLE, DESCRIPTION, and GUID. New items should be added automatically to the existing XML file, so we should collect values for title, description, and GUID from the person entering this information. A simple HTML form designed to support this process is shown here:

```
<html>
<head>
<title> Joe&apos;s Test News Page: Add News </title>
</head>
<body>
<form method="post" action="update_rss_xml.php">
<table>
<tr>
      <td>News Title:</td>
      <td><input type="text" name="title" size="50"></td>
</tr>
```

```
<tr>
     <td>Description:</td>
     <td><input type="text" name="description" size="50"></td>
</tr>
<tr>
     <td>GUID:</td>
     <td><input type="text" name="guid" size="50"></td>
</tr>
</table>
<br />
<input type="submit" value="Add News">
</form>
</body>
</html>
```

Name this file "update_news.html" and save it in the same directory you have been working in throughout this example project. Figure 6.10 shows what this HTML code produces in a browser.

After the HTML form is designed, we need to build a script that will take data from the form's text input fields and store these values as variables. As you may recall from the beginning of this chapter, even the earliest version of PHP, PHP-FI, contained built-in support for forms, so PHP is designed for precisely this type of activity. Since our form is using the POST method to submit its data, the process of mapping the text fields to variables is rather simple. We define the variables we wish to use and then access an associative array, or a special type of array that uses strings for keys rather than numbers, by providing the text field names as indices. The associative array for variables submitted using the POST method is named $_POST, which

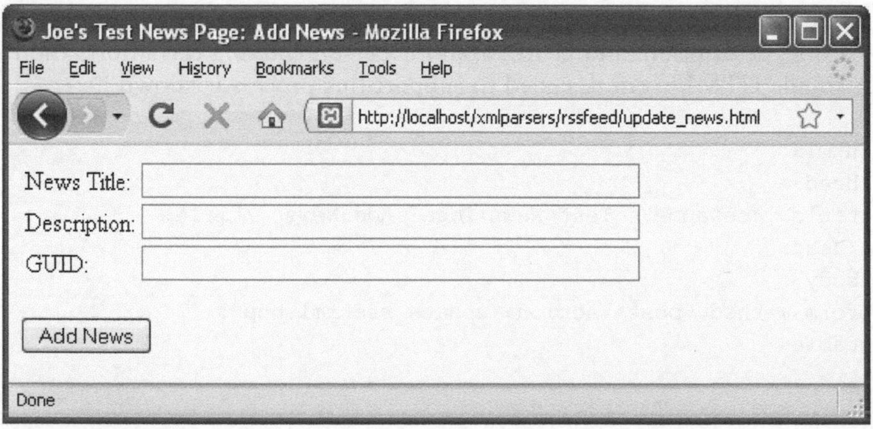

Figure 6.10 HTML Add News Form

is a shortcut for $HTTP_POST_VARS. The code to access individual values from the form fields looks like this:

```
$title = $_POST["title"];
$description = $_POST["description"];
$guid = $_POST["guid"];
```

There are two types of methods one can use when submitting forms: GET or POST. Both are highly useful and the experienced Web designer or information architect will often find themselves using both. In the GET protocol, the variables sent from a form are appended to the end of a URL string. Example URLs which have been submitted from forms using the GET method look something like this:

```
www.rhetoricalxml.com?var1=radish&var2=20&var3=a
```

In this system, the question mark symbol is used to indicate the first variable name and the equal symbol links that first variable name to a specific value. Subsequent variables names are appended to the URL using the ampersand, with their values continuing to use the equal sign for mapping purposes. So, the previous URL would enable the following variables and values to be parsed from its textual address string:

```
var1 (containing the value "radish")
var2 (containing the value 20)
var3 (containing the value 'a')
```

We can access GET variables in the same fashion as POST variables, but we use $HTTP_GET_VARS or $_GET as the associative array names instead of $HTTP_POST_VARS or $_POST. To assign the three values above to PHP variables, we could therefore use the following code:

```
$var1 = $_GET["var1"]; // $var1 now contains "radish"
$var2 = $_GET["var2"]; // $var2 now contains the number 20
$var3 = $_GET["var3"]; // $var3 now contains the character 'a'
```

If one uses the POST method, these variables are not appended to the URL and are instead passed to the script behind the scenes. GET is handy when the inner page of a website needs to be bookmarked for later use, or when the variables being passed to an XML parser should be transparent. Since the values are clearly visible in the URL string, the entire data set is clearly visible and able to be bookmarked. This could be a potential security problem since the designer might not want all of these variables to be viewable by any user of the Web page. POST is useful when the page should not be able to be bookmarked, or when large amounts of data need to be passed from a form to a script or parser. This is because the amount of data GET can append to the URL is limited to the maximum size of a URL string, which varies from browser to browser.

The full code that is used to take the information from our simple form and build new elements into our news story XML file is shown here:

```
<html>
 <head>
  <title> Joe&apos;s Test News Page: Results Page </title>
 </head>
<?php
/* This code will get the form&apos;s data variables */
$title = $_POST["title"];
$description = $_POST["description"];
$guid = $_POST["guid"];

/* This variable holds the name of the XML source file */
$xmlFileName = "rss_feed_sample.xml";

/* This function will read the contents of an XML file */
/* and store those contents as a string */
function getXMLFileContentsAsString($filename)
      {
      $xmlStringData = file_get_contents($filename);
      return($xmlStringData);
      }
/* This function will write the new XML structure */
/* into a permanent file on the hard disk drive */
function writeNewXMLFile($filename,$content)
      {
      $handle = fopen($filename, "w");
      if ($handle)
            {
            fwrite($handle,$content->asXML());
            }
    if (fclose($handle))
            return TRUE; // successful!
      else
            return FALSE; // not successful!
      }
/* Call the function to get the XML content as a string */
$xmlString = getXMLFileContentsAsString($xmlFileName);

/* Create a new SimpleXMLElement object */
$xml = new SimpleXMLElement($xmlString);
```

```
/* Add our new child element to this object */
$newItem = $xml->channel->addChild('item');
$newItem->addChild("title", $title);
$newItem->addChild("description", $description);
$newItem->addChild("guid", $guid);

echo "<body>";

/* Now, we need to write this structure back out to our XML file
*/
if (writeNewXmlFile($xmlFileName, $xml))
      {
      echo "<p>Your news article has been added.</p>";
      echo "<p>Click ";
      echo "<a href=\"displayNewsPreviews.php\">";
      echo "here</a> to visit the main page.</p>";
      }
else
      {
      echo "<p>Sorry, but there was a problem adding your
      article.</p>";
      }

echo "</body>";
echo "</html>";
?>
```

Figures 6.11–6.13 show how this script works in terms of the user interface. First, in Figure 6.11, a user enters the title, description, and GUID data into the allocated text fields using the HTML form. Next, in Figure 6.12, the script returns a page indicating the data has been successfully added to the XML file. Finally, Figure 6.13 shows the new data after it has been inserted into the XML file and re-parsed by our original news display script.

Our new XML file after this additional news item has been added looks like this:

```
<?xml version="1.0" encoding="utf-8"?>
<rss version="2.0">
<channel>
<title>Joe&apos;s Test Newsfeeds File</title>
<description>Local news from your community.</description>
    <item>
          <title>Bear Loose on Freeway</title>
          <description>A 750-pound grizzly bear was found
          wandering around aimlessly on Interstate 4 this
          afternoon.  Officials are attempting to goad the
```

```
        bear into an Animal Services&apos; truck but have
        not had any luck with this so far.</description>
    <guid> www.rhetoricalxml.com/rssparser/news.
    php?id=1</guid>
    </item>
    <item>
        <title>Stock Market Plunges</title>
        <description>The stock market has taken another
        hit due to rising prices in agriculture (corn and
        wheat) and in the oil industry.  Nasdaq has dropped
        600 points and the S&amp;P500 dropped 300
        points.</description>
    <guid> www.rhetoricalxml.com/rssparser/news.
    php?id=2</guid>
    </item>
    <item>
        <title>Toddler Saves Pet Turtle from 10-foot
        Gator</title>
        <description>A three year old boy saved his pet
        turtle, named Harold, from a 400-pound alligator
        on Tuesday.  The turtle had wandered down from the
        boy&apos;s home into a nearby swamp
        region.</description>
    <guid> www.rhetoricalxml.com/rssparser/news.
    php?id=3</guid>
    </item>
<item><title>Internet Radio is in Trouble</title><description>
The Internet radio industry is facing serious problems
due to the new advertising structure proposed by the
FCC.</description><guid> www.rhetoricalxml.com/rssparser/news.
php?id=4</guid></item></channel>
</rss>
```

Note that the new ITEM element and its associated children elements are now added to the bottom of the file. Unfortunately, there is no easy way to preserve formatting using the SimpleXML API. We have much greater control of formatting when we directly write to a text file, as we do in the content management example and single sourcing examples discussed in the next section, but, unfortunately, this process is more complex and involves using character sequences such as \n (new line) and \t (tab) to build our XML documents line by line.

At this point, our discussion of the first parser example is complete. There are several additional modifications that could be made to improve this script. In particular, a timestamp and date element, which RSS does indeed support, would be a helpful addition. It would also be useful for the script to automatically create the GUID tag so that the user would not need to

Joe's Test News Page: Add News - Mozilla Firefox
File Edit View History Bookmarks Tools Help
http://localhost/xmlparsers/rssfeed/update_news.html
News Title: Internet Radio is in Trouble
Description: The Internet radio industry is facing serious problems c
GUID: http://www.rhetoricalxml.com/rssparser/news.php?id=4
Add News
Done

Figure 6.11 Add News Form (With Sample Data)

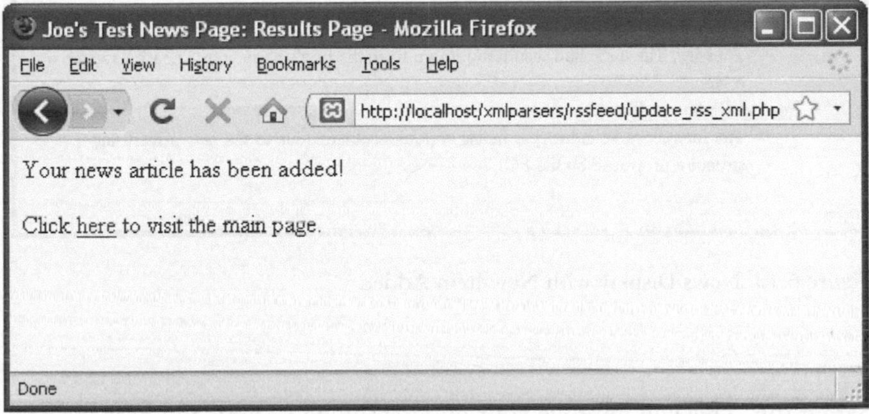

Figure 6.12 Article Added Message

look up the previous item's GUID in the XML file. If this were a real document designer working with a real client, there would likely be an additional follow up meeting where the beta product was demonstrated to the client and additional modifications were negotiated after that initial meeting. As it stands, however, Joe can rest easy knowing that his rhetorical analysis was sufficient to satisfy the needs of his hypothetical boss, Ms. Brown.

All of the files for the RSS newsfeed parsing project can be downloaded from the www.rhetoricalxml.com website. We encourage you to download the source files and experiment with adding functionality and customizing the XML file to meet the needs of your own projects. If you do not currently have a project to work on, or if you are simply wishing to learn more about SimpleXML, try making the modifications mentioned in the previous

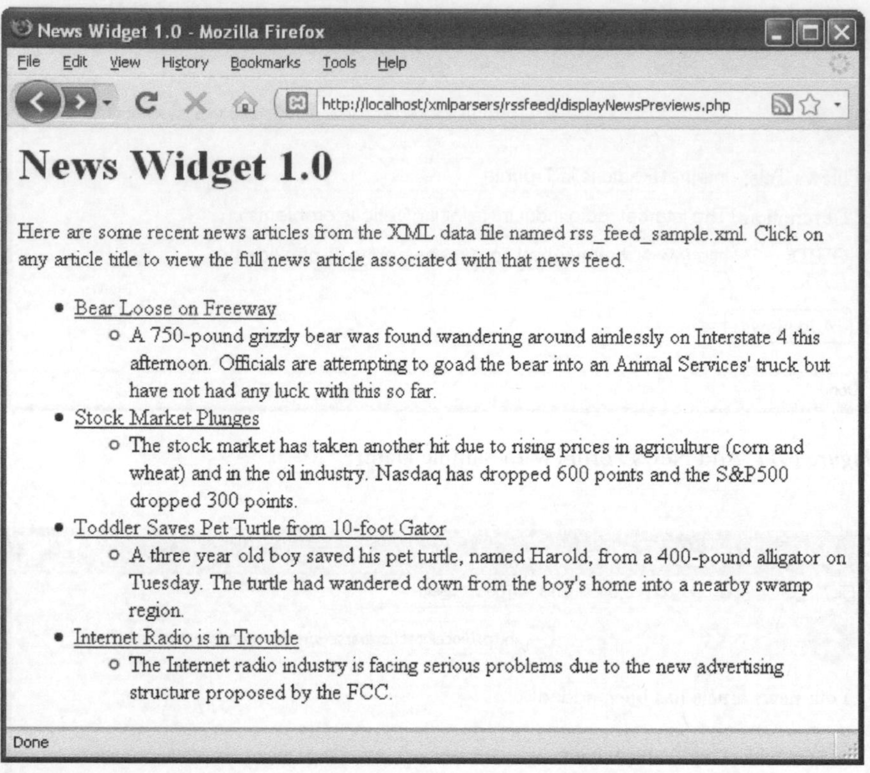

Figure 6.13 News Display with New Item Added

paragraph. This will improve your understanding of PHP and perhaps help you to better recognize this scripting language's potential use as a tool for building customized XML parsers.

For our next project, we consider a parser with more functionality and more complexity: a CMS. Parts of the CMS will then be reused for our final parsing example, which is a basic single-sourcing application.

Project 2: CMS

Now that we have discussed the simplest approach to both rhetorical analysis and the technical implementation of an XML parser, we will move on to an example that is slightly more challenging in both domains. A CMS is a software program that allows content to be collected from one or more users and then distributed to one or more audiences in its original form or in a reconfigured form. Andreas Mauthe and Peter Thomas describe a CMS as one that handles both the essence of the data, or the raw data itself, and the metadata that describes that content (4–5). The idea here is to try and capture the gist of the content by considering the items and deciding upon a reasonable set of representative characteristics to encode and then associate with them using metadata. This can be done automatically, when certain descriptive characteristics can be extracted from digital files, or manually, when a person considers an item individually and determines which facets should be represented in the metadata. Oftentimes, CMSs will combine human knowledge with technological support systems. That is the approach we follow in this example.

Typically, CMSs pair a graphical user interface with a database. This structure allows selected individuals with certain privileges to insert information into the database while other users with other types of privileges can extract information from the database. While many online CMSs use relational databases for this task, which we discuss in Chapter 3, we will be creating a system that uses XML as the primary data source. When we move into our technical discussion of how to build our simple CMS, we will be focusing on the SAX API for PHP. SAX supports XML parsing in a streaming fashion, parsing linearly from the top of a document to the bottom.

The particular type of CMS we will build for this project is a digital asset management system. An asset is a digital file that has value in a particular informational context. For example, if an information designer creates a glossy brochure to advertise a new software product, that person might need to create new image files in Adobe Photoshop® and then import them into a program like Adobe PageMaker® or FrameMaker®. Once the image files are added to the brochure, the document creator may store these images on her own hard disk drive for perpetuity. If this particular employee ever leaves the company or forgets about these files, the potential information represented by these data files is lost.

This is a problem of durability, as we discussed in Chapter 1; at the individual level of knowledge, information is not very durable and may in fact be considered volatile as any group and organizational knowledge built atop this foundation may come crumbling down. For instance, if the Web development team later decided to create a new promotional website based on this brochure, they might assume that all of the original layered Photoshop files were available somewhere in the company archives. Without the critical employee being available to hand over these files from their hard disk, the Web team would quickly realize that the plan of work and the various milestones they had planned for were neither plausible nor achievable. The repercussions of this simple loss of individual knowledge would be felt from the Web team, to management, to eventually the customers that would be purchasing this new product.

Our goal, then, is to build a simple media asset system for this type of organization. This system can be implemented in a distributed corporate setting. Rather than storing files separately on employee computers, individuals working on various projects can use such a system to store all assets in a centralized location with meaningful metadata to facilitate location and retrieval. The metadata can also be used to allow various groups within the organization to recognize the different types of information produced by each group and to better understand ways in which information can be exchanged between units.

The example we will create for this project could be used in a small organization that uses document types such as plain text, Microsoft Word®, Adobe Acrobat® PDF, and image types such as GIFs and JPEGs. With some minor customization, it could be extended to handle more complicated types of assets such as layered Photoshop® files or audio/video recordings.

Rhetorical Analysis

For our rhetorical analysis of this project, we could use a variety of different approaches, including the use of the ad hoc rhetorical analysis document mentioned in the previous example, to define the parameters of our parser. Or, we could choose to adopt a more formal rhetorical model.

For the sake of discussion, let us consider one such formal model that was created by the famous rhetorician Kenneth Burke. Burke's dramatist rhetoric studies the rhetorical situation from five different dimensions: act, scene, agent, agency, and purpose. Collectively, these dimensions are known as Burke's pentad, or, sometimes, as the dramatist pentad. From a dramatic perspective, this delineation makes perfect sense. When watching a play or reading an immersive novel, we find ourselves caught up in the relationships the protagonist has with the world around him. These relationships may engender within the reader feelings of adventurousness, humor, sadness, or excitement.

While Burke's rhetorical ideas are influential and important, his pentad is probably not the best tool for our task at hand: building a well-designed CMS. Burke's system was designed to study the motives of human agents, from the heroes acting inside the boundaries of fiction to the political orators asking an audience for an explicit call to action. While XML certainly may involve notions of act, scene, agent, agency, and purpose, a CMS will generally not have much dramatic substance with which to apply these Burkian strategies. We will, however, see how XML can be paired with dramatist rhetoric and technology when we discuss one of the real world XML examples in Chapter 7. One of our interviewees uses XML as a tool to help train cast members for interactive, dramatic performances.

In this particular CMS, the persuasive appeal of our technology is somewhat more subdued than what we might see in an analysis using Burke's pentad. This does not lessen the usefulness of this instrument as a tool for the construction and organization of knowledge, however. On the contrary, we find that Burke's pentad, when applied to a different problem solving context, may be particularly useful for thinking about the process of metadata design and distribution. Although we will not use it to guide the construction of our CMS here, it may be very useful in other contexts in which dramatic content is more predominant. We will return to this theme in Chapter 7.

So how do we choose a more appropriate rhetorical model to guide the information design process for this project? For the purposes of building a CMS, we can apply a model that has been specifically developed for information design. Saul Carliner's physical, cognitive, and affective framework is well-known for breaking information design problems down into three dimensions (45):

- The physical dimension, which involves specific things like page design, screen layout, and production decisions;
- The cognitive dimension, which is concerned with the thinking users do as they engage with an information source; and,
- The affective dimension, which deals with emotional and motivational elements and how they influence the communication process.

Carliner's framework is useful here because it can be used in a generative fashion to produce rhetorical questions related to the ways in which users physically interact with, think about, and feel about information. These questions can then be used as guides to help a designer make informed decisions about how the XML parser will function. Rather than arbitrarily linking together scripting code, metadata, and GUI components, a designer can plan out an effective information delivery system by carefully considering the questions an information seeker may have when looking for information. Beginning a project by looking at these types of questions rather than thinking about the raw metadata first is known as a "top-down" approach (Morville and Rosenfeld 44). This type of design process is largely concerned with trying to anticipate and cater to the informational needs of expected users.

Physical Design Dimension

Using Carliner's model, we will attempt to predict our users' informational needs by separating the three dimensions of information design into sets of associated questions. We can begin with the physical dimension. Since the physical dimension is concerned with characteristics such as page layout and design, this dimension is mostly focused on the ways in which the user will move through the asset management system. Here are some paired questions and answers that can help us with physical interface design:

- How will users upload assets? We will provide a Web-based form that allows users to upload and store files.

- Which types of assets will be supported? Our CMS will support five file types: plain text (.txt) files, Microsoft Word® (.doc) files, Adobe Acrobat® (.pdf) files, and selected image (.gif and .jpg) files.
- How will users search and find previously uploaded assets? We will display assets and associated metadata in a tabular fashion and provide hyperlinks to existing assets.
- How will users move from one section of the CMS to the next? We will provide a guided series of steps leading users through each procedure.
- How will users recognize their current location in the asset management system? We will provide clear labels to show the steps associated with each procedure.
- How will users know which types of metadata to supply for each uploaded asset? We will use drop down lists and selection boxes to provide appropriate metadata tags for each asset. Assets will be classified by their user-given name and description, filename, category, a list of years during which they were used, and their media type.

Cognitive Design Dimension

We can now consider the cognitive dimension, which is often the most difficult and comprehensive design exercise because it involves getting inside our users' heads, or trying to anticipate what they will be thinking about when they are using our product. Any page or document that requires too much cognitive processing from the user is inherently less usable than one with an intuitive and familiar interface.

In this particular project, we need to think about the cognitive implications of an XML parser. Here, paying careful attention to best practices from usability research will help us to create a more intuitive and less cognitively demanding CMS. For example, in his well known book *Don't Make Me Think: A Common Sense Approach to Web Usability*, Steve Krug writes that his first law of usability is "don't make me think": websites should be "self-evident. Obvious. Self-explanatory" (11). Krug justifies this law by noting three behaviors of Internet users that have surfaced from his own observations, as well as the research of usability experts such as Jakob Nielsen. Specifically, he mentions these three facts:

1. Readers scan Web pages rather than reading them fully. Because we are often in a hurry or otherwise distracted by a myriad of virtual (e.g., advertisements or instant messages are loading on the screen) or real world (e.g., the phone is ringing) cues, we quickly scan pages in order to try and find familiar elements based upon recognizable, physical design cues or snippets of familiar text (22–3). Our CMS interface should then use standardized labels and familiar interface components in order to support this scanning behavior and reduce cognitive load.

2. Readers do not make optimal Web browsing choices. Instead, they choose the first reasonable choice they come across, a behavior known as "satisficing" (24). This means that the typical user will not weigh and measure each interface item presented by our CMS in order to make the best possible decision about how it should be used. Instead, she will quickly scan the interface for the first reasonable option that seems like it might lead her to an acceptable outcome.
3. Readers do not figure out how things work before they use them. They tinker with them and muddle through as best they can without fully understanding them or reading instructions (26). This suggests that our CMS should have built-in mechanisms to deal with improper input or other interface decisions that might lead a user astray.

The cognitive dimension helps us to refine our initial ideas about physical interface design into a format that is easy for our audience to use. We do not want to ask them to think deeply about their every decision to click a button or access a dropdown menu. Carliner suggests that the cognitive design process follows a series of five steps (48–52). These steps are outlined as follows:

First, designers must analyze needs (48). We can imagine a hypothetical person to help us analyze needs, as we did when we created Joe Smith for the first example in this chapter, or we could develop use case diagrams, which are behavioral diagrams that show the relationship between hypothetical users and goals in a system. For our CMS, we will use a generalized use case diagram, as shown in Figure 6.14. Here, we see that there are two different types of users, or "actors," as they are known in use case parlance. The first type of actor, labeled "production," represents those users working on production tasks such as document creation. The second type of actor, labeled "management," represents users more concerned with locating existing assets and less concerned with the creation and annotation process. Our use case diagram therefore provides a detailed and unambiguous sketch of which tasks are likely to be performed by which users. This diagram tells us that producers will need access to all functionality, while managers will only need access to select and view tasks.

For your own projects, you can create use case diagrams from a variety of software programs such as Microsoft Visio® or SmartDraw, or by using online diagramming software such as the suite offered by Gliffy.com (see additional online resources). You can also draw use case diagrams by hand. In this case, you can sketch out the actors and tasks on a sheet of paper and then use arrows to show the relationships between the two entities. The use case diagram shown in this chapter was generated by a program named SmartDraw.

Carliner's next step in the cognitive design process is to set specific goals for the project. These goals can be business or content related and should include an evaluation component to ensure that these goals are being met

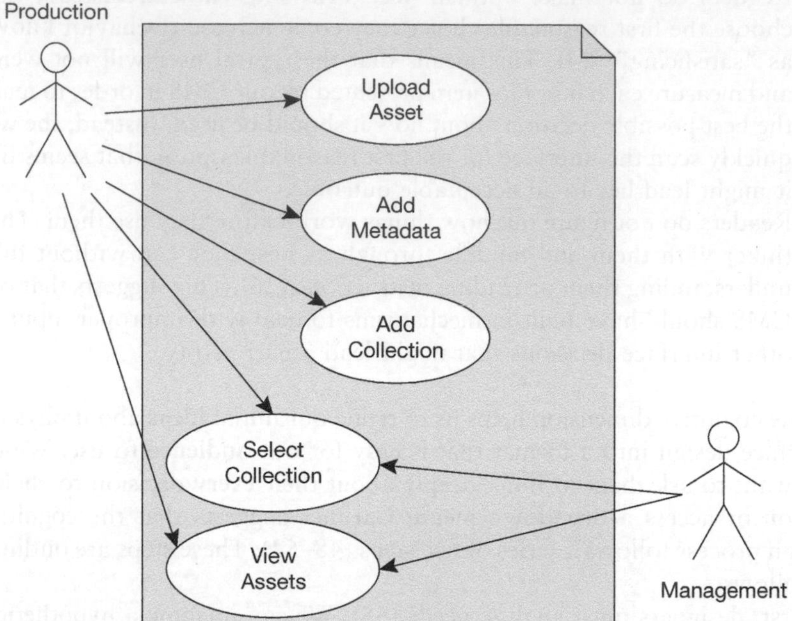

Figure 6.14 Use Case Diagram

through the information system. For our CMS, the fundamental business goal is to improve the efficiency and productivity of different organizational groups by sharing assets and reducing redundancy and duplicity. In other words, we want the assets produced by one group to be available to other groups who may need to use the same files or documents for another type of project. This process can be facilitated by knowledge managers in the manner that we described in Chapter 3.

Content related goals describe tasks a reader should be able to complete after interacting with the system (Carliner 49). Our primary content related goal for the CMS is to build a system that users can use to upload and search for assets. The evaluation of such a system would need to be implemented in a longitudinal fashion and would depend largely on the institutional context and the number of departments or divisions using the CMS.

After goals have been set, we must choose the form of our communication project. Form is related to genre. In rhetorical terms, genre refers to the typical and repeated responses that emerge over time and through space (Carliner 49). This is intrinsically related to the prototype theory we discuss in Chapter 3. Although genre is something traditionally associated with literary canons or stylistic conventions, we can also have genres associated with locations or even products. For example, the typical patterns of behavior of someone using an ATM machine will be much different than the behaviors of someone using their bank's online website to check their finances from home. Similarly, groups of people eating at a restaurant will exhibit similar normalized responses and expectations based on social conventions and the particular type of restaurant in which one is dining. By making our genre clear to users, we reduce their cognitive loads because they have prior expectations and familiarity with the process according to that defined genre. In this project, our genre for the CMS is an interactive, form-driven website; users will expect to see things like submit buttons, text fields, and procedural instructions and labels on such a site.

The next step in the cognitive design process is to prepare the design of our communication project. Depending on the complexity of a project and its specific genre, the design documents can vary to a great degree. For the CMS, we will devise an information map that shows the relationship between different files and the audiences that will need to use these files. Later, after we have considered affective design elements and a few additional cognitive design parameters, we will create simple algorithms that illustrate how the XML parser and the XML writer's modules will work.

To prepare our information map, we should consider our audiences. Since our content producers will be the individuals who produce unique assets, this audience will be tasked with the uploading and tagging of content. Other organizational groups, such as management, will then need access to the CMS as "readers" so that they can find appropriate assets and distribute these assets to other groups that may need them. In addition, all producers should have access to these files for those situations in which management is not directly involved in the process. We can therefore divide our overall project into two main chunks, or units. The first unit will be composed of functionality to add and tag assets. The next unit will be composed of files that can search existing assets, parse these assets, and then display associated metadata information. Producers will need access to both units while management will only need access to the second unit.

Finally, we must set project and product guidelines. Carliner notes that product guidelines include editorial guidelines, production specifications, and technical specifications, while project guidelines include questions of schedule, budget, and staff (51). We will include several of the product guidelines in our communication design information map. These items will include technical details such as file names and file types as well as rhetorical details such as audience and purpose. Our information map is presented a bit later in Table 6.1.

Affective Design Dimension

Lastly, we will consider the affective dimension of information design in order to produce a final set of rhetorical questions. Affective in this context is a psychological term that refers to the experience of feeling or emotion.

Affective design elements deal with issues such as motivation, attention, and satisfaction. In other words, even if the information is available and accessible, will users feel like using it?

To some extent, affective design can be linked to product aesthetics and the visual dimension of information as we discussed in Chapter 4. For example, Donald Norman is well known for exploring these types of affective and aesthetic issues in regards to the psychological implications of everyday objects. In his 2005 book *Emotional Design: Why We Love (or Hate) Everyday Things*, he writes about Japanese researchers Masaaki Kurosu and Kaori Kashimura, who found that despite different types of ATM machines having identical buttons and functionality, the Japanese ranked the ATM machines with more attractive buttons and screens as being easier to use than the ATM machines with less attractive layouts (17). These findings were then replicated, with even stronger results, when Israeli scientist Noam Tractinsky repeated this experiment by translating the Japanese instructions into Hebrew (18). Norman relates the aesthetic dimension of design into the affective dimension by claiming that since attractive things make us feel good, an attractive product will make us feel good and allow us to think more creatively (19). Creative thinking is important for a user when she needs to react to unexpected problems that may come up during a product's use, or when she needs to extend or customize tools in order to use them in ways the original developers did not anticipate.

Returning to our efforts to build an affectively usable CMS, we need to create some guiding questions that will help us to better plan our parser. Affective rhetorical questions related to our CMS can be formulated as follows:

- How can we deal with issues such as attention and motivation?
- How can we keep our users focused on the task at hand?
- What can we do to make the interaction process as pleasurable as possible?
- How can we reduce anxiety through our informational decisions?
- How can we create a simple interface that contains enough information for its users, without being overwhelming, and that is at the same time pleasing and familiar?

In order to address these affective issues, we should design our interface to provide clear instructions. We should clearly indicate how many steps are required to complete each operation. This helps reduce anxiety and gives our users an idea of how much investment is required to accomplish a task using our CMS. Additionally, when our parser eventually sorts through our XML file and creates a navigable list of assets, it should clearly demonstrate the value added by having such a content repository. This will create feelings of satisfaction for the users who find this material and realize they will not have to reinvent the wheel for their own project. We can expect that the most pleasurable moment for users will be found in the instant they discover that someone else has already done the work they were charged with. Upon realizing this, they will recognize that they need only to download an existing asset and make minor modifications to it in order to solve their problem.

These issues of consistency, labeling, and organization are important for us to address and undoubtedly help us to improve the moods and feelings of our users. Additional improvements can always be made with the help of professional graphic designers. For real world applications of this system, graphic designers need to be involved early in the design process. These skilled artists provide valuable expertise in aesthetics and design and contribute meaningfully to the overall affective usability of an information product.

Preproduction Design Tasks

Since we are building a working CMS to store typical production assets, we need to move from an abstract and conceptual idea of our CMS to a more applied blueprint by using the information from our rhetorical analysis. We will use this information to construct a rudimentary Web form that allows us to gather data from our production archivists. To do this, we need to apply the rhetorical and technical information we have collected so far to build our initial design documents.

Using the information collected in our analysis, we know that we have several initial facets to our data: name, filename, category, media type, years used, and description. In addition, we should have some unique identifier associated with each asset to differentiate items from one another. As in the prior example, we can fashion a rough algorithm to help us with the sequencing and design of our XML writer and parser. First, we will use the following algorithm for the writer:

```
While (new assets exist to be added)
      Find the asset ID number of the last asset added to the
      system.
      Display an upload form to allow for the storage of the
      file.
      After the file has been uploaded, display an additional
      form to collect the metadata for the file.
      Assign the asset ID number to allow for unique
      identification of each asset in the collection.
      Increment the current asset ID number to the next
      available asset ID number in case more assets need to be
      added.
End While
After all assets have been added and annotated with metadata,
provide a mechanism for writing all temporary data to an XML
document.
```

Our reader algorithm is even simpler. This series of instructions specifies how our parser will iterate through our XML database and display relevant information about the asset collection. This algorithm looks like this:

```
If (one or more asset collections exist)
    Provide the user with a form to select the desired XML
    file containing the asset collection.
    Open and parse the XML file.
    Display each asset contained in the XML file along with
    its associated metadata.
End If
```

Now that we have precisely defined the steps needed for the storage and retrieval of assets, we can concentrate on additional preproduction tasks by specifying how various pages will be constructed and making production decisions concerning our file and directory structure. Here we will combine guidelines for the physical dimension with the cognitive dimension by crafting a table which specifies parameters for the project. Specific columns for data concerning audience and primary purpose, which directly relate to cognitive interactions, can be combined with information about directory structure, file type, and file name, which relate to physical interactions. One way of constructing such a grouping is shown in Table 6.1. These files are listed in roughly the same order in which they will be processed by first the XML writer and then the XML parser.

Based on this table, we see that we will have a total of eleven files, three of which are dynamically generated. We can now add logic to these files by writing our scripts in PHP. The source code for each of these files is shown in Appendix C. You can also download these files from our accompanying website. If you would like to follow along with this example on your own computer, you should perform the following steps:

1. Make sure that XAMPP is installed and in working order. Instructions for installing XAMPP are given earlier in this chapter (p. 222).
2. Create a new directory named "cms" within your xmlparsers directory in the XAMPP program directory. By default, this directory would be created in "c:\xampp\htdocs\xmlparsers\cms".
3. Create empty files with the filenames shown in Table 6.1. Type in the source code shown in Appendix C that corresponds to each file name. Alternately, you can simply download the Project 2 zip file from our website and extract all of the files into your newly created CMS directory.
   a. In the root directory, you should have six PHP files: "add_asset_metadata.php", "finalize_assets.php", "parse_xml.php","process_asset_metadata.php", "select_xml.php", and "upload_asset.php."
   b. You should also have an HTML file named "upload_asset.html" in the root directory.

Table 6.1 Parser Design File Structure

File Name	*Purpose*	*File Type*	*Audience(s)*
upload_ asset.html	This document allows users to upload files from their computer using a Web-based form.	HTML document	Production
upload_ asset.php	This file checks that the asset is a supported type. It handles the reading and writing of asset IDs. It also performs basic error checking and handles the physical process of moving uploaded files into the appropriate directory.	PHP script	Production
asset_ buffer.txt	This temporary file stores the unique id number of the most recently added asset. "Asset_buffer.txt" is the default file name, but this can be changed.	Plain text document	None (works behind the scenes)
add_ asset_ metadata.php	This script provides users with a second form for assigning metadata to the uploaded asset.	PHP script	Production
process_ asset_ metadata.php	This script checks the metadata for any errors and then writes the metadata to a temporary file. It provides links which allow the user to either add more individual assets or to write all existing asset metadata to an XML file.	PHP script	Production
metadata_ buffer.txt	This temporary file stores the metadata information for each asset in a temporary file. After all assets have been uploaded, this file will be parsed in order to create the final XML file. "Metadata_buffer.txt" is the default filename, but this filename can be changed.	Plain text document	None (works behind the scenes)
finalize_ assets.php	This script loops through the temporary metadata file and obtains the asset identification and metadata information associated with each asset. These items are then transferred to a valid XML file.	PHP script	Production
asset_ collection_ mmmDD yyyy.xml	This dynamically generated XML document houses the entire asset collection in one encapsulated unit. Each asset will be wrapped in individual ASSET element tags with associated metadata embedded within.	XML document	None (works behind the scenes)
select_xml.php	This script displays any available XML files that exist in the default document directory. It displays a dropdown list that enables a user to select any of these available files for parsing.	PHP script	Production, Manage- ment
parse_xml.php	This script parses the selected XML file using the SAX extension for PHP. Each asset found in the XML file will be listed along with its corresponding metadata.	PHP script	Production, Manage- ment
asset_ collection_ transform.xsl	This is an XSL stylesheet file that is used to display the XML file in HTML format. It will loop through the XML file and display each HTML item in tabular format.	XSL document	None (works behind the scenes)

c. There will be three directories in the root directory as well: "assets", "metadata", and "sample assets". The XSL transformation sheet, "asset_collection_transform.xsl", should be in the "metadata" directory. If you are creating your files manually, you should also create the assets directory and the metadata directory manually and place the "asset_collection_transform.xsl" file in the metadata directory as described above.

4. From the root CMS directory, open the "upload_asset.html" file in your browser by navigating to "http://localhost/xmlparsers/cms/upload_asset.html". This is the first file you will need to access when uploading files to the CMS. Follow the instructions to add assets and view your asset collection using the built-in XML parser.

Building the Interface

Recall that the first task in our writer algorithm involved providing an upload form to the user. This form allows the user to select a file for storage in the CMS. Our initial Web form for this task needs to look something like the page shown in Figure 6.15. We are not concerned much with aesthetics at this early stage—though, as Donald Norman reminds us, such issues certainly need attention at some point. At this moment, however, we are more interested in building basic functionality and establishing an easy to use interface. We are also careful to indicate the number of steps necessary to complete this process as we planned for this in our rhetorical analysis. This will help with both the affective and cognitive dimensions of our user interactions since the user will know precisely how much investment is required of her.

In order to create this form in HTML, we need to use the code for "upload_asset.html" that is listed in Appendix C. Note the use of the HTML comments beginning with <!— and ending with —> in order to explain different portions of the code. This format for comments is different from the PHP conventions, but is the same format used for XML comments.

Although this CMS could be used to harvest many different types of assets, we specified during our preproduction phase that we would only accept PDF, DOC, GIF, JPEG, and TXT documents. So, we provide these instructions to the user as well. Figure 6.16 shows the bottom portion of this form after a user has selected an asset to upload. In this case, she is uploading a file named "sample_brochure.doc".

After the file has been uploaded and processed by this initial script, the user is provided with information concerning the asset's new name and recognized file type. She is then given the option to proceed to Step 2, as shown in Figure 6.17.

In Step 2, the user is taken to the "add_asset_metadata.php" page and given the opportunity to provide metadata for the uploaded asset. Here, several pieces of information are added automatically by the CMS in order to reduce cognitive and affective strain on the user. As Figure 6.18 reveals,

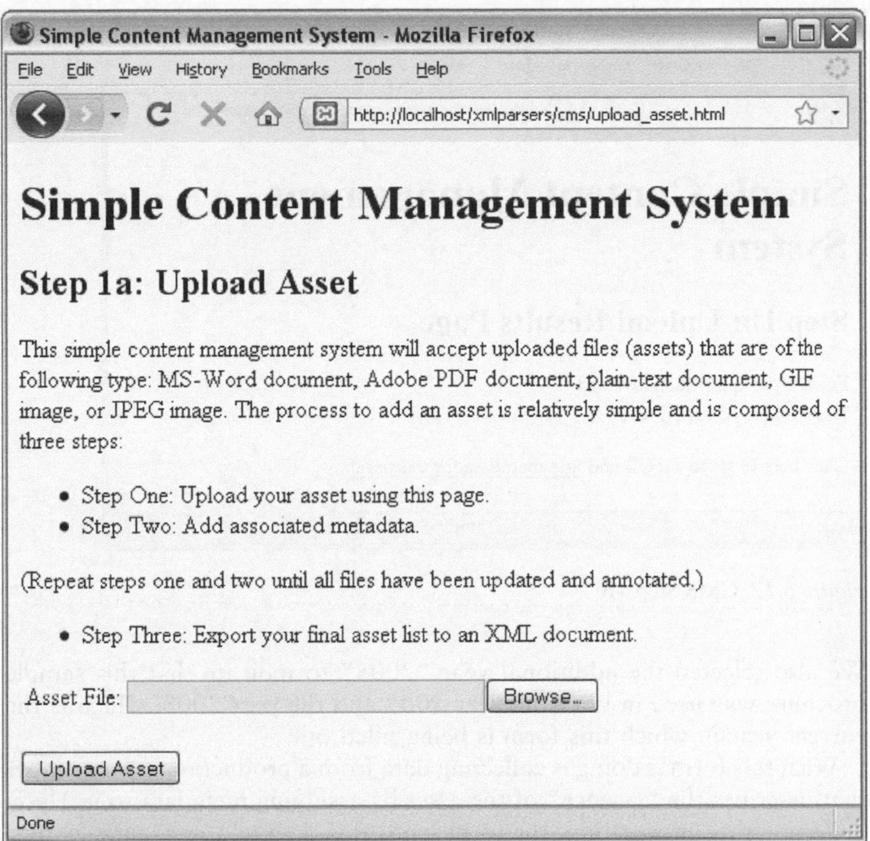

Figure 6.15 CMS Step 1a

Asset File: ıs\sample assets\sample brochure.doc Browse...

Figure 6.16 Asset Upload Screen

the asset's id, media type, and filename are automatically added by the system. In addition, a default asset name is provided, though this can be changed by the user if she so desires. Finally, the "Years Used" selection will default to the current year. The user is only required to provide categories, additional years, and an asset description. For this example, we have typed in a brief description and selected "New Media", "Print", and "Epsilon Project" as sample categories. These categories could represent different divisions of an organization or different internal project names.

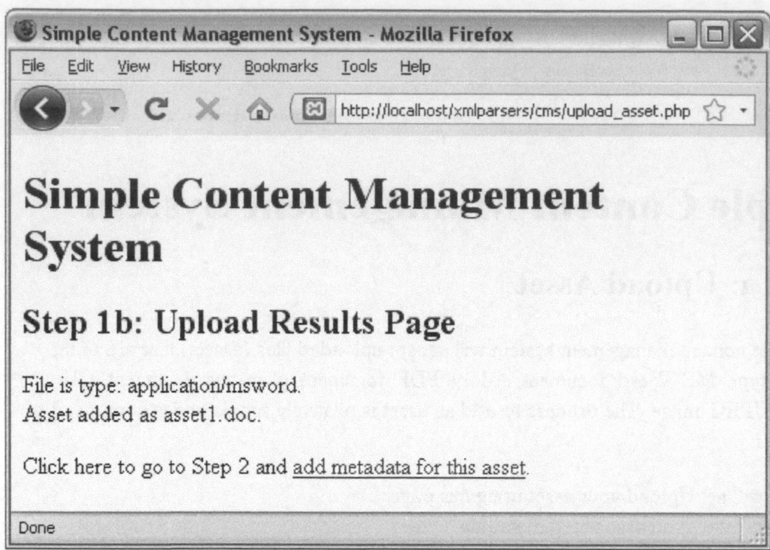

Figure 6.17 CMS Step 1b

We also selected the additional year "2005" to indicate that this sample brochure was used in both the year 2005 and the year 2008, which is the current year in which this form is being filled out.

What this form is doing is collecting data from a production team member that describes the "essence" of the asset by assigning metadata to it. Upon collecting this information, the form sends this data to a new page, named "add_asset_metadata.php". This page takes the data points and translates them into variables, then it glues them together into a single string entity separated by the sequence ***. Commas are used to separate multiple values that are related to the same entities. This notation is important because it will be used to construct the XML file later and the script will need to know where one element ends and the next begins. We will use the same convention in our single sourcing project later in the chapter. The idea is to construct a running buffer file that can then be accessed incrementally when it is time to create the final XML document.

After this process is finished, the "add_ asset_metadata.php" script attempts to initiate a file connection and write this data to a temporary file. Using our example data, the temporary buffer file contains a single line that looks like this:

```
1***asset1.doc***asset1***New Media, Print, Epsilon
Project***2008, 2005***MS-Word Document***This is a sample
brochure idea that was shown to clients before the new Epsilon
2.0 software was released. It needs a bit more work before it
can be sent to the printer.
```

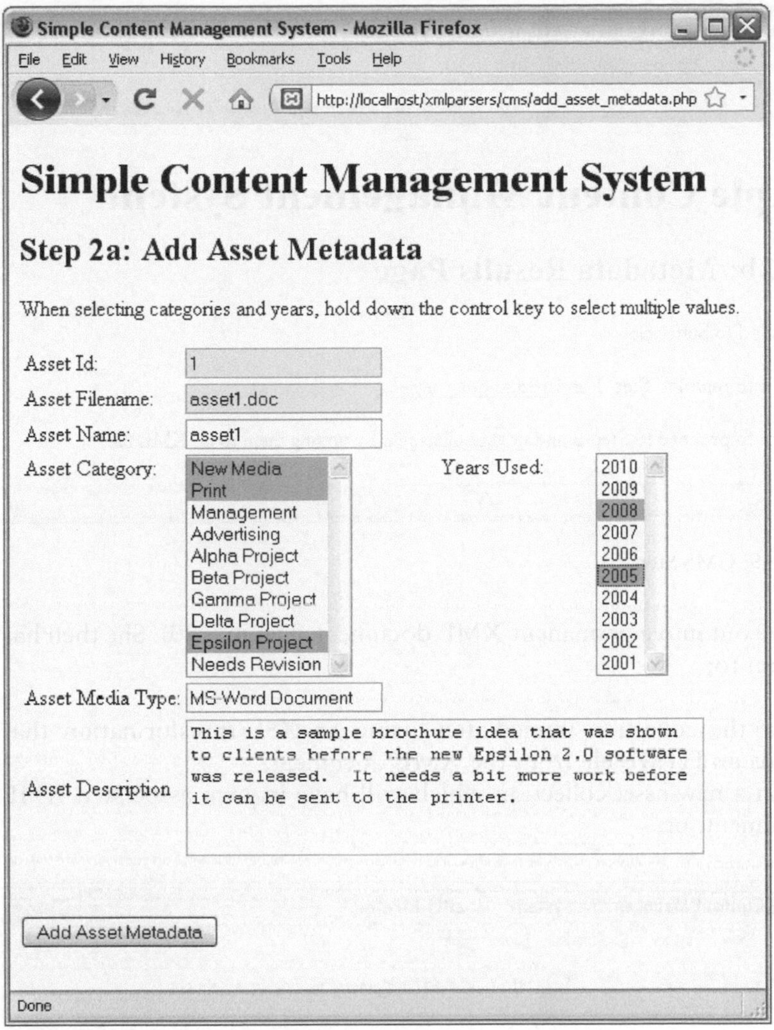

Figure 6.18 CMS Step 2a

After all assets have been added, the user is given the option to click on a link to "finalize all assets" and finish the process (Figure 6.19). Or, she may choose to return to Step 1 and continue adding assets for this particular collection. Once all assets have been uploaded, she can click on the "finalize all assets" link to produce the XML file. What this final step will do is take the information from the temporary buffer file and construct the XML document containing all assets.

After all assets have been uploaded and annotated with metadata, the user is given the option to finalize her collection and write the temporary

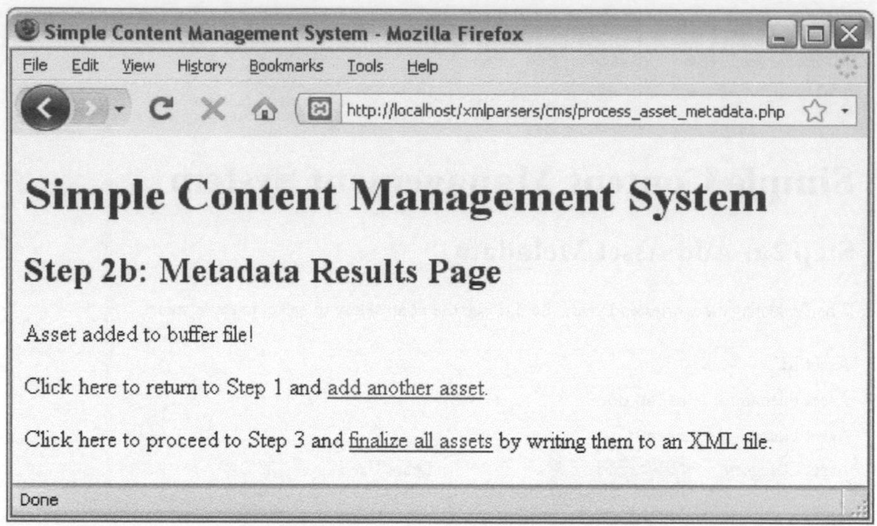

Figure 6.19 CMS Step 2b

buffer file out into a permanent XML document (Figure 6.20). She then has the option to:

a. View the collection immediately using an XSL transformation that builds an HTML file from the XML document,
b. Begin a new asset collection which will have its own associated XML document, or,

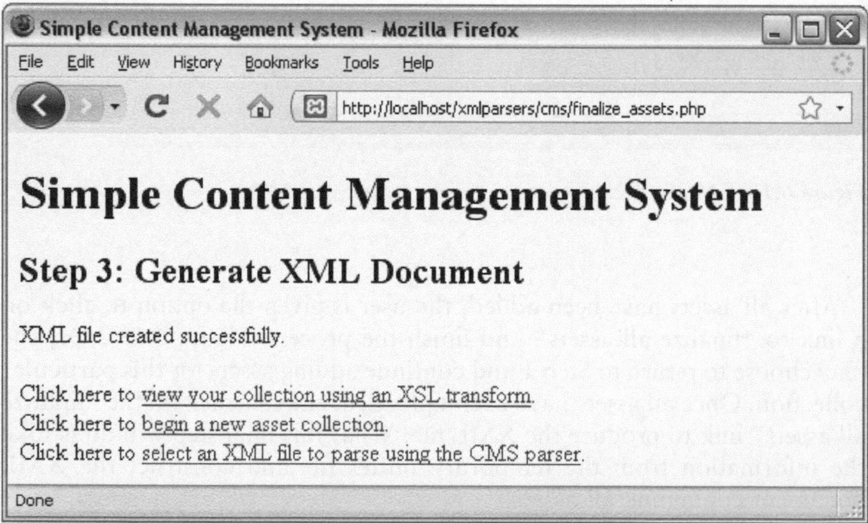

Figure 6.20 CMS Step 3

c. Select an XML document to parse using the built-in CMS parser, which is a more sophisticated version of the table presented by the XSL transformation.

After the asset collection is finalized, an XML document is dynamically created from within the CMS system. An example XML file created by the system is shown next. You will see that many of the form elements from Figure 6.18 have been directly translated into XML elements and attributes in our database. This example file also contains several additional assets that were entered using the CMS system.

```
<?xml version="1.0" encoding="utf-8"?>
<!– XML Asset Collection Document –>
<!– Automatically generated by Content Management System –>
<!– Created on Aug192008 –>
<?xml-stylesheet type="text/xsl" href="asset_collection_
transform.xsl"?>
<!DOCTYPE asset_collection [
      <!ELEMENT asset_collection (asset)*>
      <!ELEMENT asset (id,filename,name,category,years,
      media,description)>
      <!ELEMENT id (#PCDATA)>
      <!ELEMENT filename (#PCDATA)>
      <!ELEMENT name (#PCDATA)>
      <!ELEMENT category (#PCDATA)>
      <!ELEMENT years (#PCDATA)>
      <!ELEMENT media (#PCDATA)>
      <!ELEMENT description (#PCDATA)>
]>

<asset_collection>
            <asset>
                  <id>1</id>
                  <filename>asset1.doc</filename>
                  <name>asset1</name>
                  <category>New Media, Print, Epsilon Project
                  </category>
                  <years>2008, 2005</years>
                  <media>MS-Word Document</media>
                  <description>This is a sample brochure idea
                  that was shown to clients before the new
                  Epsilon 2.0 software was released.  It needs
                  a bit more work before it can be sent to the
                  printer.</description>
            </asset>
```

```
<asset>
      <id>2</id>
      <filename>asset2.pdf</filename>
      <name>asset2</name>
      <category>New Media, Print, Epsilon Project
      </category>
      <years>2008, 2005</years>
      <media>Adobe PDF Document</media>
      <description>This is the PDF version of the
      Epsilon product brochure.</description>
</asset>
<asset>
      <id>3</id>
      <filename>asset3.txt</filename>
      <name>asset3</name>
      <category>New Media, Print, Gamma Project,
      Epsilon Project</category>
      <years>2008, 2005</years>
      <media>Plain text Document</media>
      <description>These are some printing notes
      from the Gamma Project that might also be
      useful for the Epsilon project.</description>
</asset>
<asset>
      <id>4</id>
      <filename>asset4.jpg</filename>
      <name>asset4</name>
      <category>Epsilon Project</category>
      <years>2007</years>
      <media>Image File</media>
      <description>JPEG logo for the Epsilon
      project.  Red text with drop
      shadow.</description>
</asset>
<asset>
      <id>5</id>
      <filename>asset5.jpg</filename>
      <name>asset5</name>
      <category>New Media, Print, Advertising
      </category>
      <years>2007</years>
      <media>Image File</media>
      <description>This is a JPEG of the logo for
      the new Epsilon 2.0 software.</description>
</asset>
</asset_collection>
```

Although we do include a DTD file in our dynamically generated XML document, our SAX parser does not have the ability to validate our document using this internal DTD. With the DTD added, however, we are making our document more extensible by specifying the pattern that should be used for adding new data or elements to the file should this ever need to be done manually. In addition, if this asset document ever needs to be exchanged with another computer or merged into another XML file, the DTDs of this document and the receiving system could better understand one another based on this information.

After it is created, this XML file can then be directly transformed using an XSL transformation (Figure 6.21) or parsed using our more robust CMS parser (Figure 6.22).

At first glance, the custom-designed parser in Figure 6.22 looks almost identical to the HTML listing we see in Figure 6.21 when we load the XML file directly and initiate its XSL transformation. In fact, it is useful to know that much of the hard work involved with building a customized parser can be minimized by using an XSL transformation as we do here. There are, however, subtle differences in the amount of flexibility we have available to us when parsing the document. The primary advantage of using our own custom parser in this instance is that we can take advantage of PHP's built-in functions to perform additional error checking and validation on our

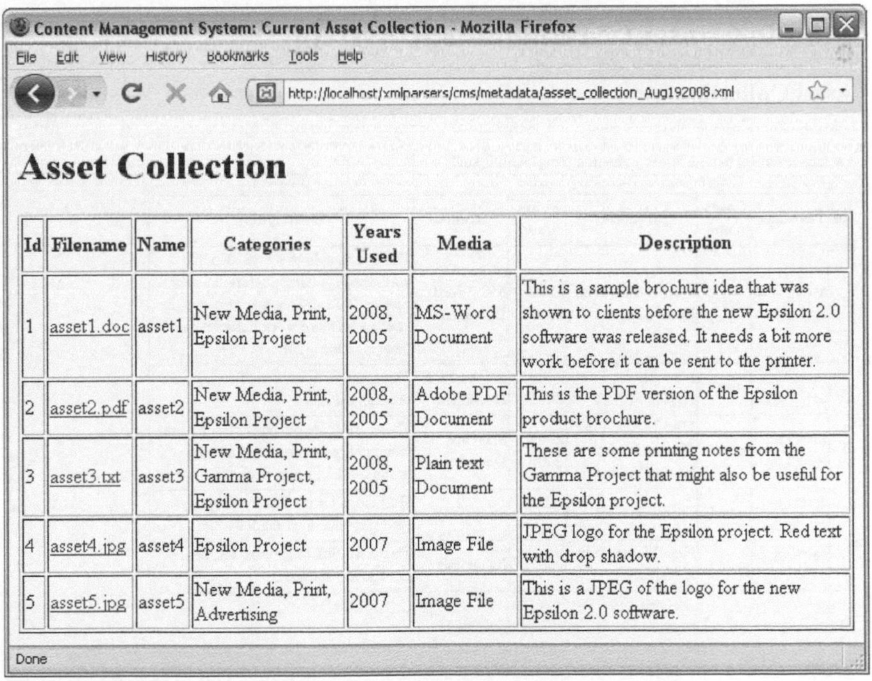

Asset Collection

Id	Filename	Name	Categories	Years Used	Media	Description
1	asset1.doc	asset1	New Media, Print, Epsilon Project	2008, 2005	MS-Word Document	This is a sample brochure idea that was shown to clients before the new Epsilon 2.0 software was released. It needs a bit more work before it can be sent to the printer.
2	asset2.pdf	asset2	New Media, Print, Epsilon Project	2008, 2005	Adobe PDF Document	This is the PDF version of the Epsilon product brochure.
3	asset3.txt	asset3	New Media, Print, Gamma Project, Epsilon Project	2008, 2005	Plain text Document	These are some printing notes from the Gamma Project that might also be useful for the Epsilon project.
4	asset4.jpg	asset4	Epsilon Project	2007	Image File	JPEG logo for the Epsilon project. Red text with drop shadow.
5	asset5.jpg	asset5	New Media, Print, Advertising	2007	Image File	This is a JPEG of the logo for the new Epsilon 2.0 software.

Figure 6.21 XSL Transformation

data. For instance, using our customized script, the CMS parser here will check and verify that a file exists before providing a link to the asset. The XSL transformation does not perform this check and could potentially lead a user to a broken link.

For example, if we were to remove the files “asset4.jpg” and “asset5.jpg” from our assets directory, we would create a situation in which our users might encounter broken links, or links to resources that no longer exist. Compare the two versions of our parser shown in Figures 6.23 and 6.24. In Figure 6.23, we see that links for assets 4 and 5 exist even though the files they are linking to are no longer accessible. In Figure 6.24, our parser has detected these anomalies and indicates to the user that these files are missing. It also removes the links to these resources so that broken links are no longer potential frustrations for the user. This contributes to a better information design along all three dimensions: physical, since the links are now more accurate; cognitive, since we improve the satisficing process by only providing meaningful links; and affective, since potential frustrations and annoyances are minimized or even outright eliminated.

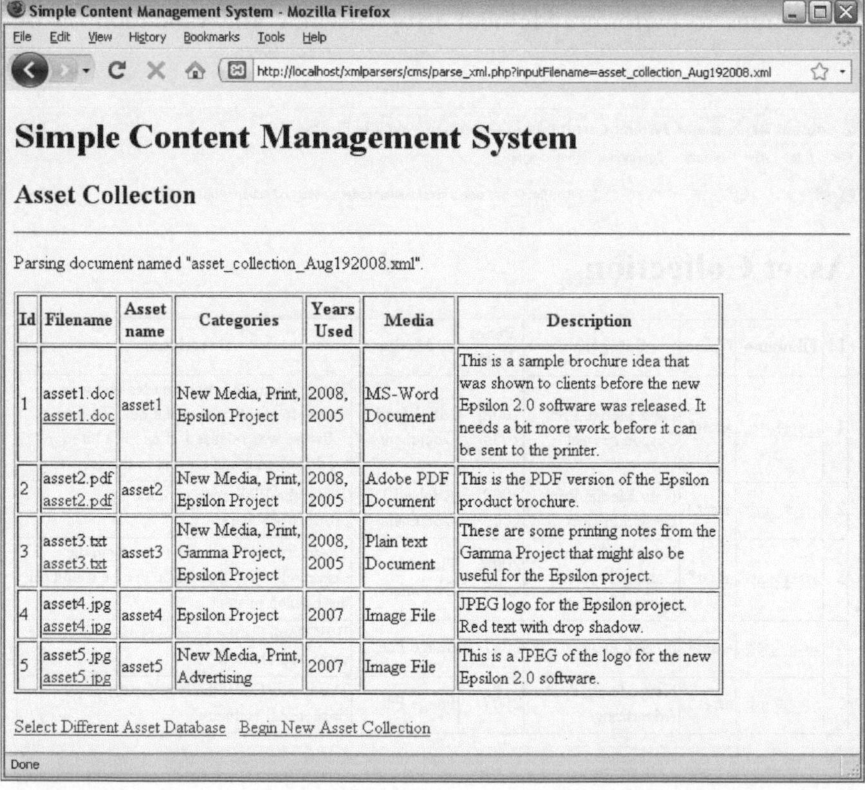

Figure 6.22 Custom Parser View

Id	Filename	Name	Categories
1	asset1.doc	asset1	New Media, Print, Epsilon Project
2	asset2.pdf	asset2	New Media, Print, Epsilon Project
3	asset3.txt	asset3	New Media, Print, Gamma Project, Epsilon Project
4	asset4.jpg	asset4	Epsilon Project
5	asset5.jpg	asset5	New Media, Print, Advertising

Figure 6.23 4-Column XSLT View

Id	Filename	Asset name	Categories
1	asset1.doc asset1.doc	asset1	New Media, Print, Epsilon Project
2	asset2.pdf asset2.pdf	asset2	New Media, Print, Epsilon Project
3	asset3.txt asset3.txt	asset3	New Media, Print, Gamma Project, Epsilon Project
4	asset4.jpg (missing)	asset4	Epsilon Project
5	asset5.jpg (missing)	asset5	New Media, Print, Advertising

Figure 6.24 4-Column View using Custom Parser

During maintenance periods, a quick scan of the asset list allows a user to determine which files have been removed or relocated and then to either remove those assets from the system (which would need to be done manually with our current CMS) or otherwise track down the assets and restore them to the assets directory.

Like the RSS feed parser we created in our first project, this CMS project could also be improved upon. For example, for large asset collections, there would need to be some mechanism for adding pagination so that only a small number of records are shown on each page. Also, it would be helpful to have an automated editing function to revise or add to existing metadata, as well as an automated delete function to remove assets no longer in the system. Lastly, it would be useful to perform a postproduction rhetorical analysis on this system to ensure that the physical, cognitive, and affective design parameters were effectively achieved as outlined during the planning phase of this project. Additionally, in a real production environment, it is always a good idea to perform some usability testing involving real human participants rather than relying too much on hypothetical personas and use case diagrams.

Fortunately, some of the work we have done for this large project can carry forward into our final project, which is a single sourcing demonstration. Project 3 is designed to show one way of creating a very basic level three single sourcing system that can modularize and repurpose XML content based on self-selected user skill levels.

Project 3: Single Sourcing System

While we discussed single sourcing technologies in Chapter 3, we will review a few of the important points about single sourcing here before building our third and final XML parser. First, we need to agree upon a common meaning for single sourcing so that we can build an XML parser that is in line with this conceptual idea. We will use the definition of single sourcing that is offered by the Society for Technical Communication's Single Sourcing Special Interest Group. This definition is cited by Joe D. Williams in his 2003 literature review of the single sourcing literature. Williams writes that single sourcing is "using a single document source to generate multiple types of document outputs; workflows for creating multiple outputs from a document or database" (321).

Next, we need to understand why single sourcing is useful to technical communicators and other professionals working with large volumes of textual data. We discussed many of these reasons in Chapter 3, but we will mention them again here. Ann Rockley notes that single sourcing technologies, or structured writing methodologies in general, offer the following benefits (350):

- The ability to unify content and ensure that content can be used (and reused) for a variety of informational needs;

- A standardized system for working in collaborative authoring environments;
- A framework for delivering electronic documents through automation, using information processing technologies; and
- The potential for cost and time savings through the use of more efficient techniques in terms of both the authoring and reuse of documents.

Like Rockley, Locke Carter agrees that single sourcing can provide benefits in terms of cutting costs, boosting revenue, creating more efficient means of production and distribution, and adding flexibility to the document design process. He also notes that single sourcing can result in an increase in the quality of products and a "faster responsiveness to a constantly changing marketplace" (317). Both the improvements in quality and market reaction time can be largely linked to the fact that content can be updated from a central location and then streamed to ancillary media using single sourcing technologies.

On the other hand, Carter also warns that document designers must be cautious of single sourcing technology because this process disrupts the traditional craftsman process of designing documents individually, for a specific context and audience, from start to finish. In Carter's words, single sourcing "puts pressure on the seemingly stable constructs of the writer and the document in ways that many previous innovations have not" (318). Because of this pressure, we must be careful when considering the complex rhetorical spaces formed by document developers, media, and potential audiences. In situations where writers and document designers are comfortable with using single sourcing and the same content is generally reused among various distribution media, single sourcing may be a good process to consider. When circumstances call for subtle and distinct voices to emerge from these various documents, however, or when content is highly specialized and idiosyncratic, single sourcing may not be such a good idea. In these types of situations, a general CMS may be a better choice if you are aiming for the automation and electronic distribution of information.

If single sourcing is a strategy well-suited for your own organizational culture and its associated information ecology, then you can use XML as a major part of your implementation. XML makes it simple to precisely define which modules of data can be reused across documents and how that data should appear within each individual document. Our primary goal in single sourcing is to write our source content once, then provide access to different configurations of this content using customized views. Building a parser for this type of task requires an analysis of our different user categories and their associated viewpoints, as discussed in the next section.

Now that we have considered some important caveats related to the design of single sourcing systems, we can begin this project's rhetorical analysis. Since single sourcing technologies are commonly used to develop software documentation, we will be extending the example of our content management system from Project 2 by adding some documentation to the project. Rather than using the Web-based version of that project that we discussed in Project 2, we will be using a small Microsoft Windows® version of the application named Asset Management System (AMS) 1.0 to better distinguish our online documentation from the product itself. The default AMS 1.0 interface is captured in Figure 6.25.

Though AMS 1.0 is a working application that we developed using the Microsoft Visual Basic® programming language, we will not go into any of the code details here as we intend to use this only as a demonstration of a software program that could be documented with an online single sourcing system. Online documentation continues to be a good option for software developers because it is cheaper to develop. It requires no media and there is usually only a small expense associated with Web hosting. This type of documentation is also easier to maintain, requiring only updates to the online documentation files rather than the shipping out of new CD documentation every time a new update or software release is launched.

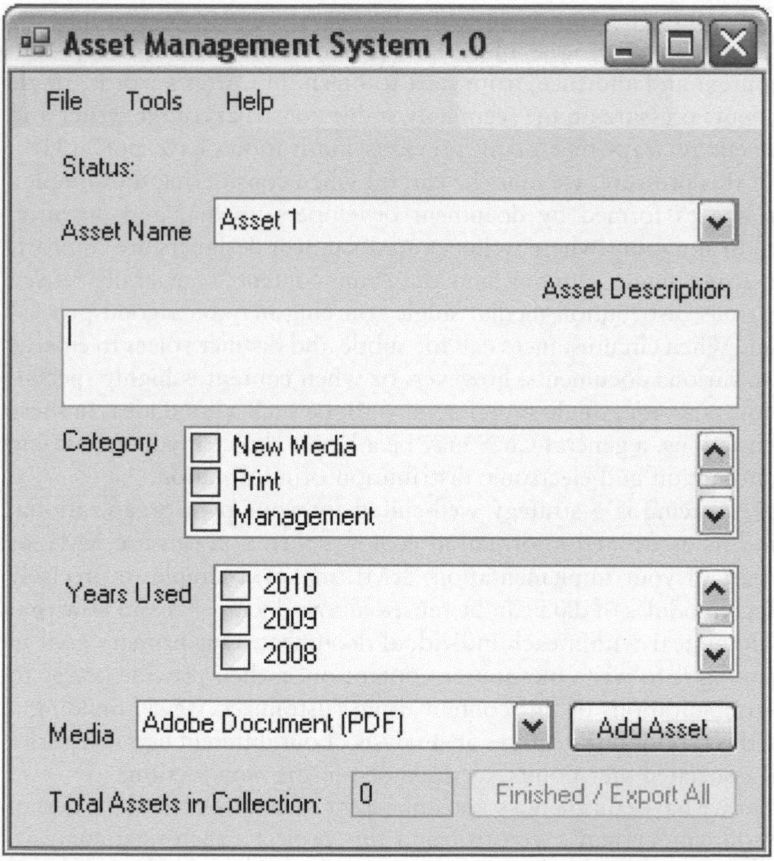

Figure 6.25 Asset Management System (AMS) 1.0

This application mimics the functionality of the CMS we created in Project 2 with the exception that it is designed to generate the XML metadata only. In other words, it will handle the XML file creation, but it does not manage and store the assets themselves. Using this application, a person can add asset metadata one item at a time by providing details about the assets, their categories, the years during which these assets were used, and their media types. AMS 1.0 is designed with several toolstrip menus and submenus to better reflect the typical GUI applications that have different feature levels for different types of users (see Figure 6.26). You can download this sample application from the text's website and enable the default configuration by using these toolstrip menus and navigating to Tools, Options, Load All Defaults. Or, you can enable these same defaults by pressing the control key, the alt key, and the letter "a" key all at the same time when the application is active on your screen. Choose "Yes" when you are asked to erase all existing categories.

In a fully developed single sourcing system, we would document each of these features in separate modules and then pull in only those feature descriptions deemed appropriate for a particular type of user and the common tasks associated with that type of user. In this project, we will add single sourcing functionality for an introductory help module.

In terms of functionality, the AMS 1.0 software imitates the Web interface we designed for Project 2 and produces the same type of well-formed XML.

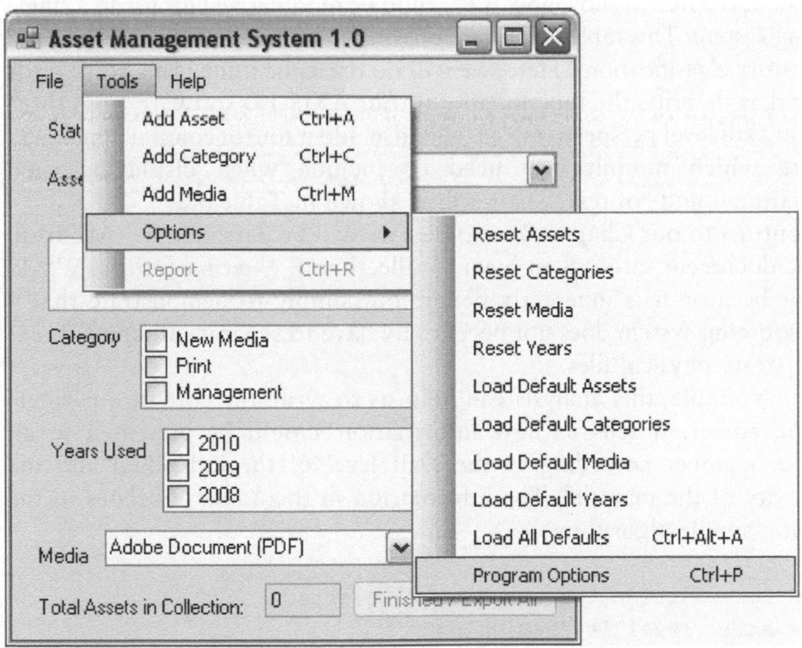

Figure 6.26 AMS 1.0 Sample Menus

It generates an identical XML structure with two exceptions. First, it does not include a DTD, since the application itself does not do any validation. (Our single sourcing system, however, does support validation since the new DOM API we are using allows us to validate against an internal DTD.) Also, the AMS 1.0 application does not include the filename tag as this is used solely for asset storage purposes, which is not a feature of this particular product. Rather, this type of application could be used when access to Apache or PHP was not available to an archivist. Of course there is a tradeoff in that the application was designed for the Windows operating system and other operating systems are not supported. In addition, the software requires .NET runtime libraries. These are specialized collections of code which users running older version of Windows may not have installed.

Now that we have an idea of the functionality of the software and the module that we will be documenting, we can try yet another type of rhetorical anlaysis, which is a bottom-up approach. This is in contrast to the top-down approach we used for Project 2. In a bottom-up approach, we start with the data rather than the predicted informational needs of our audience. This type of rhetorical analysis is more concerned with finding the appropriate level of granularity with which to surround a unit of text and finding the means of combining and repurposing these textual units in a manner that is compatible with our informational needs.

In Chapter 3, we used a simple table to divide the XML entities used for the DOE's ACHRE to show how they could be modularized for use in a single sourcing system. This table separated information into semantic, generic, and XML entity classifications. Here, we will do the same thing using some units of text that describe the functionality of our AMS 1.0 software from three different skill level perspectives. We will also add a fourth column which will indicate which modules we need to include when displaying each informational node of text. The result is shown in Table 6.2.

In contrast to our Chapter 3 example, we will be parsing our XML from a single document rather than from a collection of external entities. We do this not because it is inherently better, but simply to demonstrate that a single sourcing system does not necessarily have to separate all units of text into separate physical files.

Though simple, this analysis can help us to write the core of our single sourcing parser. It tells us how information should be presented to an audience member according to the skill level of the individual and the complexity of the material. The information in this table translates to the following simple algorithm:

```
If the Introduction topic has been selected
If the skill level is "beginning"
Display module "ib"
Else if the skill level is "intermediate"
```

```
Display module "ib" and module "ii"
Else if the skill level is "advanced"
Display all three modules: "ib", "ii", and "ia"
End If
```

With this algorithm now developed, we can design our parser's file structure and borrow portions of the writing component from Project 2. Table 6.3 shows the files that we will be using for the single sourcing demonstration. Like the CMS from Project 2, we have a mixture of HTML, PHP, XML, TXT, and XSL files being used in this system. Since we are automatically creating the ID based on what we developed earlier in Table 6.1, we can also reduce the number of files we need by two and simplify the writer component for this new parser.

Table 6.2 Bottom-Up Analysis for Single Sourcing System

Semantic Unit	*Generic Unit*	*XML ID Name*	*Modules to Include*
Introduction module written at the beginner's level	module	introduction_beginner: ib	ib
Introduction module written at the intermediate user's level	module	introduction_intermediate: ii	ib, ii
Introduction module written at the advanced user's level	module	introduction_advanced: ia	ib, ii, ia

Here are the steps to follow if you would like to recreate this project on your own. Interestingly enough, these instructions that you are reading now can be considered single sourced from our prior CMS example since they are describing the same process and only modifying a few units of text here and there in order to be compatible with Project 3. This is the type of reuse, combination, and repurposing that makes the single sourcing methodology so powerful for situations in which certain units of information are used in multiple places.

First, you will need to create a new directory named "ss" that will hold the third parser. If you are using the default XAMPP settings on a Windows computer, the full path to this directory will most likely be "c:\xampp\htdocs\xmlparsers\ss". Next, you should follow the steps below to install the single sourcing demo on your own computer.

1. Make sure that XAMPP is installed and in working order. Instructions for installing XAMPP are given earlier in this chapter.

Table 6.3 Single Sourcing File Structure

File Name	*Purpose*	*File Type*	*Audience(s)*
add_module.html	This document allows users to write documentation modules for particular software features.	HTML document	Writers
process_module.php	This file does some basic error checking and adds the documentation module to a temporary buffer file. It then provides the writer with links to add new modules or to finalize the XML documentation file.	PHP script	Writers
ss_buffer.txt	This temporary file stores the unique id number, title, skill level, and documentation text for each software documentation module.	Plain text document	None (works behind the scenes)
finalize_modules.php	This script loops through the temporary buffer file and obtains the documentation information for each module. These items are then transferred to a valid XML file.	PHP script	Writers
documentation.xml	This dynamically generated document houses the entire documentation collection in one encapsulated unit. It includes an internal DTD for validation with the DOM parser.	XML document	None (works behind the scenes)
parse_xml.php	This script parses the selected XML file by skill level to display the appropriate documentation modules for the given user type. It applies a simple algorithm to provide additional advanced information to skilled users.	PHP script	End Users (Production) and Writers (Testing)
ss_allmodules_transform.xsl	This is an XSL stylesheet file that is used to display all documentation modules that currently exist in the XML file in HTML format. It displays a hierarchical structure using unordered lists.	XSL document	Writers

2. Create a new directory named “ss” within your xmlparsers directory in the XAMPP program directory. By default, this directory would need to be created in “c:\xampp\htdocs\xmlparsers\ss”.
3. Create empty files with the filenames shown in Table 6.3. Type in the source code shown in Appendix D that corresponds to each file name. Alternately, you can simply download the Project 3 zip file from our website and extract all of the files into your newly created single sourcing directory.
   a. In the root directory, you should have three PHP files: “process_module.php”, “finalize_modules.php”, and “parse_xml.php”.
   b. You should also have an HTML file named “add_module.html”.
   c. There should be a single directory named “metadata” that exists in the root directory. The XSL transformation sheet, “ss_allmodules_transform.xsl”, should be in the “metadata” directory. If you are creating your files manually, you should create the metadata directory manually and place the “ss_allmodules_transform.xsl” file in the metadata directory as described above.
4. From the root single sourcing directory, open the “add_module.html” file in your browser. This is the first file you will need to access when using the single sourcing demo. Follow the instructions to add documentation modules and later to view your documentation using the built-in XML parser.

As we mentioned previously, since we have already designed a perfectly good writer module for Project 2, we can adapt this model to our current needs without much trouble. If you review the source code for the single sourcing demonstration that is provided in Appendix D and compare it to the code in Appendix C, you will note that the writer component is very similar to the one we used for the CMS. In fact, as we note in Chapter 3, a single sourcing system is really a specialized instance of a CMS anyway, so this makes sense. Our modified parser is what will distinguish our single sourcing system from the AMS we built in Project 2. As with the CMS example, the first step in actually using the system involves typing in some data and submitting a Web form (Figure 6.27).

The second step of the writing process is also very similar to the mechanism from our CMS example. The user is provided with an option to continue adding documentation modules or to finalize the master documentation database by writing it to an XML file (Figure 6.28).

After the user has returned to Step 1 and proceeded to add two more modules (intermediate and advanced) for our “Introduction” feature, she is rewarded with this XML file being generated by our custom single sourcing application:

```
<?xml version="1.0" encoding="utf-8"?>
<!— XML Single Sourcing Documentation File —>
```

```
<!— Automatically generated by Single Sourcing Demo —>
<!— Created on Aug192008 —>
<?xml-stylesheet type="text/xsl" href="documentation_
transform.xsl"?>
<!DOCTYPE documentation_modules [
    <!ELEMENT documentation_modules (module)*>
    <!ELEMENT module (id,feature,skill_level, documentation)>
    <!ELEMENT id (#PCDATA)>
    <!ELEMENT feature (#PCDATA)>
    <!ELEMENT skill_level (#PCDATA)>
    <!ELEMENT documentation (#PCDATA)>
]>
<documentation_modules>
<module>
<id>introduction_beginner</id>
<feature>introduction</feature>
<skill_level>beginner</skill_level>
<documentation>Asset Management System 1.0 is a tool for
organizing and labeling collections of organizational
documents.  You can use this system to add asset labels for
categories and media and to assign a list of years during which
that asset may have been used.  In addition, you can provide a
detailed description of the asset and why it may have long term
importance for your organization.  To get started with AMS 1.0,
click the \"Tutorial\" link from this help file.
</documentation>
</module>
<module>
<id>introduction_intermediate</id>
<feature>introduction</feature>
<skill_level>intermediate</skill_level>
<documentation>This program supports many features such as
user-configurable fields for categories, year ranges, and media
types.  In addition, the program includes options for clearing
all fields or for restoring defaults using the (Tools, Options,
Reset) menus from the Tools menubar.  Messages concerning
program operations are displayed in the Status: field.
This field is directly above the Asset Name dropdown menu in the
AMS interface. </documentation>
</module>
<module>
<id>introduction_advanced</id>
<feature>introduction</feature>
<skill_level>advanced</skill_level>
<documentation>Advanced features include the ability to use
keyboard shortcuts and to import and export settings using XML
```

```
files.  Keyboard shortcuts for common commands are found in the
Keyboard Shortcuts section of this help system. Program options
default to verbose mode for extra warnings and customized
messaging from the AMS system.  To disable verbose mode, see
Disabling Verbose Mode in the list of help
topics.</documentation>
</module>
</documentation_modules>
```

We now have three different documentation modules that we can use for our AMS software. According to the rubric developed during our rhetorical analysis, we will assume the beginners will need the beginner text only. We can then repurpose the beginner's text module by combining it with the intermediate and advanced users' modules and repurpose the intermediate user's module by combining it with the advanced user's module. The final outcome in terms of document delivery will then look like the coding shown on the next page.

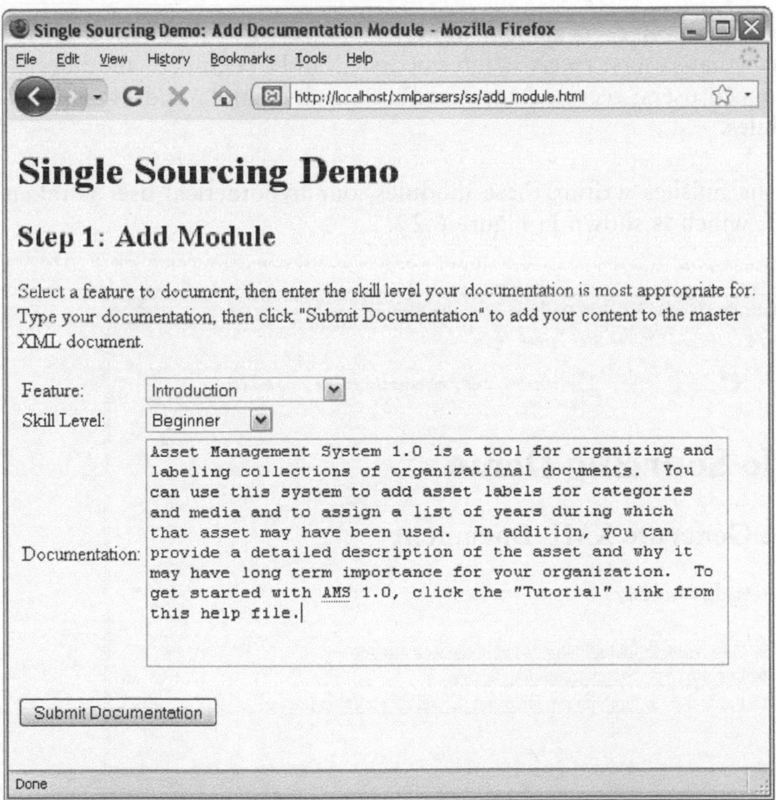

Figure 6.27 Single Sourcing Step 1

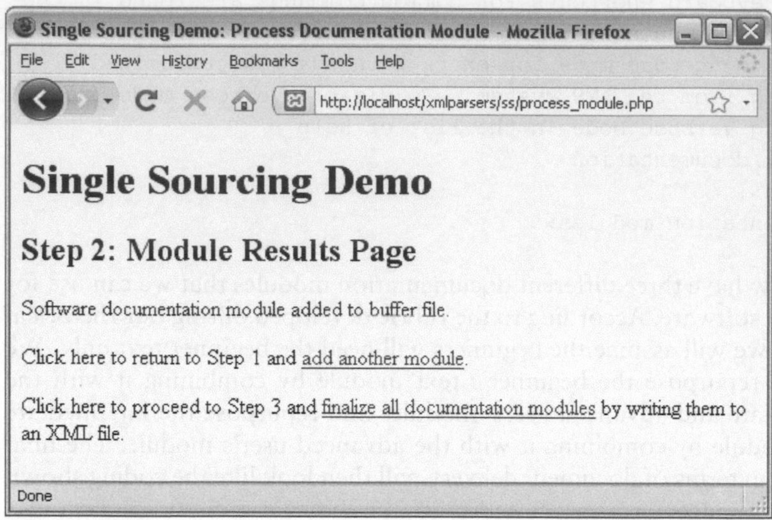

Figure 6.28 Single Sourcing Step 2

- Beginners: receive "beginner" module;
- Intermediate users: receive "intermediate" and "beginner" modules;
- Advanced users: receive "advanced", "intermediate", and "beginner" modules.

After she finishes writing these modules, our hypothetical user is taken to Step 3, which is shown in Figure 6.29.

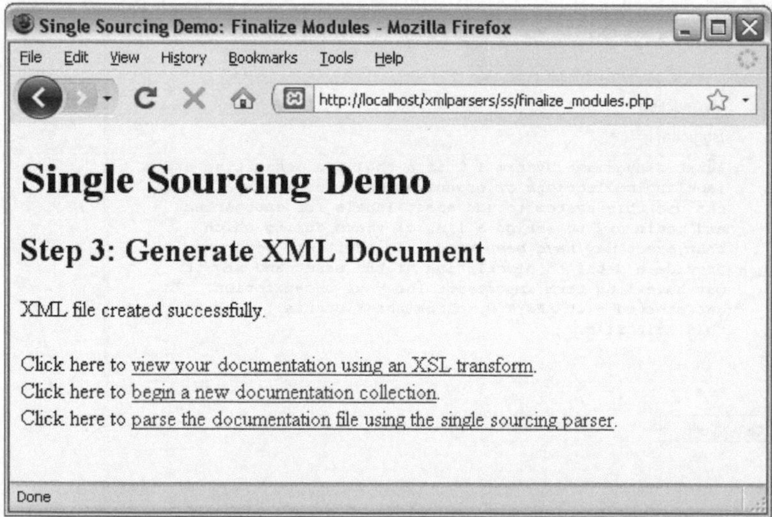

Figure 6.29 Single Sourcing Step 3

At this point, the author has the choice of viewing the data through an XSL transformation, starting a new documentation collection from scratch, or parsing the documentation file using a customized single sourcing parser. The parser for this application is found in the "parse_xml.php" file in Appendix D. Note how the script moves the user through the authoring process in an iterative and recursive fashion, allowing her to make selections and then return back to the same script any number of times until the documentation is complete. It then performs some simple XML parsing using the XML DOM model by looking at the data stored within each <skill_level> element and determining if it matches the current informational needs at the skill level self-selected by the user. The PHP code to access a particular DOM node looks like this:

```
/* First, get any skill level nodes (there is actually only
one) associated with each module */
$skillLevels = $module->getElementsByTagName("skill_level");
/* Now, get the value associated with that node.  This will be
"beginner", "intermediate", or "advanced." */
$skillLevel = $skillLevels->item(0)->nodeValue;
```

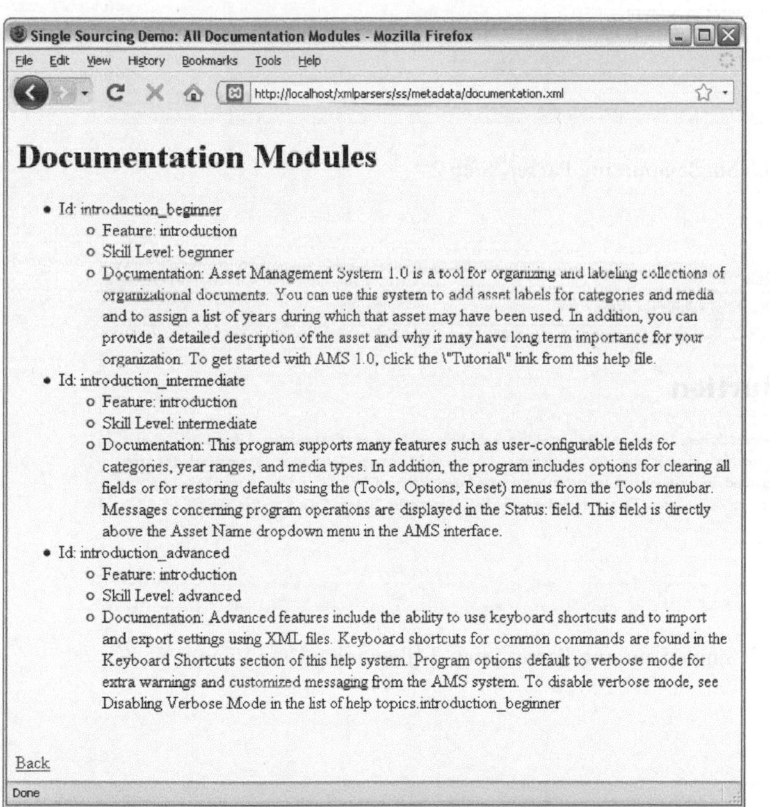

Figure 6.30 Single Sourcing After XSL-T

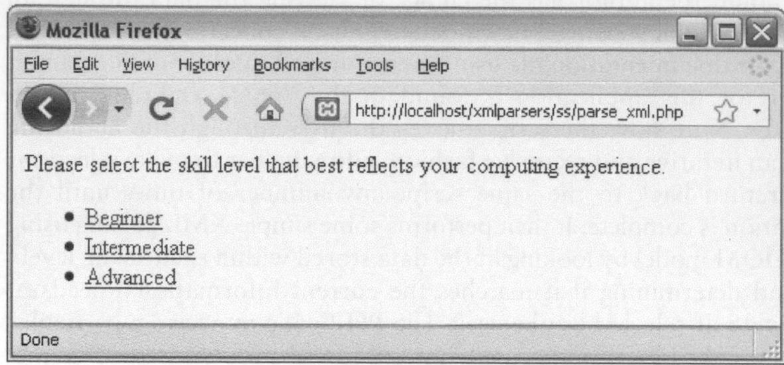

Figure 6.31 Single Sourcing Parser, Step 1

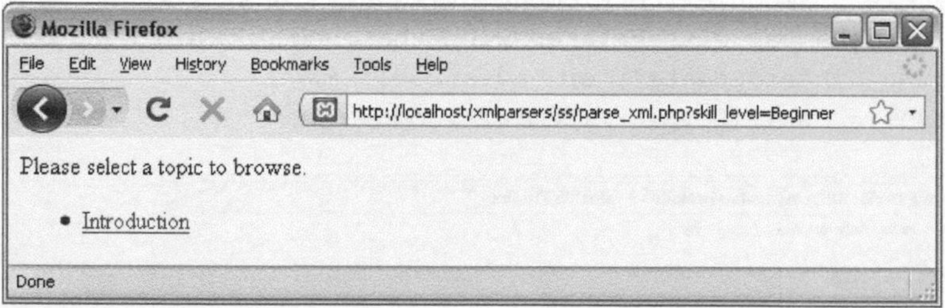

Figure 6.32 Single Sourcing Parser, Step 2

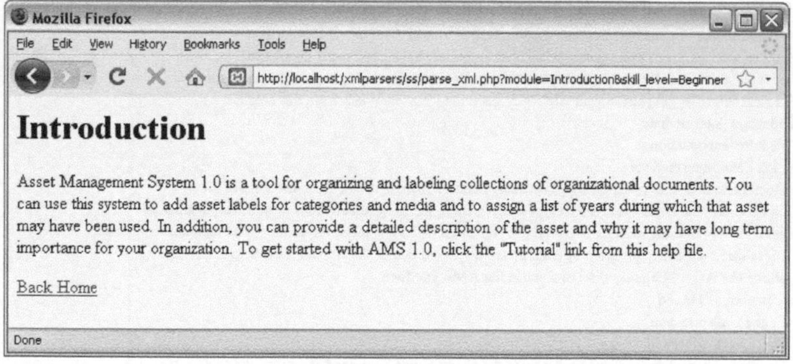

Figure 6.33 Single Sourcing Parser, Step 3 (Beginning User View)

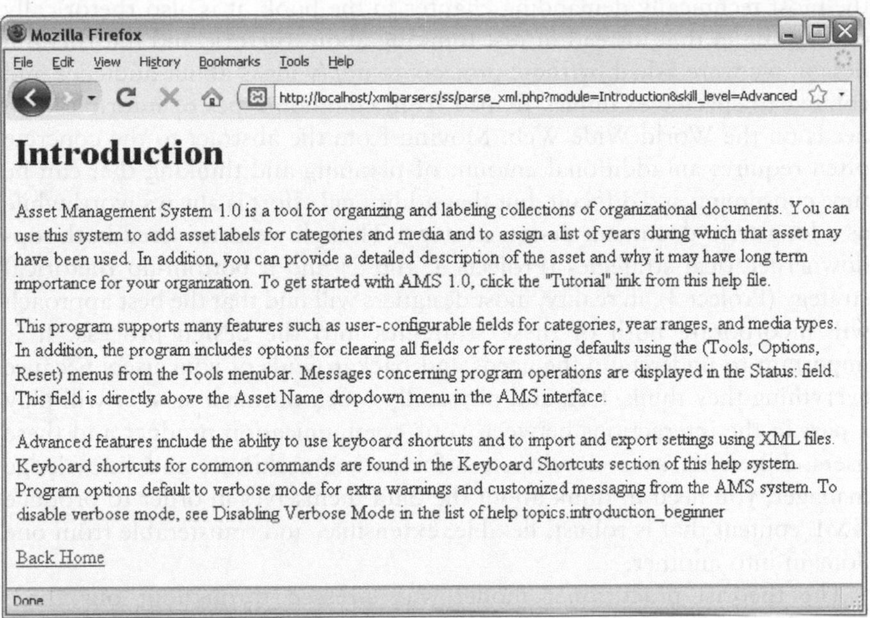

Figure 6.34 Single Sourcing Parser, Step 3 (Advanced View)

Final output from this system can be seen using the XSL transformation (Figure 6.30) which shows *all* documentation modules, or by using the customized single sourcing parser, which shows a beginner's navigational sequence in Figures 6.31–6.33. Note how the informational needs of a beginning user (Figure 6.33) are different from those of an advanced user (Figure 6.34) and how the system attempts to anticipate and meet the needs of both users according to our bottom-up rhetorical analysis.

Although this is a very simple example of a single sourcing system, it does demonstrate the fundamental idea behind repurposing textual nodes. It also shows how an XML parser can manipulate multiple content modules that are written once and then combined using different permutations in order to support different user views or perspectives. Studying the code in Appendix D will give you a better idea of the basic mechanics behind a primitive level three single sourcing system.

Chapter Summary

This chapter introduced the PHP scripting language and demonstrated how PHP and its associated XML libraries can be leveraged in order to build custom XML parsers. By building these custom parsers, we were able to put to use both the ideas about knowledge management and rhetoric we have been discussing in the first half of the book as well as some of the technical skills we discussed in the second half. Although this is probably the most technically demanding chapter in the book, it is also rhetorically demanding in that instead of just thinking about rhetoric and information design, we were asked by these projects to *apply* ideas about audience and context to specific examples shaped from common types of informational needs on the World Wide Web. Moving from the abstract to the concrete often requires an additional amount of planning and thinking that can be time consuming and difficult, but that additional effort is always worthwhile as a long term document design strategy. Though we considered both top-down rhetorical strategies (Projects 1 and 2) and a bottom-up rhetorical strategy (Project 3), in reality, most designers will find that the best approach will incorporate both of these strategies into the design process. It is important to understand the needs and backgrounds of your users because everything they think, feel, and physically touch during a session will play a part in the interactions between your communication product and these users. Likewise, as a developer, information architect, and knowledge manager, you need to think about the data themselves in order to produce XML content that is robust, flexible, extensible, and transferable from one domain into another.

The theorist–practitioner model was stressed throughout our three examples because it is so important to the professional communicator working with XML technologies. On the theoretical side, one must recognize that the humanistic elements of information management and design are often overlooked for the sake of technical efficiency or simplicity. We need to be aware of this and be ready to consider the persuasive characteristics of information design just as we consider the technological issues. Humans are emotional, fallible, and oftentimes unpredictable, and our software programs need to be designed to take these factors into account.

On the practitioner side, we need to recognize that by relying on pre-existing parsers, our creative potential and expressive capacities are limited by the designs of other companies or other individuals. Only by immersing ourselves in the low-level programming of XML parsers can we truly design an interactive system for dealing with XML code in exactly the way we want. Existing parsers work wonderfully for validating code and communicating standardized information over networks, but to truly explore the capacity of XML as an expressive language will inevitably require a greater degree of control and flexibility that general parsers can provide. Similarly, the process of building customized CMSs or single sourcing systems that flex and bend with the organic movements of a distributed organization will necessitate some amount of technical creativity on the part of the parser designers.

Although this chapter is heavy in programming code and syntactical complexity, we want to conclude by reminding our readers that the XML parser design process is not just about programming or coding scripts. The skilled professionals who build such systems must be aware of these technical details, but as we have stressed in this chapter, they also must pay attention to the larger rhetorical context of the information transfer process. For instance, what groupings will emerge in terms of audiences, skills, and information-seeking patterns? How will these groups differ on both sides of the parsing process, where one side is concerned with adding data to the repository and the other side is focused on retrieving useful information from this data store? Is a certain type of information being privileged at the expense of other types, and, if so, what are the implications of such a strategy? By considering these elements as well as building skills in information architecture and scripting, one can supplement the technical with the humanistic, the logical with the artistic. The document designer or information architect with skills in audience analysis and other forms of rhetorical acumen will inevitably find themselves in greater demand as new Internet technologies increasingly push us closer to Berners-Lee's vision of a distributed, yet integrated, Semantic Web.

In Chapter 7, our final chapter, we will explore some case studies focused on XML in the real world and discuss how XML plays a role in various types of careers. We will see how researchers and industry experts use XML to structure, manage, and communicate data that has meaning within specific professional contexts.

Discussion Questions

1. This discussion question contains two parts:
   a. In a small group, discuss the specific information that is gathered through the RAX instrument. What rhetorical considerations are well-represented, and which areas are lacking?
   b. Assume you are part of a team that is called upon to build an XML database for an environmental agency. The database will be used to keep track of particularly heinous industrial business practices that may have severe environmental consequences. This XML file needs to contain internal information useful to the environmentalists as well as information that will be distributed to the public by way of the agency's website. Discuss how you will need to modify the RAX survey to better suit the needs of this particular project.
2. Compare Projects 1 and 2 from this chapter. How do CMSs and single sourcing systems relate to one another? What are the similarities and differences? What other examples of CMSs do you use in your own information-seeking habits?
3. Under what conditions might it be better to present more information to beginning users rather than less in a single sourcing system? In Project 3, why did we choose to reverse this convention and present more information to users that had a higher level of skill?
4. Revisit Chapter 3 and read about Rockley's four levels of single sourcing. Our example in this chapter was a level three single sourcing system. What changes would we need to make to this application to move it

towards a level four system, or an EPSS? Discuss the advantages and disadvantages of such a system over a dynamic customized content (level three) implementation.

Activities

1. Download the sample XML file and PHP files for Project 1 from the book's website. Using XAMPP, test the news syndication system and see if you can get it working. Experiment with adding additional news stories to the XML file and try adding the file as a live bookmark in Mozilla Firefox®. What happens when you remove or rename the elements in the newsfeed XML file? Experiment with the XML file and try adding some additional elements and attributes from the RSS specification page at www.rssboard.org/rss-specification.
2. Design an HTML form with two text fields and a submit button. Send the form data to a PHP script named "process.php". In the "process.php" file, write some code to display the variables back to the user in different formats (heading1, from within a table, using italics, and so on). Experiment with changing the form method from GET to POST and see what changes occur in your script. Once you have mastered this, try to use SimpleXML to add the information from your PHP form to an XML file (hint: check the appendices for some sample scripts that deal with writing to text files).
3. Download the CMS Project 2 example files from this text's website. Research "include files" in the PHP documentation and modify this project to use an include file to display the redundant information that appears in the various project files. For example, you might choose to include the HTML tags that begin a new document in a "header.inc" file and the HTML tags that end a new document in a "footer.inc" file. Consider how the process of using include files forces you to think about information on a larger scale rather than focusing on the "document level" of granularity. How does the use of include files help to create a more efficient CMS?
4. Download the AMS 1.0 software from our website. Experiment with writing documentation modules for some of the features that were not discussed in this chapter. What design considerations are you asked to make when writing content that you know will be reused in multiple contexts for multiple users?
5. Using the documentation produced in the previous question, incorporate this new module into the single sourcing demo project. Discuss how the single sourcing demo project could be improved based on the ideas of Rockley, Carter, and Williams that were discussed in this chapter.

Sample Group Project: Hi-Tek Inc.

Assume you are working for Hi-Tek Inc., a large international organization that manufactures cellular phone accessories. Hi-Tek's primary manufacturing plant is located in Chicago, but your group has recently been asked to create a substantial amount of written material explaining Hi-Tek's plan for a partial relocation overseas. The purpose of this material will be to explain how this relocation will affect Hi-Tek employees in participating (relocating) divisions.

In order to produce versions of this material rapidly in several different formats, your group will need to use a single sourcing methodology. As a class, determine which information needs to be included in order to meet the needs of this audience of Hi-Tek employees. Next, define some standard XML elements to use for each type of information as it is stored in its central location. For instance, you might choose to use an <international_employee> element to wrap around each affected employee and a <domestic_employee> element to indicate non-affected personnel.

In smaller groups of 4–5 students, each group then needs to consider a specific type of delivery document from the following list:

a. a printed memorandum,
b. an e-mail,
c. a page for Hi-Tek's standard operating procedures,
d. a Web page,
e. a brochure page, or
f. a prerecorded telephone message.

As you think about your selected format, consider the following questions:

- Does this format require any additional XML tags to be stored in the central information repository? If so, what are they?
- How would a DTD or schema be helpful in the collaborative XML creation process?
- How can a database table be used to single source your information? Could a database be used instead of XML? In conjunction with XML?
- What XSL transformations would be useful for these various informational needs?

After initial procedures have been created, discuss as a class the best way to facilitate the transfer of XML data from a single, central repository to each group's selected format.

References

Albers, Michael J. "Introduction." *Content & Complexity: Information Design in Technical Communication*. Michael J. Albers and Beth Mazur, eds. Mahwah, NJ: Lawrence Erlbaum Associates, 2003. 1–13.

—— "Complex Problem Solving and Content Analysis." *Content & Complexity: Information Design in Technical Communication.* Michael J. Albers and Beth Mazur, eds. Mahwah, NJ: Lawrence Erlbaum Associates, 2003. 263–83.

Burke, Kenneth. *A Grammar of Motives.* Berkeley: University of California Press, 1969.

Carliner, Saul. "Physical, Cognitive, and Affective: A Three-Part Framework for Information Design." *Content & Complexity: Information Design in Technical Communication.* Michael J. Albers and Beth Mazur, eds. Mahwah, NJ: Lawrence Erlbaum Associates, 2003. 39–58.

Carter, Locke. "The Implications of Single Sourcing for Writers and Writing." *Technical Communication* 50.3 (2003): 317–20.

Krug, Steve. *Don't Make Me Think! A Common Sense Approach to Web Usability.* 2nd ed. Berkeley, CA: New Riders, 2006.

Mauthe, Andreas, and Peter Thomas. *Professional Content Management Systems.* Chichester, West Sussex: Wiley, 2004.

Morville, Peter, and Louis Rosenfeld. *Information Architecture for the World Wide Web.* Sebastopol, CA: O'Reilly, 2007.

Norman, Donald A. *Emotional Design: Why We Love (or Hate) Everyday Things.* New York, NY: Basic Books, 2005.

PHP.net. "Php: Xml Parser Functions." 2007. November 11, 2007. <http://us2.php.net/xml>.

Rockley, Ann. "Single Sourcing: It's About People, Not Just Technology." *Technical Communication* 50.3 (2003): 350–4.

Williams, Joe D. "The Implications of Single Sourcing for Technical Communicators." *Technical Communication* 50.3 (2003): 321–7.

Additional Online Resources

1. EditPlus Download: www.editplus.com
2. Gliffy.com: Online Diagramming Software: www.gliffy.com
3. PHP.net History: http://us.php.net/history
4. PHP.net Introductory Tutorial: http://us2.php.net/tut.php
5. PHP.net SimpleXML Functions: http://us2.php.net/simple_xml
6. PHP.net website: www.php.net
7. RSS 2.0 Specification: www.rssboard.org/rss-specification
8. XAMPP Download: www.apachefriends.org/en/xampp.html
9. XML in PHP5: What's New? http://devzone.zend.com/node/view/id/1713#Heading13

7 XML and Your Career

XML and Knowledge Management at Work in Interdisciplinary Contexts

Chapter Overview

The first section of this final chapter serves as a review and an overall summary of the book's first six chapters. In this section, we review the concepts from our earlier chapters and then discuss how these ideas translate to skills that are used in various careers in industry and research. Each chapter is briefly summarized and its key ideas are revisited here.

In the second section of this chapter, we suggest that XML is an important technology for many types of professionals—but especially document designers and professional communicators working in digital environments—to be familiar with in the workplace. In particular, we look at careers in technical communication, technical editing, digital media, library science, and interdisciplinary research. Professionals working in these fields are likely to see a continued growth in the use of XML and its associated technologies for solving problems in these types of domains.

The last portion of this chapter includes the results of an interview session we conducted with five professional individuals who use specific XML documents for various tasks in research and industry. As we discuss these interviews, we weave in ideas from the first six chapters of the book in order to discuss the unique ways that XML can be used to frame and represent data about the world. We hope that this final chapter helps readers to synthesize some of the theoretical content and apply the practical ideas of this book to see how XML solves real world, practical types of information design problems. To aid in this transition from theory to practice, we will summarize the first six chapters before moving on to our interview and discussion.

Chapter Reviews

In Chapter 1, we discussed several reasons why it is important for technical communicators to be familiar with XML. In particular, we noted the ideas of Johnson-Eilola, Michael Hughes, Corey Wick, and others who suggest that expertise in knowledge management technologies empowers these professionals to exercise a greater role in their industries and advance both themselves and the profession of technical communication. The ability of communication professionals to add, delete, change, and select meaning from units of information, as Slack, Miller, and Doak suggest, is an integral part of what employees in a knowledge economy are asked to do.

Chapter 1 also defined additional ways in which what we might call metaknowledge management (knowledge about knowledge management) and XML acumen can be critical skills in a variety of different fields and for a variety of different rhetorical purposes. We discussed the ways in which disciplines frame knowledge and the importance of histories, contexts, and social relationships in representing knowledge and facilitating information exchange. In studying examples such as the original Eureka project and Eureka II that were used by the Xerox corporation, we observed the combination of applied techniques from IT and theoretical concepts from social constructionism. These examples were provided to show how seemingly incompatible ideas from theory and practice can coexist to create a usable knowledge management system. We also discussed how a bottom-up approach that focused on large amounts of warehoused data turned out to be more useful to these technicians than even the existing documentation compiled by the corporation.

Chapter 2 introduced XML and explained how this markup language is different from its more ubiquitous cousin, HTML. We discussed how XML gives knowledge workers power by allowing them to provide semantic and ontological metadata that describes what the data *is*, rather than simply how it should appear on the computer screen. This distinction is critical and is reinforced by the definition of metadata itself: data that is about, or describes, other data. The idea of XML as an object-oriented language was touched upon and proved to be an important concept later in the book when we discussed the implications of information modularity, encapsulation, and reuse.

The remainder of Chapter 2 focused on the building blocks of XML: declarations, elements, attributes, entities, and DTDs. It included a discussion of the proper syntactical techniques for using these components in the document design process. We explained that well-formed XML is XML that is syntactically correct and meets the specifications designed by the World Wide Web consortium, and that valid XML is XML that meets the requirements for positioning, sequencing, and naming elements and attributes that are specified by an internal or external DTD. For beginners without much experience in coding, or for those readers with Web design experience in HTML but not in XML, this chapter explained how to construct XML documents using these rules of well-formedness and validity. Several examples were then provided which showed XML at work in real documents such as the Final Report of the ACHRE. By breaking this document down into modules, paragraphs, lists, and titles, we saw the implications of using XML to modularize content so that it can be packaged and repurposed for modern electronic communication systems.

In Chapter 3, we wove in additional rhetorical threads as we discussed the Semantic Web and how XML technology can be used to create more intelligent networks of humans and computers. Classical classification schemes from Aristotle and Cicero were used to show how the same types of strategies at play in dividing and describing XML information have long been used in other contexts because human cultures have always been interested in describing the ways in which things are similar and dissimilar to one another. We also employed the work of scholars such as Berners-Lee, Foucault, Kuhn, Bowker, and Star to illustrate the rhetorical implications of naming and classifying objects. Following this, we demonstrated how XML is a technology that draws on these rhetorical phenomena.

We also introduced prototype theory in Chapter 3 to explain how individuals use cognitive shortcuts to help store new experiences. These shortcuts facilitate the comparison of old experiences to new situations based on similarities and common patterns. Cognitive psychologists have studied this process carefully and use terms like scripts, frames, and schemas to further explore this type of experiential encoding. We applied some of these ideas from this chapter later in Chapter 6 when we designed our custom parsers. Finally, in Chapter 3 we spent some time talking about single sourcing systems and the importance of this methodology to certain types of distributed writing situations. Many of these ideas about single sourcing also carried over to Chapter 6 when we moved into additional applied examples of parser design.

In the beginning of Chapter 4, we discussed the rhetorical implications of separating content from its form and discussed why this is such an important concept for the XML document designer to understand. We associated the presentation of a document with its style and explained how traditional rhetors often separated rhetoric from philosophy by looking at the former as "elegant" speaking and the latter as "wise" speaking (Whitburn 45). With XML, we must be careful to remember that the format itself is just a set of guidelines for building structured data sources. The ways in which we format and display that content for an audience to read has significant implications for the credibility of the message itself, as well as for the credibility or ethos of any associated authors. No matter how wise our message might be, without an elegant presentation, it often loses its impact and effectiveness.

Chapter 4 then picked up where Chapter 2 left off and continued a discussion of some of the technical facets of XML. We learned ways to format and display XML that allow us to take advantage of the rhetorical power of carefully designed visualization techniques and aesthetic designs. In particular, we encountered two different techniques that we can use to format and display XML content: CSS and the XSL. We walked through several examples that were formatted using these techniques and the chapter concluded with an introduction to style sheet transformations, which enable document designers to transform an XML document into other types of presentation formats. These formats include other types of XML documents, HTML or XHTML documents, PDF documents, SVG documents, VRML documents, and many others.

Some advanced topics related to XML were discussed in Chapter 5. We returned to the importance of namespaces, which we briefly introduced in Chapter 2, by explaining how these encapsulating units are designed to deal with problems of collision and recognition. These problems occur when multiple data sources or elements are combined into single documents, leading to the possibility of ambiguity or redundancy with named entities. Schemas were introduced as an alternate means for validating XML documents using a syntax that is itself also written in XML.

The XPath and XLink languages were also covered in Chapter 5. These technologies are used to link to or access data embedded within XML documents using a standard notation. Given the importance of XML as a core technology in what Berners-Lee has referred to as the Semantic Web, these languages will continue to grow in popularity as they enable different databases to connect over networks and negotiate meaning at precise node locations using virtual roadmaps. For instance, an XML parser charged with compiling two XML libraries into a single collection might be faced with a recognition problem if one document used the element name ZIP_CODE and another used the element name POSTAL_CODE. To resolve this issue, a simple XPath expression can be used by the parser to identify the offending elements from deep within the documents' hierarchies, then XLink can be used to refer the reader to a link or style guide with preferred nomenclature, which might be listed in the namespace definition of the original XML documents. Additionally, XLink can direct the parser to a customized URL with a script designed to handle these types of collision issues automatically.

In the previous chapter, Chapter 6, we assembled our knowledge of rhetoric, XML, and knowledge management and set to work building some applied, real world examples of parser systems. We wrote about how parsers are often not given the attention they deserve because they are seen as unsexy or irrelevant. Parsers are in fact very important to the transactional and distributional acts of communication that occur using XML data, but they are so wide ranging and diverse that they are hard to characterize using a simple set of parameters. To demonstrate the wide variety of parsers that can be used with XML and explore the rhetorical implications of parser design, we designed three different types of Web-based parsers and used three different rhetorical strategies to guide their design.

For our first parser project, we used personas to develop an information context in which an individual was tasked with developing a system to handle streaming news information using RSS feeds. Using an ad hoc rhetorical analysis and the SimpleXML extension for PHP, we constructed a parser and an XML writer program without using a great deal of computational overhead. The second and third examples, which were parsers designed to function as CMSs, were slightly more sophisticated. We used a top-down design process following Saul Carliner's physical, cognitive, and affective design dimensions for the second project. Then, we followed a bottom-up approach for the third project by conceptualizing the data nodes as a heuristic for guiding the interface design and gradually adding layers of complexity to our parser. The general CMS required stepping up to the SAX parser and designing a more robust writer module, while the third project took advantage of the full-featured DOM extension for PHP.

XML and Your Career

Throughout this book, we have discussed examples of XML and rhetoric taken from a variety of disciplines, from technical communication, to library science, to digital media and IT, to the biological and health sciences. Here, we are more explicit as we elaborate on just a few of the many professions in which we believe XML is likely to play an ever important role:

1. The *technical communicator*. While not all technical communicators will find themselves working directly with XML, the newest software products used in the field of technical communication, from Adobe FrameMaker® to the latest Microsoft® Office suite, are increasingly using XML to structure internal file formats and facilitate network exchange using the Internet. In addition, distributed teams of writers continue to use technologies such as CMSs and single sourcing methodologies to overcome the rhetorical and technical challenges imposed by the absence of a common location. We saw in Chapter 6 that these technologies can be driven by XML data sources.
2. The *technical editor*. Although technical editing is obviously a subset of technical communication, the technical editor is one professional in particular who may find herself using XML to a greater extent. As Michael J. Albers notes, technical editors routinely find themselves editing content that has been contributed by multiple authors and written from multiple locations (191). Editing these documents requires particular skill in understanding how to create coherent and consistent documents from these multiple source materials (192). It is therefore important for technical editors to understand XML for two reasons. First, it is likely that the materials being assembled from multiple authors may be encoded in XML format, and an editor will need to know how to enforce validity within these XML files in order to ensure consistency and coherence from these authors. Second, many of the skills that are associated with XML content creation, such as knowledge representation/encoding and the ability to define an appropriate level of granularity for chunked information, are also important for the technical editor to possess. These skills are useful for assembling a single document from multiple sources.

3. The *digital media practitioner*. Digital media professionals will continue to find themselves on the cutting edge of technology, utilizing the latest software to produce media assets for industries as wide ranging as film and animation, Web development, sound and audio recording, and modeling and simulation. As several of our interviewees note later in this chapter, XML has some definitive advantages as a data source for these types of projects over other technologies such as relational database management systems. From working with Web 2.0 projects using AJAX methodologies to building new innovative frameworks for managing filming locations in preproduction film tasks, digital media practitioners will continue to have plenty of opportunities to use XML and associated XML technologies.
4. The *library scientist*. Even before the Dewey Decimal System was invented by Melvil Dewey in the late nineteenth century, library scientists were pioneering innovative new methods for classifying data and creating taxonomical structures of information. Better methods for storage and retrieval are important issues for librarians, particularly as the amount of electronic information stored in digital files continues to grow exponentially. While XML is not the only metadata system used to classify and annotate information, it is a widely used and popular format. Experience with XML is an important skill for these professionals to have or to work towards obtaining.
5. The *interdisciplinary professional* or *researcher*. Because of its focus on standardization and its ability to easily carry from one computing system to another, interdisciplinary researchers are likely to continue using XML as a lightweight and portable scripting solution to annotate or standardize projects and papers.

Interview Overview

Over a two month period, we collected interviews from five professionals working in various fields. We interviewed a college professor, two technical communicators, a software engineer, and one individual who is both a professor and a software developer for a large video game development company. Each of these individuals varies in their background education, familiarity with XML, and occupation, but all have used XML to some degree in various types of projects. We specifically sought a diverse pool of experts to form our interview panel in order to better reflect the wide variety of uses XML has in research and industry.

The ten interview questions asked respondents to reply with information about ways in which they have personally used XML to solve problems for particular types of tasks. In addition to standard questions about their backgrounds and job descriptions, we asked these experts to respond to specific technological questions about validation, transformation, and parsing. We also asked them to tell us why they chose XML to solve these problems. Though the interviews were conducted individually, we reprint the responses to each question here one after another in order to show the differences between these experts and their informational needs and to show how they each chose to apply XML technologies in unique ways. After the responses, we include a brief commentary and holistic summary of the interview process.

Participant Biographies

Our first interview participant, Bill Albing (BA in our interview), works as a knowledge developer for FarPoint Technologies in Morrisville, North Carolina. He is co-founder and editor-in-chief of KeyContent.org, which is an online portal with articles about XML, single sourcing, and other innovative technologies used in the field of technical communication. Bill has over fifteen years' experience in engineering and technical writing and enjoys applying his expertise to the design of wikis, XML, and other aspects of Web 2.0 to his work at FarPoint. He has presented on the topic of XML numerous times at a number of venues.

Our second participant, Sherry Steward (SS), has been working with markup languages since the early days of SGML. Sherry is currently Director of Applied Research and Life Cycle Support for a simulation and training company in Orlando, Florida. Sherry is tasked with managing specialty engineering disciplines, integrated logistics support services, technical documentation development, and simulation and training projects for military acquisitions. Her specialty is Interactive Electronic Technical Manuals, legacy data conversion, and intelligent technical documentation. She has a Ph.D. in Texts and Technology from the University of Central Florida.

Our third participant, Professor J. Michael Moshell (JMM), has a background in digital media and computer science, which he taught from 1975 through 2000. He served as the founding Program Director for Digital Media at the University of Central Florida in 2002, and as Division Head from 2002 to 2005. Michael's interests converge on the use of graphics, games and the Internet for the creation of situations where students learn from one another. Michael received his Ph.D. in Computer Science from Ohio State University in 1975.

Before joining the Florida Interactive Entertainment Academy (FIEA) as a faculty member, our fourth interviewee, Michael Gourlay (MG), worked as a Senior Software Engineer at Electronic Arts (EA) as the lead graphics programmer on the popular football video game Madden '06. He was also lead programmer on the NASCAR game series for several years and received a patent for algorithms he developed for interactive, high-bandwidth online applications. Prior to joining EA, he performed scientific research primarily using computational fluid dynamics and the world's largest massively parallel supercomputers. His previous research includes work with nonlinear dynamics in quantum mechanical systems and atomic, molecular, and optical physics. Michael received his degrees in physics and philosophy from Georgia Tech and the University of Colorado at Boulder.

Our final interviewee, Thomas Gorence (TG), works for Integrity Arts & Technology/i.d.e.a.s., an innovative spin-off company that was originally owned by Disney-MGM Studios. I.d.e.a.s. stands for imagery, design, editorial, art, and sound, and the company employs an interdisciplinary blend of musicians, writers, filmmakers, programmers, designers, editors, teachers, engineers, and artists (Integrity Arts & Technology, Inc. online). After working for the United States Air Force as a computer programmer, Thomas attended Full Sail for digital media design. Thomas has been programming computers since the age of thirteen, when he taught himself the language QBASIC, and has also worked at a recording studio as an audio engineer and producer.

Interview Questions

Question 1: Briefly describe your job, your day to day duties, and your role within your organization.

BA: At FarPoint, I am responsible for the product documentation of a range of software components that we sell to application developers, mostly who use Microsoft Visual Studio® as a development environment; this means I'm responsible for production and maintenance of the deliverables as well as the development of the original content. FarPoint is a small software components manufacturer and our best selling product is a .NET spreadsheet component that gives software applications spreadsheet and grid capability.

For about a half dozen products, much of the product documentation, including an API reference (or class library documentation) is automatically generated from source code comments (which is XML tagged) and the actual structure of the software. There are also sets of procedural (tasks) documentation and tutorials that I develop to explain the range of uses of the product. Beyond the strictly technical documentation, I also develop case studies, technical white papers, and a range of marketing literature from data sheets to e-mail newsletters to blog entries. For each product there are two forms of compiled help, one of which integrates with the Microsoft Visual Studio® development environment. We also produce a set of PDF files (for users to print, since we no longer provide hard copy user manuals), a Read Me, and Web-viewable versions of all this content. In my spare time, I am editor-in-chief of a groupware-based website that is the basis for a professional association, KeyContent.org.

SS: I am the director of applied research and life cycle support for a simulator company where I am responsible for the overall supportability of training devices and simulators designed and manufactured by DEI. Several products are necessary to support and operate these simulators throughout the product life cycle, and as such, I direct the functional areas responsible for the reliability, maintainability, and availability of electronic components. I also handle logistics support analysis and provisioning, system safety, technical documentation, and training.

The technical documentation we develop is used by our customers to operate and maintain the training devices and to buy spare parts as needed. The operation and maintenance data is delivered in a traditional technical manual print format; however, it is also delivered as an Electronic Technical Manual or an Interactive Electronic Technical Manual. When we deliver this format, manual data is tagged in XML or SGML so that it can be manipulated and used in mobile environments. I also direct the applied research efforts, most of which are focused on supportability process improvement or expanding simulator capability.

JMM: I work as a Professor of Digital Media. In this capacity I design and teach courses, conduct research projects, write grant proposals, develop new curricula, and serve on University committees.

MG: I have two jobs: I am a research associate at the University of Central Florida in the graduate program FIEA. Here, I create curriculum for the program and teach graduate students how to make interactive media (such as video games and military simulations). I also do research. My other job is as a Senior Software Engineer for EA where I write software for video games including Madden NFL, NASCAR, and several other games. Among many other things, I wrote the visual effects system used by many EA Sports titles including Madden, NCAA, NASCAR, NFL Tour, Fight Night Big (a.k.a. Pummel), Euro Cup, and FIFA Street. I also wrote the network application layer architecture used by Madden and NASCAR.

TG: I am the lead developer and programmer in the graphics department for i.d.e.a.s./Integrity Arts. I am involved with creating technical specifications, project maintenance, documentation, testing, and deployment. My day to day duties usually involve Adobe Flash® development (ActionScript programming), PHP/MySQL development, and graphic design.

Question 2: Please describe a project in which you have used XML.

- ***What was the purpose of this project, and who was the audience? Were there multiple authors?***
- ***What was your design process like?***
- ***Did you use any brainstorming tools or audience analysis techniques to help structure your XML hierarchy or did you "code from scratch"?***

- ***Did the structure of your XML document (e.g., its elements and attributes and the relationships between them) remain the same throughout the duration of the project, or were revisions necessary?***

BA: Our product documentation and the automation of its generation is based on the use of XML. From the source code comments in C# (Authors' Note: C# is the "C Sharp" programming language) code files, which are done in XML tags, to the import of snippet example code, which is kept in XML tagged text files, to the transformation of FrameMaker® authored content from XML to HTML, XML is key to the product documentation process. Much of the XML process is standardized and defined. The source code comments are Microsoft® conventions. Some of the XML was developed by me and somewhat resembles the organization of types of documentation that is now seen in DITA with Tasks, Reference, and Topic types. We regularly extend the XML format with an eye towards having the new parser continue to accept older files. Most of the time, we add to the existing XML format and rarely change something so drastically that old files fail to work with the new parser.

SS: We use XML for the development of an electronic tech manual using Microsoft Internet Explorer® as the browser. Our process begins with requirements analysis and planning, design of the XML structure and interactive features, XML tagging and DTD development, test, and implementation. We generally have a technical writer who collects and analyzes the data for the technical manual, but we may have another person set up the XML structure and actually tag the content. Most of the time, this process is driven by schedule, cost, and complexity. Our design and audience approach is based mostly on military specifications and the user environment. Sometimes, we do make adjustments to the structure after the document is tested by the user, but these adjustments are rarely anything substantial.

JMM: The Cast Member Performance Management (CMPM) Project involves developing a theory and practice for managing the behavior of actors in learning-oriented multi-player online role playing games (MORPGs). A central tool of CMPM is the Cast Performance Management System (CPMS), a distributed Java and PHP application that coordinates the activities of online actors. The purpose is to provide new ways of using computer game technology for learning in K-12 and university education. Its ultimate audience will be the students in schools and universities around the world, we hope. My graduate student and I are authoring these materials, with help from "seasonal labor" (semester-long student projects). We coded from scratch, but the structure is based on a formal model that we call the Scenario Segment Guide. The structure of our documents (there are several) are constantly evolving as we implement the Java and PHP code and test the resulting system.

MG: At FIEA and EA we use XML in nearly all games to describe many forms of data including visual effects. The purpose of the projects is to entertain, and the audience includes anybody who plays or develops video games. Video game software (e.g., the "engine" and asset conditioning software) effectively has four sets of users, each with different requirements: End users who play the games or potentially modify them, artists who supply artistic content ("assets") used in the game, producers and level designers who supply various other kinds of content such as layout and high-level descriptions of autonomous behavior, and other programmers who author everything else. Typically a modern video game for a current console has a team of anywhere from 30 to 150 people.

The design process is somewhat formal and includes several phases, starting with the game pitch. After the pitch is approved then preproduction starts, which includes research, design, and prototyping. We have employed various tactics including traditional and more recent methods. We used to spend about six to eight weeks writing design documents, then spent the rest of the year writing software. More recently we have started to adopt techniques resembling "SCRUM," where we alternate between design and production. For other parts of the project we have used brainstorming tools such as MindMap, but the XML formats usually derive from higher level requirements of the software design. Often the XML layout has a direct analogy to the classes used to implement the software.

More recently, we have had a push towards designing the XML separately from the implementation so that we could in principle have file formats that outlive and span across any given implementation so that even if we have multiple competing engine implementations then at least they could share their file formats. In practice, we have not adopted that practice yet. We regularly extend the XML format with an eye towards having the new parser continue to accept older files. Most of the time, we add to the existing XML format and rarely change something so drastically that old files fail to work with the new parser.

TG: A recent example project involving XML would be a trivia application. Basically, this application is composed of a list of multiple-choice questions and feedback depending on whether the person got the question right or wrong. The purpose of the project was to create an interactive educational exhibit that could randomly select a question from an external file. This file could easily be updated by someone without knowledge of programming. The design process usually starts with planning, followed by the design of the XML file structure itself. A common tool I use for almost all XML applications is known as ECMAScript for XML, or for short, E4X. It makes parsing and searching the information within an XML file much easier. For most trivia projects, the format stays the same. The main element tags are TRIVIA (the main "wrapper" tag), QUESTION (to "wrap" each question) and ANSWER (child nodes of QUESTION tag). For each set of ANSWER tags inside a QUESTION tag, one of those elements has a "CORRECT" attribute. This format is extremely simple, and rarely needs any revisions. If a revision is needed, it would only be for a special case, such as adding time limits to certain questions.

Question 3: Why did you choose to use XML rather than another technology (such as a relational database)?

BA: For two reasons. First, the industry is moving towards more machine-independent and platform-independent, semantically meaningful ways of tagging content. Second, the tools we are using, that are generally used by our industry, are using XML and HTML. For instance, we use FrameMaker® because it's an industry-accepted authoring tool for large amounts of content. And we generate XML from there because we can easily transform it into HTML without the need for third party tools. You used to have to buy Quadralay WebWorks or other tools (and some people still do) but we found we can get around that by transforming (using XSLTs) the raw XML ourselves and making HTML. Another reason is that Microsoft Visual Studio® (the development environment often used by software developers) uses XML in the source code for putting in comments. So, we can use tools such as Innovasys Document! X to generate documentation automatically, grab the XML, and generate HTML. So it is natural for us to work with XML since it is part of how software developers do their code comments.

SS: Our choice is based on customer requirements and standards.

JMM: First, XML is uniquely low-cost. The concept can be explained in minutes, and you can immediately begin to build documents with an ordinary word processor. Second, tools exist in Java, in PHP, and indeed in most major programming environments that make it easy to transform XML into data structures that are used within the telecommunication and graphical display elements of our programs.

MG: We do use relational databases for other kinds of information, especially where we need to execute sophisticated relational queries on the data. XML works well when we need some text format which describes some asset, usually in some static way. We like XML because parsers for it are readily available; the format is familiar to most programmers and some non-technical people so humans can modify the file format in the absence of a specialized authoring tool.

TG: XML tends to be more of a "standard" format, in that an XML file has a definitive structure and a definitive set of data, no matter what is accessing the XML file (whether it be a website, Adobe Flash® animation, software, text document, etc.). Using the E4X function library I mentioned previously, I can also bring many of the benefits of a relational database to XML (specifically, search queries and linking multiple entries to each other). While XML will not replace relational databases, it can emulate their function. So, in most cases, I choose to go with XML, unless there are performance issues.

Question 4: What process did you use to author your XML content? Did you use any tools or software programs (including software that auto-generates XML code)?

BA: We use Adobe FrameMaker® to author a good deal of documentation, mostly User's Guide topics and procedural tasks. This allows us to output XML which we then transform into HTML. That is half of it. The other half is automatically generated API Reference (class library) documentation. If your software product has a public interface (called an API) then you have to publish information about that interface.

Often if the software is big, you do not want to write it by hand and there are tools for generating it automatically. This happens when the tool looks at the source code comments and the structure of the software itself and builds pages of documentation. So with these two parts, the User's Guide and the API Reference, we have a complete document set. XML is used throughout. The final output of both parts is a set of HTML pages which are then combined into a compiled help file. Also, from Adobe FrameMaker® we can make printable PDFs of the User's Guide.

SS: We use Notepad for most of our XML tagging. Sometimes we use the built-in editors that come with the authoring environment, but it is easier to use Notepad. I know that many of the tools have all the bells and whistles; however, I can work faster in Notepad.

JMM: There are three principal XML documents in our system: the Scenario Model, multiple Session Records, and Session Logs. A Scenario Model is constructed manually. Currently, we use word processing, but we could use a syntax-directed tool to automatically build well-formed and valid code. We just have not done so yet. Session Records are automatically constructed from the Scenario Model by the Java application. The Session Record contains the particulars of a particular "run" of the game: who played what roles, which goals and learning objectives have been met, what reconfigurations of the user interface have been made (e.g., boxes dragged around the screen, etc.). Session Logs are also automatically generated by the Java application. They record all changes to the system's state, as well as all dialogue that occurred through the built-in text chat system.

MG: We use multiple methods to author the XML. Initially, the systems had their data described procedurally, at which point we made the decision to read the descriptions from a file. Instead of authoring the files from scratch we wrote serialization methods to output the existing descriptions in XML form. Subsequently, people modified the XML files using regular text editors. Later, people developed several specialized authoring tools to modify and author XML files either directly or indirectly. It is common for games to have in-game "tweakers" which update the data structures directly after which the classes containing the data can serialize the data out to XML.

TG: All of my XML is coded with a regular text editor, such as Notepad. There are many tools to make coding easier via color-coded syntax and auto-completing phrases; however, I have not found anything that works much better than the traditional copy and paste workflow.

Question 5: What type of parser (SAX, DOM, etc.) did you use for your project?

- *Was this parser commercially available, or did you build it?*
- *Why did you choose this parser, and were there any particular advantages or disadvantages to it?*

BA: We use the SAXON processor with our home-grown XSLTs to convert our raw XML into either finished HTML files or text files (snippets) that are brought into generating HTML files later in the process. We also use Innovasys Document! X to automatically generate a lot of HTML documentation of the product. We then combine the automatically generated HTML along with the XML-to-HTML files of human-authored content to make a completely compiled online help system for our customers.

SS: Our authoring environment is furnished by the government. The environment includes a built-in editor, parser, reader, and style sheet editor. These tools are basic; they don't include many of the convenient functions found in high-end commercial authoring environments. They work well, but the author needs to know the document markup language and structure fairly well in order to make the tools efficient.

JMM: I wrote my own ad hoc XML processor in Java, based on DOM4J. It is open source. I chose it because it easily integrates with Java in the Eclipse integrated development environment (IDE).

MG: We usually use SAX (for in-game use), but sometimes DOM (usually for tools used internally). At EA, we use a parser written internally because we have to use software developed by EA employees or for which we have obtained licenses. The process of obtaining a license is cumbersome and we generally need to modify source code so that we can control the memory footprint since most games ship on fixed memory systems. At FIEA, we use an open source parser called Expat. Here we have more limited resources and Expat serves our needs adequately.

TG: Being an Adobe Flash® project, there are a few different ways to parse the data. Adobe Flash® 8 and earlier (ActionScript 2.0) has an awkward way of parsing XML information, so I normally rely on a custom class called XMLConstruct made for ActionScript 2.0. Adobe Flash® 9 (ActionScript 3.0), however, has built-in functions to parse XML, including the E4X library. The aforementioned XMLConstruct class was created by a group of developers at Indivision.net. The ActionScript 3.0 parser is built into the new Adobe Flash® CS3 commercial software.

Using the XMLConstruct parser is extremely advantageous in that it allows you to refer to a specific XML element (also known a "node") by name as opposed to by number. For example, without the XMLConstruct class, to refer to an element, it would look something like this:

```
myNode = myXML.childNode[0].childNode[1].
nextSibling;
```

Using the XMLConstruct class, the same result could be attained using this syntax:

```
myNode = myXML.trivia[0].question[1].answer;
```

Question 6: Did you use style sheets or transformations with your XML documents? If so, which types?

BA: Yes, we wrote our own DTD to handle a small set of elements in our procedural documentation (User's Guide) and our own XSLT that transforms the XML of our procedural documentation into HTML. We also use a separate XSLT to transform many snippets which are in XML into individual text files that are pulled into a larger process that inserts the text (code snippets) into pages of automatically generated documentation. (Authors' note: see our online website for some examples of the DTDs and the XSLTs that Bill's company uses.)

SS: Yes, we use both DTDs and style sheets in the documentation. Any formatting or manipulation is done using the style sheet editor.

JMM: Not yet. We are evolving the DTD for our CMPM dialect (we call it cpXML) but as yet the parsing is ad hoc, driven by the architecture of our Java application.

MG: Not usually.

TG: Typically, I do not use style sheets or transformations with my XML documents. Any formatting or manipulation is done after the information has been parsed by the parent application (in most cases, this is Adobe Flash®).

Question 7: How did you validate your XML documents, if at all? Did you use DTDs, Schema, or another method?

BA: The XML that we output from Adobe FrameMaker® is valid XML. We do not have a separate step in our process of validating the XML. We use a DTD that we wrote in-house, and FrameMaker® does any validating that is needed. The authoring process indicates invalid XML when we author, so by the time we get to production, it is all valid XML.

SS: We use a DTD that is based on the military standard most of the time, but we also use the built-in parser. Both seem to work well.

JMM: Again, we handle this by producing error messages within our Java code if the DOM4j system objects to any well-formedness issues. In this way, we report any validation issues when our internal scanner fails to understand a structure.

MG: At EA, the tools teams sometimes employ validation facilities built into the DOM parser.

TG: When using Adobe Flash®, the only validation needed is standard error checking. If the application works, and is bug-free, the XML is considered valid. When dealing with XML for websites (for example, RSS feeds), then many different tools are used for validation, my favorite being the W3C online validator.

Question 8: In your mind, what are the primary strengths and weaknesses of XML as a communications technology?

BA: XML, as we use it, is great for human use: it is a tagging language that makes sense. And it is scalable. We only need a few elements, so we can use a small DTD that we wrote in-house. We do not need a behemoth like DocBook. XML is flexible and easy to work with. We wrote our own XSLTs and avoided using third party software that would have cost the company lots of money. XML is about putting content in containers with meaningful names. A topic contains a subtopic that contains a paragraph. Simple enough.

SS: The strengths are flexibility, portability, and the ability to use it without buying expensive tools.

JMM: The strengths are broad acceptance and simplicity. The weakness is that there is no systematic way to find relevant namespaces. It sometimes seems like luck and accident when one finds good ontologies.

MG: The primary strength I see is that customizing a parser for XML is extremely easy, especially when using a SAX-based parser which mostly entails writing callbacks and a simple push-down automata. Usually, we can get away with finite-state automata and occasionally employ a stack to handle recursive elements, which are rare. A primary weakness is that it is verbose, so although editing by hand is straightforward, it takes a much larger number of keystrokes. Also, reading XML is slightly harder than reading other declarative languages. Writing procedural phrases in XML is especially cumbersome.

TG: The strengths of XML include its open text format, its ability to be edited using standard text editors, and the ease of reading and editing the information. The weaknesses are directly related to its strengths: being a physical file format, it can create issues with performance, especially if many different applications are trying to read from the same file at once. This is where a standard database would be ideal.

Question 9: What skills and competencies are important to technical professionals wishing to learn more about XML and XML-related technologies? Aside from learning the subject matter itself, would you recommend any additional fields of study or particular coursework?

BA: The challenge with using XML and content is to know how to structure your content. What is each piece of content, and in what container does it belong? Once you have figured out that, then you can reuse content, you can organize the content, and you can transform and filter the content. Whether you have a database, a CMS, or just raw files, you can work with XML in a myriad of ways. But, being familiar enough with the content to know how to structure it into containers: that is the essential skill needed for working with XML. The tools will grow and will help you with that task, so there are no specific tools you need to know up front.

SS: In the case of designing electronic manuals, users are looking for user-friendly interfaces and useful information. A working knowledge of document markup languages is extremely useful, including knowing how to design and develop DTDs, style sheets, interface prompts, and dialogue. An XML or SGML electronic technical manual is a technical information database that stores data in modules and provides access to numerous media and external databases. It would be beneficial for students to learn how to develop structural diagrams so that they can see the hierarchy and sequence and nested elements visually. I believe database design or basic IT principles would be very helpful to students who are interested in doing more than just tagging and manipulating text in XML.

JMM: In our case, computer science has been very useful. We haven't tried to use tools that might be appropriate for non-programmers, since our principal objective has been to build our own software. For many or most XML users, programming should not be a requirement. The tools appropriate to these audiences are not yet part of our repertoire.

MG: I teach (or review with) my students the basics of formal language theory and the Chomsky hierarchy, which applies to parsing in general. I also teach my students how to use a SAX-based parser and they are required to implement a parser that they use in a game engine that they write. People dealing with any parser should understand the fundamentals of automata, specifically finite-state and push-down.

TG: A good understanding of the concepts of database redundancy and database design are essential in being efficient with XML, or any database at all, for that matter. A working knowledge of HTML would also be extremely helpful, since XML could be easily considered a regular HTML page, with custom tags as opposed to predefined ones. XML has so many uses that additional fields of study are too numerous to count; however, I will emphasize some of the particulars I brought up earlier in the interview. E4X is one of the most powerful tools for dealing with XML data, and learning how to use it should be a top priority for anyone involved with XML. XML also allows you to emulate any number of database designs, including single-linked-lists, double-linked-lists (also known as graphs) and many others. Understanding the differences between various database structures is the best recommendation I could give. Sometimes database design (or, in this case, XML file structure) is such a blank sheet of paper, it is hard to know where to start.

Question 10: Are there any other comments or thoughts about your project you would like to share?

BA: XML is the foundation technology, the underlying tool, the Tupperware containers in which to put information, and, by doing so, free it from previous constraints in delivery, presentation, and maintenance. It is the basis of content management and information reuse and single sourcing—or it can be. The concept of putting information (content) in containers is one of the key concepts of using XML. There are so many uses of XML today. The uses of e-commerce and electronic transactions are growing and probably driving the use of XML more than our meager efforts to contain technical content in documentation. The new buzzwords are informatics and analytics (really just information and analysis, with machine-reading thrown in), which are made possible by XML and XML-related technologies. Here are some exciting new applications of XML:

- Small companies are using XML databases behind their websites and manufacturers are publishing their catalog of millions of parts on the Web using XML.

- CMSs offer complete control to an organization by allowing authors and SMEs to check out content and revise it.
- RSS is a way for newspapers and other organizations to share their news.
- Portals within an organization, the next generation of Intranet, allow employees to access company information throughout an organization.
- The genome project and ongoing DNA research is an example of the huge amounts of information that are being tagged. Medline is a comprehensive literature database of life sciences and biomedical information.

SS: In the case of electronic technical manuals, I think XML is better suited for technical document collections or technical manuals that require constant revisions rather than one time developments.

JMM: XML has proven to be much easier to work with than I anticipated. The tools such as DOM4j and PHP's SimpleXML just seem to work reliably and on the first try. This makes projects move much faster than some other technologies.

MG: In game development, it is useful to have compact file formats that parse very quickly. Since XML is a text format, it is fundamentally slow to parse. Some XML parsers have subsystems that partially pre-parse and generate a binary file format that has a direct relationship to XML. It is important for interactive applications to have access to tools such as that.

Interview Discussion

As we can see from these interviews, XML is used for a variety of different applications in industry and research. Our respondents used it for purposes ranging from encoding materials for interactive electronic technical manuals to guiding a metadata system in order to direct the work of interactive performance cast members and dramatic performers in a research setting. Though the interviews stand fairly well on their own, we do want to highlight ten emergent themes that have come out of these responses:

1. There are many different tools, software programs, and technical protocols that are associated with XML technology or that support or extend this technology in some capacity. Several of these technologies have been mentioned earlier in the book, but some have not. From the interviews, we see professionals working with XML using Microsoft Visual Studio.NET®, C#, Java, Eclipse, Adobe Flash®, Adobe FrameMaker®, and with several other custom programs or authoring environments. The sheer number of possibilities available for both authoring tools and parsing tools suggests that this markup language has a promising future ahead.

2. DITA is a popular XML-based framework for authoring and distributing technical information. Like DocBook, a markup language for technical documentation, it can be used to modularize content using a set of preexisting procedures and tags that are built into the DTD. Both DITA and DocBook are very popular and will likely continue to grow in popularity as content management and single sourcing become even more common in software documentation processes. We discuss DITA and DocBook in more detail in Chapter 5.

3. XML can be used to allow programmers and technical communicators to work more efficiently with one another. As Bill noted, some of the software he uses automatically generates comments in XML format. These comments can then be harvested using special software and then be incorporated as part of the documentation. When both programmers and technical communicators speak the same language (i.e., XML), there is bound to be a better working environment for both groups.

4. Despite the best laid plans of the document designer, outside forces will sometimes determine the actual course of action for projects in the real world. Sherry stated this most succinctly in the interview when she stated, "our choice is based on customer requirements and standards." Here we could perhaps use quotation marks around the word "choice" to show that this is not really a choice in many instances, but rather a set of parameters that one must adhere to in order to win a contract or actually be hired to complete the work. She later stated that even her authoring environment was imposed by the government! Whether these requirements are imposed by the customer, by the client, or by the capabilities of your team, it is an unusual case in industry when one can simply proceed doing things "the way they should be done." Being aware of these political forces, and being prepared to defend and justify one's design decisions, is another important rhetorical skill that one should be aware of in industrial settings.

5. Not all XML authors use validation, but the option is nice to have. Several of our respondents indicated that they did use an internal DTD to validate their XML documents, but others indicated that they did not do any type of validation, or that they simply made sure the data was good enough to work in whatever program they were using to parse it. This shows that despite the inherent ability of XML to be validated against a DTD or schema document, it is not always necessary, nor practical, to do so for all types of applications.

6. The biggest perceived benefit of using XML was the language's low overhead in terms of its minimal cost, shallow learning curve, and its ability to run on multiple platforms under multiple operating systems and hardware configurations. Michael Gourlay also noted that parsers are readily available

for XML, which is an important consideration. Even though one can certainly design their own custom parsers, as J. Michael Moshell indicates that he did in his interview (and as we write about in some detail in Chapter 6), the resources (time, money, and experience) for such a task are not always readily available.

7. Despite the availability of expensive authoring tools and other sophisticated suites for handling XML, many authors profiled in this interview preferred to use old-fashioned text editors for authoring their XML content. We agree that text editors are useful for the authoring component, though it is nice to have support for validation and transformation testing for more sophisticated types of documents. We provide some links to useful XML authoring tools, of both the plain text and validating variety, on our website.

8. In terms of education and skill development, many interviewees noted that computer science skills, particularly in database design, are important when working with XML parsers. Michael Gourlay also mentions the importance of understanding automata theory and formal language hierarchies, which are common topics in computer science. For less technically demanding types of operations, the key skill noted is the ability to visualize how information should be broken down into modular units and the experience to know which elements should be used for which types of data. Additionally, expertise from technical communication and library science courses should provide skills and competencies in learning how to achieve proper levels of granularity within an XML document. Many of these skills will often come with experience in working on applied projects.

9. Though we did not explicitly ask any questions about knowledge management, it is clear that XML is being used by these individuals in varying capacities to deal with this very issue. From the explicit knowledge model used by Thomas in his trivia application, which directly asked users whether or not they knew an answer to a question, to the more implicit and tacit model used by J. Michael Moshell in his CMPM system, these XML projects all functioned under the assumption that the XML data would help to facilitate and transfer knowledge among an organizational group and its members. As both Bill Albing and Michael Gourlay noted, XML helps to improve the durability and scope of knowledge by standardizing and encapsulating communication and organizational tasks in distributed teams. This could be done by bringing programmers and technical communicators together using a shared language, as Bill noted. Or, it might be done by helping to facilitate the design documents used in the "SCRUM" sessions that alternate between design and production tasks in a game design cycle, as Michael Gourlay discussed in regards to modern game design techniques.

10. The future of XML looks very bright. As we see the beginnings of Berners-Lee's Semantic Web on the horizon, it is clear that XML will be the driving force and harbinger of many next generation Internet and network technologies. Technologies such as AJAX will continue to emerge and drive innovation towards more user-centered and reader-centered types of communication on the World Wide Web. In addition to its implications for the Semantic Web, XML's use in mobile technology also reveals that this language is gaining in popularity and becoming a more mature and stable platform upon which to build distributed applications. Though XML continues to experience some trouble imposed by political battles over proprietary formats, it seems clear that these issues will eventually be worked out and the language will continue to grow and evolve in a meaningful fashion.

Chapter Summary

This chapter presented the results of an interview which captured information about how XML is used by professionals in different fields and occupations. It also summarized the key concepts from this book and explained how knowledge management, XML, and rhetoric can function as synergistic elements and building blocks for modern technical communication.

Having discussed knowledge management, XML, and rhetoric throughout this book, we can now take a step back and see how these various components work together (and sometimes against one another) in the professional life of a practitioner working in the Information Age. We have stressed two primary themes throughout this book:

1. Students and scholars who wish to deliver usable and effective communication products in our digital economy must be theorist-practitioners as well as symbolic-analysts (we discuss this more in Chapter 1). This involves understanding both technological issues as well as the larger social, cultural, and critical contexts that inform and surround these technologies. In the case of XML, it involves both understanding the nuts and bolts of how this markup language functions as well as understanding the inherent tension between the way knowledge is *actually* constructed in the real world (through messy, socially mediated, and often unquantifiable processes), and how our information systems *lead us to believe* knowledge is created (through the acquisition of a greater number of neat, packaged information units). In other words, we can never manipulate pure units of knowledge, but only the imperfectly defined and socially constructed symbols that represent aspects or facets of this knowledge. These symbols are defined from a particular shared perspective or a particular set of social conditions with its own idiosyncratic list of linguistic limitations, affordances, and parameters. Despite the limitations of this practice, the ability to work with and apply

such semantic symbols is extremely important, since it is all we have. In addition, understanding the tension between these two models of thought about knowledge transfer—social constructionism and logical positivism—allows us to build better networks for knowledge dissemination that consider both the social and technological dimensions of tacit and explicit knowledge transfer.

2. Rhetoric is a powerful tool and technique for improving our symbolic-analytic skills. By its very nature, rhetoric asks us to think about the communication process in a formal way, by considering our audience, our communicative intent, and the potential sources of noise that may disrupt the integrity of our messages. In digital environments, this noise may take the shape of bad usability decisions or it may even be produced by improperly developed classification schemas. Understanding how to leverage rhetorical tools to our advantage helps us to improve our knowledge management systems, even when such systems are driven by IT and digital computers—tools that were far in the future when classical rhetors debated science and philosophy in the streets of Greece and Rome.

As this is the concluding chapter, we invite readers to continue this dialogue using the online resources associated with this text. These include our companion website at www.rhetoricalxml.com, our Wiki site with user-contributed and editable content (including these interview questions), and the "contact us" mechanisms listed online or in this book. We would be pleased to hear about how you are using XML in your own studies and careers and we hope that reading this book is only the beginning of your own quest to learn more about knowledge management, rhetoric, and technical content creation using modern Internet technologies.

Discussion Questions

1. Besides the ways we mention in this chapter, what other applied situations can you describe that involve the convergence of knowledge management strategies and rhetoric? Do these two elements apply to *any* social or organizational interaction? How can XML assist in capturing these types of interactions?
2. Carefully read through the survey questions that interviewees were asked to respond to in this chapter. What is the overall rhetorical intent of this survey? Now, based on the ideas you have read about in this book and your own project work and experience, think about how you would revise the interview above to include additional questions related to XML, or how you would extend the interview to also gather explicit data about the knowledge management practices used in an organization. Does this change the survey's rhetorical context? If so, how has the rhetorical intent changed based on your modifications?

3. Aside from our own discussion of the XML interview, what other trends or ideas emerge from this collection of ideas from XML practitioners? How do these interviews illustrate the similarities or differences between these different careers? Are there any potential problem areas that might emerge if these different professionals were tasked to work on the same type of problem together? What disciplinary boundaries are at play here? If necessary, refer back to Chapter 1 and read about Thomas Kuhn's ideas about "normal science" to help guide your answers here.

Activities

1. Find a professional working in your anticipated occupational field that uses XML somehow in his or her career. Conduct a brief interview with this person using the original questions from this chapter or the questions you have written in response to Discussion Question 2. Bring your questions in to share with your classmates.
2. Using your knowledge about XML and the rhetorical design process, write a short proposal for a project in which you could use XML in your own studies or line of work. Include rough physical design parameters for your project including a total number of anticipated files, elements, and Web pages. Also, include an audience analysis and a summary of any cognitive or affective design issues that you may need to deal with during the project. Your total proposal should be between five and seven pages and should include a timeline for completion as well as other traditional proposal elements (any budget issues, technical resources, and so forth).

References

Albers, Michael J. "The Technical Editor and Document Databases: What the Future May Hold." *Technical Communication Quarterly* 9.2 (2000): 191–206.

Albing, Bill. Personal interview. December 19, 2007.

Carliner, Saul. "Physical, Cognitive, and Affective: A Three-Part Framework for Information Design." *Content & Complexity: Information Design in Technical Communication*. Michael J. Albers and Beth Mazur, eds. Mahwah, NJ: Lawrence Erlbaum Associates, 2003. 39–58.

Gorence, Thomas. Personal interview. November 27, 2007.

Gourlay, Michael. Personal interview. November 27, 2007.

Integrity Arts & Technology, inc. "I.D.E.A.S." 2006. December 24, 2007. www.integrityarts.com/index.php

Moshell, J. Michael. Personal interview. November 21, 2007.

Steward, Sherry. Personal interview. December 21, 2007.

Appendix A
ACHRE—Executive Summary

Publication Information

The Final Report of the ACHRE (stock number 061-000-00848-9), the supplemental volumes to the Final Report (stock numbers 061-000-00850-1, 061-000-00851-9, and 061-000-00852-7), and additional copies of this Executive Summary (stock number 061-000-00849-7) may be purchased from the Superintendent of Documents, U.S. Government Printing Office.

All telephone orders should be directed to:
Superintendent of Documents
U.S. Government Printing Office
Washington, D.C. 20402
(202) 512-1800
Fax (202) 512-2250
8 a.m. to 4 p.m., Eastern time, M-F

All mail orders should be directed to:
U.S. Government Printing Office
P.O. Box 371954
Pittsburgh, PA 15250-7954

An Internet site containing ACHRE information (replicating the Advisory Committee's original gopher) will be available at George Washington University. The site contains complete records of Advisory Committee actions as approved; complete descriptions of the primary research materials discovered and analyzed; complete descriptions of the print and non-print secondary resources used by the Advisory Committee; a copy of the Interim Report of October 21, 1994, and other information. The address is www.seas.gwu.edu/nsarchive/radiation. The site will be maintained by the National Security Archive at GWU.

Printed in the United States of America

The Creation of the Advisory Committee

On January 15, 1994, President Clinton appointed the ACHRE. The President created the Committee to investigate reports of possibly unethical experiments funded by the government decades ago.

The members of the Advisory Committee were fourteen private citizens from around the country: a representative of the general public and thirteen experts in bioethics, radiation oncology and biology, nuclear medicine, epidemiology and biostatistics, public health, history of science and medicine, and law.

President Clinton asked us to deliver our recommendations to a Cabinet-level group, the Human Radiation Interagency Working Group, whose members are the Secretaries of Defense, Energy, Health and Human Services, and Veterans Affairs; the Attorney General; the Administrator of the National Aeronautics and Space Administration; the Director of Central Intelligence; and the Director of the Office of Management and Budget. Some of the experiments the Committee was asked to investigate, and particularly a series that included the injection of plutonium into unsuspecting hospital patients, were of special concern to Secretary of Energy Hazel O'Leary. Her department had its origins in the federal agencies that had sponsored the plutonium experiments. These agencies were responsible for the development of nuclear weapons and during the Cold War their activities had been shrouded in secrecy. But now the Cold War was over.

The controversy surrounding the plutonium experiments and others like them brought basic questions to the fore: How many experiments were conducted or sponsored by the government, and why? How many were secret? Was anyone harmed? What was disclosed to those subjected to risk, and what opportunity did they have for consent? By what rules should the past be judged? What remedies are due those who were wronged or harmed by the government in the past? How well do federal rules that today govern human experimentation work? What lessons can be learned for application to the future? Our Final Report provides the details of the Committee's answers to these questions. This Executive Summary presents an overview of the work done by the Committee, our findings and recommendations, and the contents of the Final Report.

The President's Charge

The President directed the Advisory Committee to uncover the history of human radiation experiments during the period 1944 through 1974. It was in 1944 that the first known human radiation experiment of interest was planned, and in 1974 that the Department of Health, Education, and Welfare adopted regulations governing the conduct of human research, a watershed event in the history of federal protections for human subjects.

In addition to asking us to investigate human radiation experiments, the President directed us to examine cases in which the government had intentionally released radiation into the environment for research purposes. He further charged us with identifying the ethical and scientific standards for evaluating these events, and with making recommendations to ensure that whatever wrongdoing may have occurred in the past cannot be repeated.

We were asked to address human experiments and intentional releases that involved radiation. The ethical issues we addressed and the moral framework we developed are, however, applicable to all research involving human subjects.

The breadth of the Committee's charge was remarkable. We were called on to review government programs that spanned administrations from Franklin Roosevelt to Gerald Ford. As an independent advisory committee, we were free to pursue our charge as we saw fit. The decisions we reached regarding the course of our inquiry and the nature of our findings and recommendations were entirely our own.

The Committee's Approach

At our first meeting, we immediately realized that we were embarking on an intense and challenging investigation of an important aspect of our nation's past and present, a task that required new insights and difficult judgments about ethical questions that persist even today.

Between April 1994 and July 1995, the Advisory Committee held sixteen public meetings, most in Washington D.C. In addition, subsets of Committee members presided over public forums in cities throughout the country. The Committee heard from more than two hundred witnesses and interviewed dozens of professionals who were familiar with experiments involving radiation. A special effort, called the Ethics Oral History Project, was undertaken to learn from eminent physicians about how research with human subjects was conducted in the 1940s and 1950s.

We were granted unprecedented access to government documents. The President directed all the federal agencies involved to make available to the Committee any documents that might further our inquiry, wherever they might be located and whether or not they were still secret.

As we began our search into the past, we quickly discovered that it was going to be extremely difficult to piece together a coherent picture. Many critical documents had long since been forgotten and were stored in obscure locations throughout the country. Often they were buried in collections that bore no obvious connection to human radiation experiments. There was no easy way to identify how many experiments had been conducted, where they took place, and which government agencies had sponsored them. Nor was there a quick way to learn what rules applied to these experiments for the period prior to the mid-1960s. With the assistance of hundreds of federal officials and agency staff, the Committee retrieved and reviewed hundreds of thousands of government documents. Some of the most important documents were secret and were declassified at our request. Even after this extraordinary effort, the historical record remains incomplete. Some potentially important collections could not be located and were evidently lost or destroyed years ago.

Nevertheless, the documents that were recovered enabled us to identify nearly 4,000 human radiation experiments sponsored by the federal government between 1944 and 1974. In the great majority of cases, only fragmentary data was locatable; the identity of subjects and the specific radiation exposures involved were typically unavailable. Given the constraints of information, even more so than time, it was impossible for the Committee to review all these experiments, nor could we evaluate the experiences of countless individual subjects. We thus decided to focus our investigation on representative case studies reflecting eight different categories of experiments that together addressed our charge and priorities. These case studies included:

- experiments with plutonium and other atomic bomb materials
- the Atomic Energy Commission's program of radioisotope distribution
- nontherapeutic research on children
- total body irradiation
- research on prisoners
- human experimentation in connection with nuclear weapons testing
- intentional environmental releases of radiation
- observational research involving uranium miners and residents of the Marshall Islands.

In addition to assessing the ethics of human radiation experiments conducted decades ago, it was also important to explore the current conduct of human radiation research. Insofar as wrongdoing may have occurred in the past, we needed to examine the likelihood that such things could happen today. We therefore undertook three projects:

- A review of how each agency of the federal government that currently conducts or funds research involving human subjects regulates this activity and oversees it.
- An examination of the documents and consent forms of research projects that are today sponsored by the federal government in order to develop insight into the current status of protections for the rights and interests of human subjects.
- Interviews of nearly 1,900 patients receiving out-patient medical care in private hospitals and federal facilities throughout the country. We asked them whether they were currently, or had been, subjects of research, and why they had agreed to participate in research or had refused.

The Historical Context

Since its discovery one hundred years ago, radioactivity has been a basic tool of medical research and diagnosis. In addition to the many uses of the x-ray, it was soon discovered that radiation could be used to treat cancer and that the introduction of "tracer" amounts of radioisotopes into the human body could help to diagnose disease and understand bodily processes. At the same time, the perils of overexposure to radiation were becoming apparent.

During World War II the new field of radiation science was at the center of one of the most ambitious and secret research efforts the world has known—the Manhattan Project. Human radiation experiments were undertaken in secret to help understand radiation risks to workers engaged in the development of the atomic bomb.

Following the war, the new Atomic Energy Commission used facilities built to make the atomic bomb to produce radioisotopes for medical research and other peacetime uses. This highly publicized program provided the radioisotopes that were used in thousands of human experiments conducted in research facilities throughout the country and the world. This research, in turn, was part of a larger postwar transformation of biomedical research through the infusion of substantial government monies and technical support.

The intersection of government and biomedical research brought with it new roles and new ethical questions for medical researchers. Many of these researchers were also physicians who operated within a tradition of medical ethics that enjoined them to put the interests of their patients first. When the doctor was also a researcher, however, the potential for conflict emerged between the advancement of science and the advancement of the patient's wellbeing.

Other ethical issues were posed as medical researchers were called on by government officials to play new roles in the development and testing of nuclear weapons. For example, as advisers they were asked to provide human research data that could reassure officials about the effects of radiation, but as scientists they were not always convinced that human research could provide scientifically useful data. Similarly, as scientists, they came from a tradition in which research results were freely debated. In their capacity as advisers to and officials of the government, however, these researchers found that the openness of science now needed to be constrained.

None of these tensions were unique to radiation research. Radiation represents just one of several examples of the exploration of the weapons potential of new scientific discoveries during and after World War II. Similarly, the tensions between clinical research and the treatment of patients were emerging throughout medical science, and were not found only in research involving radiation. Not only were these issues not unique to radiation, but they were not unique to the 1940s and 1950s. Today society still struggles with conflicts between the openness of science and the preservation of national security, as well as with conflicts between the advancement of medical science and the rights and interests of patients.

Key Findings

Human Radiation Experiments

- Between 1944 and 1974 the federal government sponsored several thousand human radiation experiments. In the great majority of cases, the experiments were conducted to advance biomedical science; some experiments were conducted to advance national interests in defense or space exploration; and some experiments served both biomedical and defense or space exploration purposes. As noted, in the great majority of cases only fragmentary data are available.
- The majority of human radiation experiments identified by the Advisory Committee involved radioactive tracers administered in amounts that are likely to be similar to those used in research today. Most of these tracer studies involved adult subjects and are unlikely to have caused physical harm. However, in some nontherapeutic tracer studies involving children, radioisotope exposures were associated with increases in the potential lifetime risk for developing thyroid cancer that would be considered unacceptable today. The Advisory Committee also identified several studies in which patients died soon after receiving external radiation or radioisotope doses in the therapeutic range that were associated with acute radiation effects.
- Although the AEC, the Defense Department, and the National Institutes of Health recognized at an early date that research should proceed only with the consent of the human subject, there is little evidence of rules or practices of consent except in research with healthy subjects. It was commonplace during the 1940s and 1950s for physicians to use patients as subjects of research without their awareness or consent. By contrast, the government and its researchers focused with substantial success on the minimization of risk in the conduct of experiments, particularly with respect to research involving radioisotopes. But little attention was paid during this period to issues of fairness in the selection of subjects.
- Government officials and investigators are blameworthy for not having had policies and practices in place to protect the rights and interests of human subjects who were used in research from which the subjects could not possibly derive direct medical benefit. To the extent that there was reason to believe that research might provide a direct medical benefit to subjects, government officials and biomedical professionals are less blameworthy for not having had such protections and practices in place.

Intentional Releases

- During the 1944–74 period, the government conducted several hundred intentional releases of radiation into the environment for research purposes. Generally, these releases were not conducted for the purpose of studying the effects of radiation on humans. Instead they were usually conducted to test the operation of weapons, the safety of equipment, or the dispersal of radiation into the environment.

- For those intentional releases where dose reconstructions have been undertaken, it is unlikely that members of the public were directly harmed solely as a consequence of these tests. However, these releases were conducted in secret and despite continued requests from the public that stretch back well over a decade, some information about them was made public only during the life of the Advisory Committee.

Uranium Miners

- As a consequence of exposure to radon and its daughter products in underground uranium mines, at least several hundred miners died of lung cancer and surviving miners remain at elevated risk. These men, who were the subject of government study as they mined uranium for use in weapons manufacturing, were subject to radon exposures well in excess of levels known to be hazardous. The government failed to act to require the reduction of the hazard by ventilating the mines, and it failed to adequately warn the miners of the hazard to which they were being exposed.

Secrecy and the Public Trust

- The greatest harm from past experiments and intentional releases may be the legacy of distrust they created. Hundreds of intentional releases took place in secret, and remained secret for decades. Important discussion of the policies to govern human experimentation also took place in secret. Information about human experiments was kept secret out of concern for embarrassment to the government, potential legal liability, and worry that public misunderstanding would jeopardize government programs.
- In a few instances, people used as experimental subjects and their families were denied the opportunity to pursue redress for possible wrongdoing because of actions taken by the government to keep the truth from them. Where programs were legitimately kept secret for national security reasons, the government often did not create or maintain adequate records, thereby preventing the public, and those most at risk, from learning the facts in a timely and complete fashion.

Contemporary Human Subjects Research

- Human research involving radioisotopes is currently subjected to more safeguards and levels of review than most other areas of research involving human subjects. There are no apparent differences between the treatment of human subjects of radiation research and human subjects of other biomedical research.

- Based on the Advisory Committee's review, it appears that much of human subjects research poses only minimal risk of harm to subjects. In our review of research documents that bear on human subjects issues, we found no problems or only minor problems in most of the minimal-risk studies we examined.
- Our review of documents identified examples of complicated, higher-risk studies in which human subjects issues were carefully and adequately addressed and that included excellent consent forms. In our interview project, there was little evidence that patient-subjects felt coerced or pressured by investigators to participate in research. We interviewed patients who had declined offers to become research subjects, reinforcing the impression that there are often contexts in which potential research subjects have a genuine choice.
- At the same time, however, we also found evidence suggesting serious deficiencies in aspects of the current system for the protection of the rights and interests of human subjects. For example, consent forms do not always provide adequate information and may be misleading about the impact of research participation on people's lives. Some patients with serious illnesses appear to have unrealistic expectations about the benefits of being subjects in research.

Current Regulations on Secrecy in Human Research and Environmental Releases

- Human research can still be conducted in secret today, and under some conditions informed consent in secret research can be waived.
- Events that raise the same concerns as the intentional releases in the Committee's charter could take place in secret today under current environmental laws.

Other Findings

The Committee's complete findings, including findings regarding experiments conducted in conjunction with atmospheric atomic testing and other population exposures, appear in Chapter Seventeen of the Final Report.

Key Recommendations

Apologies and Compensation

The government should deliver a personal, individualized apology and provide financial compensation to those subjects of human radiation experiments, or their next of kin, in cases where:

- efforts were made by the government to keep information secret from these individuals or their families, or the public, for the purpose of avoiding embarrassment or potential legal liability, and where this secrecy had the effect of denying individuals the opportunity to pursue potential grievances;

- there was no prospect of direct medical benefit to the subjects, or interventions considered controversial at the time were presented as standard practice, and physical injury attributable to the experiment resulted.

Uranium Miners

- The Interagency Working Group, together with Congress, should give serious consideration to amending the provisions of the Radiation Exposure Compensation Act of 1990 relating to uranium miners in order to provide compensation to all miners who develop lung cancer after some minimal duration of employment underground (such as one year), without requiring a specific level of exposure. The act should also be reviewed to determine whether the documentation standards for compensation should be liberalized.

Improved Protection for Human Subjects

- The Committee found no differences between human radiation research and other areas of research with respect to human subjects issues, either in the past or the present. In comparison to the practices and policies of the 1940s and 1950s, there have been significant advances in the federal government's system for the protection of the rights and interests of human subjects. But deficiencies remain. Efforts should be undertaken on a national scale to ensure the centrality of ethics in the conduct of scientists whose research involves human subjects.
- One problem in need of immediate attention by the government and the biomedical research community is unrealistic expectations among some patients with serious illnesses about the prospect of direct medical benefit from participating in research. Also, among the consent forms we reviewed, some appear to be overly optimistic in portraying the likely benefits of research, to inadequately explain the impact of research procedures on quality of life and personal finances, and to be incomprehensible to lay people.
- A mechanism should be established to provide for continuing interpretation and application in an open and public forum of ethics rules and principles for the conduct of human subjects research. Three examples of policy issues in need of public resolution that the Advisory Committee confronted in our work are: (1) clarification of the meaning of minimal risk in research with healthy children; (2) regulations to cover the conduct of research with institutionalized children; and (3) guidelines for research with adults of questionable competence, particularly for research in which subjects are placed at more than minimal risk but are offered no prospect of direct medical benefit.

Secrecy: Balancing National Security and the Public Trust

Current policies do not adequately safeguard against the recurrence of the kinds of events we studied that fostered distrust. The Advisory Committee concludes that there may be special circumstances in which it may be necessary to conduct human research or intentional releases in secret. However, to the extent that the government conducts such activities with elements of secrecy, special protections of the rights and interests of individuals and the public are needed.

Research involving human subjects. The Advisory Committee recommends the adoption of federal policies requiring:

- the informed consent of all human subjects of classified research. This requirement should not be subject to exemption or waiver;
- that classified research involving human subjects be permitted only after the review and approval of an independent panel of appropriate nongovernmental experts and citizen representatives, all with the necessary security clearances.

Environmental releases. There must be independent review to assure that the action is needed, that risk is minimized, and that records will be kept to assure a proper accounting to the public at the earliest date consistent with legitimate national security concerns. Specifically, the Committee recommends that:

- Secret environmental releases of hazardous substances should be permitted only after the review and approval of an independent panel. This panel should consist of appropriate, nongovernmental experts and citizen representatives, all with the necessary security clearances.
- An appropriate government agency, such as the Environmental Protection Agency, should maintain a program directed at the oversight of classified programs, with suitably cleared personnel.

Other Recommendations

The Committee's complete recommendations, including recommendations regarding experiments conducted in conjunction with atmospheric atomic testing and other population exposures, appear in Chapter Eighteen of the Final Report.

What's Next: The Advisory Committee's Legacy

Interagency Working Group Review

The Interagency Working Group will review our findings and recommendations and determine the next steps to be taken.

Continued Public Right To Know

The complete records assembled by the Committee are available to the public through the National Archives. Citizens wishing to know about experiments in which they, or family members, may have taken part, will have continued access to the Committee's database of 4,000 experiments, as well as the hundreds of thousands of further documents assembled by the Committee. The Final Report contains "A Citizen's Guide to the Nation's Archives: Where the Records Are and How to Find Them." This guide explains how to find federal records, how to obtain information and services from the member agencies of the Interagency Working Group and the Nuclear Regulatory Commission, how to locate personal medical records, and how to use the Advisory Committee's collection.

Supplemental volumes to the Final Report contain supporting documents and background material as well as an exhaustive index to sources and documentation. These volumes should prove useful to citizens, scholars, and others interested in pursuing the many dimensions of this history that we could not fully explore.

Appendix B
RAX Form

Use this form as a brainstorming and planning tool when designing custom XML presentation systems or parsers.

- Who is my primary audience?

  ____

  ____

  ____

  ____

- Who are the secondary and tertiary audiences?

  ____

  ____

  ____

  ____

- What is the purpose of this metadata system?

  ____

  ____

  ____

  ____

- What are the informational needs of my audiences?

  ____

  ____

  ____

  ____

- How should the information be arranged and presented for this audience?

  ____

  ____

  ____

  ____

- What kinds of backgrounds will my audiences possess? Will they have a high literacy level? What about their level of technical literacy? What style will work well for these needs?

  ____
  ____
  ____
  ____

- What vocabularies will my audiences use to identify relevant elements and data nodes within the XML hierarchy?

  ____
  ____
  ____
  ____

Appendix C
Source Code for CMS

File 1: upload_asset.html

```
<!–
Project: Content Management System.
Component: Writer (File 1 of 5).
Filename: "upload_asset.html".
Purpose: Display a form with simple instructions and a file
upload field.
–>
<!–
Note that we are using the XHTML document type and namespace
for our HTML form page. This could just as easily be done in
normal HTML, but using XHTML is useful because it asks us to
make the same syntactical decisions (using well-formed code)
that we will need to make when writing our XML documents.
–>
<!DOCTYPE html PUBLIC "-//W3C//DTD XHTML 1.0 Transitional//EN"
"www.w3.org/TR/xhtml1/DTD/xhtml1-transitional.dtd">
<html xmlns="www.w3.org/1999/xhtml">
 <head>
  <title> Simple Content Management System </title>
 </head>

 <body>
<h1>Simple Content Management System</h1>
<h2>Step 1a: Upload Asset</h2>
<p>This simple content management system will accept uploaded
files (assets) that are of the following type: MS-Word document,
Adobe PDF document, plain-text document, GIF image, or JPEG
image.  The process to add an asset is relatively simple and is
composed of three steps:</p>
<ul>
      <li>Step One: Upload your asset using this page.</li>
      <li>Step Two: Add associated metadata.</li>
```

```
</ul>
     <p>(Repeat steps one and two until all files have been
     updated and annotated.)</p>
<ul>
     <li>Step Three: Export your final asset list to an XML
     document.</li>
</ul>
<!-
This form will "post" the data, or send it behind the scenes
to the file specified in the action attribute.  In this case, it
goes to a file named upload_asset.php. Note the enctype=
"multipart/form-data" which is a special encoding instruction
that is used to configure the form for file uploading.
->
<form enctype="multipart/form-data" method="post" action=
"upload_asset.php">
<table>
     <tr>
          <td>Asset File:</td>
          <!- This is a form filename field.  It allows us to
          upload files from our computer to a server. ->
          <td><input type="file" name="asset" size="35" />
          </td>
     </tr>
</table>
<br />
<!-
When the submit button is pressed, the filename   will be
sent to the upload_asset.php page.  Only files of size 100k
or smaller will be allowed (this is what is specified in the
value="100000").
->
<input type="hidden" name="MAX_FILE_SIZE" value="100000" />
<input type="submit" name="submit" value="Upload Asset" />
</form>
</body>
</html>
<!- End of "upload_asset.html" file ->
```

File 2: upload_asset.php

```
<!-
Project: Content Management System.
Component: Writer (File 2 of 5).
Filename: "upload_asset.php".
```

```
Purpose: Process the uploaded asset, rename it, and move it to
the assets directory.
->
<!DOCTYPE html PUBLIC "-//W3C//DTD XHTML 1.0 Transitional//EN"
"www.w3.org/TR/xhtml1/DTD/xhtml1-transitional.dtd">
<html xmlns="www.w3.org/1999/xhtml">
 <head>
  <title> Simple Content Management System </title>
 </head>
 <body>
<h1>Simple Content Management System</h1>
<h2>Step 1b: Upload Results Page</h2>
<?php
/*
The $bufferFilename variable is used to store the name of the
temporary file we will be writing to.  For this exercise, this
file needs to be in the same directory as this script, but it
could be changed by including the directory path as part of
this variable.
*/
$bufferFilename = "asset_buffer.txt";

/*
This array defines the different types of files our CMS supports.
*/
$supportedTypes = array();
$supportedTypes[] = "image/gif";
$supportedTypes[] = "image/jpeg";
$supportedTypes[] = "image/pjpeg";
$supportedTypes[] = "application/msword";
$supportedTypes[] = "text/plain";
$supportedTypes[] = "application/pdf";
/*
This array is used to map the appropriate file extensions onto
our assets.
*/
$fileSuffixes = array();
$fileSuffixes["image/gif"] = ".gif";
$fileSuffixes["image/pjpeg"] = ".jpg";
$fileSuffixes["image/jpeg"] = ".jpg";
$fileSuffixes["application/msword"] = ".doc";
$fileSuffixes["text/plain"] = ".txt";
$fileSuffixes["application/pdf"] = ".pdf";
/*
```

```
This function will open the asset "buffer" text file to obtain
the index id number of the last added asset.
*/
function getLastAssetNumber($filename)
      {
      /* open in "read only" mode.  If file doesn't exist,
      the CMS will treat this asset as the first item in the
      collection. */
      if ($fp = @fopen($filename, "r"))
            {
            /* read the last asset number */
            $lastAssetNumber = fgets($fp);
            }
      /* if this number is not available, then return zero as
      the last asset number */
      if (empty($lastAssetNumber)) $lastAssetNumber = 0;
      /* this will be either zero (new collection) or some
      number N of previously existing assets */
      return($lastAssetNumber);
      }
/*
This function will open the asset "buffer" text file and write
the index id number of the most recently added asset.
*/
function writeNewAssetNumber($filename, $newNum)
      {
      $fp = fopen($filename, "w"); // open in "overwrite" mode
      $bytes_written = fputs($fp, $newNum."\n");
      }
/*
The first thing we will do when uploading our asset file is
to access the temporary filename from the special $_FILES
array that is automatically created by PHP based on the "FILE"
form element submitted on the prior page.  This filename is
stored in an index named [x]["tmp_name"] where x is the name
of the form field specified in the form.  In this case, it is
named "asset".  We can also obtain the type of file using the
["type"] index (this value may be slightly different in
browsers other than Internet Explorer).
*/
$originalFileName = $_FILES["asset"]["name"];
$fileName = $_FILES["asset"]["tmp_name"];
$fileType = $_FILES["asset"]["type"];
/*
Here, we will get the file suffix (.doc, .pdf, etc.) which can
help to minimize problems with certain browsers not recognizing
```

```
the default file types correctly.  First, the explode function
will turn the file name into an array split by where the period
appears.  For example, if the filename was "information.doc",
then $temp[0] would contain the value "information" and
$temp[1] would contain the value "doc".  So, we store the suffix
into a variable named $fileSuffix.
*/
$temp = explode(".",$originalFileName);
$fileSuffix = $temp[1];
/*
Next, we need to make sure this variable contains data.
If not, we provide a link back to the previous page and ask
the user to enter data for the filename field.  The $errors
variable will keep track of total errors and if this variable
is greater than zero, the user will be asked to go back and
provide any missing data.
*/
if (empty($fileName))
      {
      echo "A valid filename is required!<br />";
      $errors++;
      }
/*
Here is the part where we check the total errors and redirect
the user if need be.
*/
if ($errors == 1)
      {
      echo "There was an error with your data.<br />";
      echo "Please <a href=\"upload_asset.html\">go back</a>
      to correct it!<br />";
      }
else
      {
      /*
      This code will check to make sure the file type is
      supported for our content management system (jpg, gif,
      doc, txt, or pdf)
      */
      /* If so, we need to try and process the uploaded
      file. */
      if (in_array($fileType, $supportedTypes) || ($fileSuffix ==
      "doc"))
            {
            /* find the number of the last added asset */
```

```
$lastAssetNumber = getLastAssetNumber
($bufferFilename);

/* add one to it so that it will be unique */
$currentAssetNumber = $lastAssetNumber + 1;

/* tell the user the calculated file type */
echo "File is type: $fileType.<br />";

/* find the correct file suffix to use (e.g., .pdf or
.doc) */
$newFileSuffix = $fileSuffixes[$fileType];

/* if unable to find new file suffix, use the existing
file suffix */
if (empty($newFileSuffix)) $newFileSuffix = "."
.$fileSuffix;

/* create a new filename based on assetX.Y
(X=new id number, Y=suffix) */
$newFileName = "asset".$currentAssetNumber.
$newFileSuffix;

/*
Assuming everything is okay, move the uploaded
file to the assets directory and provide a link to
step 2.
*/
if (move_uploaded_file($fileName, "assets\\
$newFileName"))
      {
      echo "Asset added as ".$newFileName.".";

      /* Call the function we defined at the top
      of this page to write the asset number to a
      text file */
      writeNewAssetNumber($bufferFilename,
      $currentAssetNumber);
      ?>

      <!— This code will display a link to the
      next step in the process, which is adding
      metadata. —>
      <p>Click here to go to Step 2 and <a href=
      "add_asset_metadata.php?asset_id=<?php echo
      $currentAssetNumber; ?>&asset_filename=
```

```
                    <?php echo $newFileName; ?>">add metadata for
                    this asset</a>.</p>
                    <?php
                    }
                else
                    {
                    /* If the file is unable to be uploaded,
                    provide an error message */
                    echo "Unable to add asset ".$fileName.".";
                    }
                }
        else
                {
                /* If there is a problem with the file type, display
                an error message for the user. */
                echo "Sorry, but $fileType is not a supported file
                type.<br />";
                }
        }
?>
</body>
</html>
<!— End of "upload_asset.php" file —>
```

File 3: add_asset_metadata.php

```
<!—
Project: Content Management System.
Component: Writer (File 3 of 5).
Filename: "add_asset_metadata.php".
Purpose: Add associated metadata for uploaded assets.
—>
<!—
We need to first obtain the asset identifier that is sent from
the upload page.  We will use this identifier in our form below
to indicate the ID for a particular asset.  Since we are
passing this information through the URI, we need to use the
GET protocol.  We will also grab the asset&apos;s filename using
the same method.
—>
<?php
$assetId = $_GET["asset_id"];
$assetFilename = $_GET["asset_filename"];
$temp = explode(".",$assetFilename);
$defaultName = $temp[0];
$suffix = $temp[1];
```

```
switch($suffix)
      {
      case "doc":
            $mediaType = "MS-Word Document";
            break;
      case "pdf":
            $mediaType = "Adobe PDF Document";
            break;
      case "gif":
      case "jpg":
            $mediaType = "Image File";
            break;
      case "txt":
            $mediaType = "Plain text Document";
            break;
      default:
            $mediaType = "Unknown Media";
            break;
      }

?>
<!DOCTYPE html PUBLIC "-//W3C//DTD XHTML 1.0 Transitional//EN"
"www.w3.org/TR/xhtml1/DTD/xhtml1-transitional.dtd">
<html xmlns="www.w3.org/1999/xhtml">
 <head>
  <title> Simple Content Management System </title>
 </head>
<body>
<h1>Simple Content Management System</h1>
<h2>Step 2a: Add Asset Metadata</h2>

<!—
This form will "post" the data, or send it behind the scenes
to the file specified in the action attribute.  In this case, it
goes to a file named process_asset_metadata.php
—>
<p>When selecting categories and years, hold down the control
key to select multiple values.</p>
<form method="post" action="process_asset_metadata.php">
<table>
      <tr>
            <td>Asset Id:</td>
            <!— This is a simple text field that is "read only".
            The value is sent from the uploading
            form. —>
```

```
        <td colspan="3"><input type="text" readonly name=
        "asset_id" value="<?php echo $assetId; ?>"></td>
    </tr>
    <tr>
        <td>Asset Filename:</td>
        <!- This is a simple text field that is "read only".
        The value is sent from the uploading
        form. ->
        <td colspan="3"><input type="text" readonly
        name="asset_filename" value="<?php echo
        $assetFilename; ?>"></td>
    </tr>
    <tr>
        <td>Asset Name:</td>
        <!- This is a simple text field ->
        <td colspan="3"><input type="text" name=
        "asset_name" value="<?php echo $defaultName;
        ?>"></td>
    </tr>
    <tr>
        <td valign="top">Asset Category:</td>
        <!- This is a drop down field ->
    <td><select name="asset_category[]" size="10" multiple>
    <option value="New Media">New Media</option>
    <option value="Print">Print</option>
    <option value="Management">Management</option>
    <option value="Advertising">Advertising</option>
    <option value="Alpha Project">Alpha Project</option>
    <option value="Beta Project">Beta Project</option>
    <option value="Gamma Project">Gamma Project</option>
    <option value="Delta Project">Delta Project</option>
    <option value="Epsilon Project">Epsilon Project</option>
    <option value="Needs Revision">Needs Revision</option>
    </select>
    </td>
        <td valign="top">Years Used:</td>
        <!- This is a drop down field ->
        <td>
        <select name="asset_years_used[]"size="10"
        multiple>
            <?php
            for ($i = 2010; $i > 2000; $i-)
                {
                if (date("Y") == $i)
        echo "<option value=\"$i\" selected>$i</option>";
                else
```

```
                        echo "<option value=\"$i\">$i
                        </option>";
                        }
                ?>
            </select>
        </td>
    </tr>
    <tr>
        <td>Asset Media Type:</td>
        <!— This is a simple text field —>
        <td colspan="3"><input type="text" name=
        "asset_media_type" value="<?php echo $mediaType;
        ?>"></td>
    </tr>
    <tr>
        <td>Asset Description</td>
        <!— This is a textarea, which provides more room
        for writing —>
        <td colspan="3"><textarea name="asset_description"
        rows="5" cols="45"></textarea></td>
    </tr>
</table>
</body>
<br /><br />
<!—
When the submit button is pressed, each of the values typed
into the text fields and the textarea form will be stored in the
names associated with each form element (asset id, asset
filename, asset_name, asset_category, asset_media_type, and
asset_description).
—>
 <input type="submit" name="submit" value="Add Asset Metadata">
 </form>
</html>
<!— End of "add_asset_metadata.php" file —>
```

File 4: process_asset_metadata.php

```
<!—
Project: Content Management System.
Component: Writer (File 4 of 5).
Filename: "process_asset_metadata.php".
Purpose: Checks metadata for errors and then writes out
metadata to a temporary file.
—>
```

```
<!DOCTYPE html PUBLIC "-//W3C//DTD XHTML 1.0 Transitional//EN"
"www.w3.org/TR/xhtml1/DTD/xhtml1-transitional.dtd">
<html xmlns="www.w3.org/1999/xhtml">
<head>
      <title> Simple Content Management System </title>
</head>
<body>
<h1>Simple Content Management System</h1>
<h2>Step 2b: Metadata Results Page</h2>
<?php
/*
The $bufferFilename variable is used to store the name of the
temporary file we will be writing to.  For this exercise, this
file needs to be in the same directory as this script, but it
could be changed by including the directory path as part of
this variable.
*/
$bufferFilename = "metadata_buffer.txt";
/*
The first thing we will do is to grab the various
characteristics of our asset from the form.
*/
$assetId = $_POST["asset_id"];
$assetFilename = $_POST["asset_filename"];
$assetName = $_POST["asset_name"];
$assetMediaType = $_POST["asset_media_type"];
$assetDescription = $_POST["asset_description"];
$assetCategory = $_POST["asset_category"][0];
$assetYearsUsed = $_POST["asset_years_used"][0];
/*
Next, we need to make sure these variables all contain data.
If not, we provide a link back to the previous page and ask
the user to enter data for all fields.  The $errors variable
will keep track of total errors and if this variable is greater
than zero, the user will be asked to go back and provide any
missing data.
*/
/* If the asset ID is missing, show an error message. */
if (empty($assetId))
      {
      echo "Asset id is missing!<br />";
      $errors++;
      }
/* If the asset name is missing, show an error message. */
if (empty($assetName))
      {
```

```
	echo "Asset name is required!<br />";
	$errors++;
	}
/* If the asset filename is missing, show an error message. */
if (empty($assetFilename))
	{
	echo "Asset filename is required!<br />";
	$errors++;
	}
/* If the asset category is missing, show an error message. */
if (empty($assetCategory))
	{
	echo "Asset category is required!<br />";
	$errors++;
	}
else
	/* Otherwise, loop through and add a comma to separate
	each category */
	{
	foreach($_POST["asset_category"] as $cat)
		{
		if (empty($temp))
			$temp = $cat;
		else
			$temp = $temp.", ".$cat;
		}
	}
$assetCategory = $temp;
/* If no years are selected, show an error message. */
if (empty($assetYearsUsed))
	{
	echo "Asset year is required!<br />";
	$errors++;
	}
else
	/* Otherwise, loop through and add a comma to separate
	each year used */
	{
	foreach($_POST["asset_years_used"] as $year)
		{
		if (empty($temp2))
			$temp2 = $year;
		else
			$temp2 .= ", ".$year;
		}
	}
```

```
$assetYearsUsed = $temp2;
/* If the asset media type is missing, show an error message.
*/
if (empty($assetMediaType))
      {
      echo "Asset media type is required!<br />";
      $errors++;
      }
/* If the asset description is missing, show an error message.
*/
if (empty($assetDescription))
      {
      echo "Asset description is required!<br />";
      $errors++;
      }
/*
Here is the part where we check the total errors and redirect
the user if need be.
*/
if ($errors > 0)
      {
      echo "There were one or more errors with your data.
      <br />";
      echo "Please <a href=\"javascript:history.back()
      \">go back</a> button and correct your input!<br />";
      }
else
      {
      /*
      What we are going to do now is to "glue" the variables
      together into a single line using the PHP concatenation
      operator, which is the period (.) In order to keep each
      variable distinct, we will use the character sequence
      *** to separate each entry.
      Also, a \n (newline escape sequence) is used to make each
      additional entry begin on a new line.
      */
      $gluedString =
$assetId."***".$assetFilename."***".$assetName."***".$assetCate
gory."***".$assetYearsUsed."***".$assetMediaType."***".$assetDe
scription."\n";
      /*
      Now, we create a connection to the second buffer text
      file; the first argument is the filename and the second
      argument is the "mode". "a" mode means "append" and new
      text content will be added to the end of the file.
```

```
$fp is a variable that holds the "link" or connection
to the file that we open and then sends this link to the
fwrite function for the purpose of writing data to the
file.
*/
if ($fp = fopen($bufferFilename,"a"))
      {
      /* This will write the glued string to the file */
      if (fwrite($fp, $gluedString))
            {
            echo "Asset added to buffer file!<br />";
            /* This closes the file after the writing
            is done */
            fclose($fp);
            }
      else
            /* This is displayed if the file cannot be
            written to */
            {
            echo "There was an error adding your asset.
            <br />";
            echo "Please check the file permissions for
            your buffer file.<br />";
            echo "The buffer filename is:
            ".$bufferFilename.".<br />";
            }
      }
else
      {
      echo "Unable to open asset metadata file!";
      }
}
?>
<p>Click here to return to Step 1 and <a href="upload_asset.
html">add another asset</a>.</p>
<p>Click here to proceed to Step 3 and <a href="finalize_
assets.php">finalize all assets</a> by writing them to an
XML file.</p>
</body>
</html>
<!– End of "process_asset_metadata.php" file –>
```

File 5: finalize_assets.php

```
<!–
Project: Content Management System.
```

```
Component: Writer (File 5 of 5).
Filename: "finalize_assets.php".
Purpose: Loops through metadata file and writes out all assets
and their associated metadata to a valid XML file.
->
<!DOCTYPE html PUBLIC "-//W3C//DTD XHTML 1.0 Transitional//EN"
"www.w3.org/TR/xhtml1/DTD/xhtml1-transitional.dtd">
<html xmlns="www.w3.org/1999/xhtml">
 <head>
  <title> Simple Content Management System </title>
 </head>
<body>
<h1>Simple Content Management System</h1>
<h2>Step 3: Generate XML Document</h2>
<?php
/*
The $bufferFilename variable is used to store the name of the
temporary file we will be reading from. For this exercise, this
file needs to be in the same directory as this script,
but it could be changed by including the directory path as
part of this variable.  The $xmlOutputFilename will be the
name of the file used to store our XML collection.  Note that
this default filename will automatically append the current
month (in the format Jan, Feb, Mar, etc.) day, and year
(in four digit format) to the latter half of the filename.
*/
$bufferFilename = "metadata_buffer.txt";
$metaDataDirectory = "metadata\\";
$xmlOutputFilename = "asset_collection_".date("MdY").".xml";
/* This function writes the beginning of the XML file using UTF-
8 (Latin) encoding */
function startXMLdocument($xmlString)
      {
      $xmlString = "<?xml version=\"1.0\" encoding=\"utf-8\"
      ?>\n";
      $xmlString .= "<!- XML Asset Collection Document ->\n";
      $xmlString .= "<!- Automatically generated by Content
      Management System ->\n";
      $xmlString .= "<!- Created on ".date("MdY")." ->\n";
      $xmlString .= "<?xml-stylesheet type=\"text/xsl\"
      href=\"asset_collection_transform.xsl\"?>\n\n";
      return($xmlString);
      }
/*
This function writes the document type definition (DTD) as well
as the top level asset_collection container to the XML file
```

```
*/
function insertDTD($xmlString)
      {
      $xmlString .= "<!DOCTYPE asset_collection [\n";
      $xmlString .= "\t<!ELEMENT asset_collection (asset)*>\n";
      $xmlString .= "\t<!ELEMENT asset (id,filename,name,
      category,years,media,description)>\n";
      $xmlString .= "\t<!ELEMENT id (#PCDATA)>\n";
      $xmlString .= "\t<!ELEMENT filename (#PCDATA)>\n";
      $xmlString .= "\t<!ELEMENT name (#PCDATA)>\n";
      $xmlString .= "\t<!ELEMENT category (#PCDATA)>\n";
      $xmlString .= "\t<!ELEMENT years (#PCDATA)>\n";
      $xmlString .= "\t<!ELEMENT media (#PCDATA)>\n";
      $xmlString .= "\t<!ELEMENT description (#PCDATA)>\n";
      $xmlString .= "]>\n\n";
      $xmlString .= "<asset_collection>\n";
      return($xmlString);
      }
/*
This function closes the asset_collection container
*/
function endXMLdocument($xmlString)
      {
      $xmlString .= "</asset_collection>\n";
      return($xmlString);
      }
/*
This function will loop through the temporary buffer file
and find each asset and its metadata.  It can tell where one
asset ends and another begins because they are each separated
by newlines (or the \n character sequence).  It can also
tell where one element ends and the next one begins by the
*** character sequence.  Each asset is converted into an XML
element with its values written out into the XML file as
character data.
*/
function addAssetCollection($filename,$xmlString)
      {
      if ($filecontents = file($filename))
            {
            $xmlString = startXMLdocument($xmlString);
            $xmlString = insertDTD($xmlString);
            foreach($filecontents as $singleLine)
                  {
                  $line = explode("\n",$singleLine);
                  $units = explode("***",$line[0]);
```

```
                $xmlString .= "\t\t<asset>\n";
                $xmlString .= "\t\t\t<id>".$units[0]."
                </id>\n";
                $xmlString .= "\t\t\t<filename>".$units[1]."
                </filename>\n";
                $xmlString .= "\t\t\t<name>".$units[2]."
                </name>\n";
                $xmlString .= "\t\t\t<category>".$units[3]."
                </category>\n";
                $xmlString .= "\t\t\t<years>".$units[4]."
                </years>\n";
                $xmlString .= "\t\t\t<media>".$units[5]."
                </media>\n";
                $xmlString .= "\t\t\t<description>".
                $units[6]."</description>\n";
                $xmlString .= "\t\t</asset>\n";
                }
            $xmlString = endXMLdocument($xmlString);
            return($xmlString);
            }
        else
            {
            echo "Unable to open file!";
            }
        }
/*
This function handles the actual writing of the XML document
to a permanent file. It will use the same metadata directory
and filename as specified by the variables at the top of this
script.
*/
function writeXMLFile($metaDataDirectory, $fileName, $xmlString)
    {
    $fp = fopen($metaDataDirectory.$fileName, "w");
    if ($bytesWrittenToFile = fwrite($fp, $xmlString))
            {
            echo "XML file created successfully!<br />";
            return TRUE;
            }
    else
            {
            echo "There was a problem writing to file.<br />";
            echo "Here is the XML content: <br /><pre>
            $xmlString</pre>";
            return FALSE;
```

```
        }
    }

/*
Here is where the addAssetCollection is "called" to create
the XML string that will be written to the permanent file.
For now, the entire XML document will be stored in a temporary
string variable named $xml.
*/
$xml = addAssetCollection($bufferFilename, $xmlString);
/*
This segment of code writes the XML string to a permanent file.
If it is unable to create the file, the error message defined in
the writeXMLFile function will be displayed.
*/
if (writeXMLFile($metaDataDirectory, $xmlOutputFilename, $xml))
    {
    echo "<br />";
    echo "Click here to <a href=\"".$metaDataDirectory.$xml
    OutputFilename."\">view your collection using an XSL
    transform</a>.<br />";
    echo "Click here to <a href=\"upload_asset.html\">begin
    a new asset collection</a>.<br />";
    echo "Click here to <a href=\"select_xml.php\">select an
    XML file to parse using the CMS parser</a>.";
    }
?>

</body>
</html>
<!- End of "finalize_assets.php" file ->
```

File 6: select_xml.php

```
<!-
Project: Content Management System.
Component: Reader/Parser (File 1 of 2).
Filename: "select_xml.php".
Purpose: Selects the XML document to send to the parser.
->
<?php
/*
This variable holds the name of the directory containing the
metadata for the content management system */
$metaDataDirectory = "metadata";
/*
```

```
This variable would need to be modified to run on non-Windows
operating systems.  This is a system command that will list the
current files with an XML extension.  The /b characters tell the
system to print this information in "bare" mode with only
filenames in the output.  This system will also work only if the
XML documents are stored in metadata directory as defined in the
$metaDataDirectory variable.
*/
$directoryListingCommand = "dir /b {$metaDataDirectory}\*.xml";
/*
This command will load the results from this system command
into an array.  Each filename will be stored in a separate index
location in the array.  Explode is a special function in PHP
that will figuratively blast apart (or tokenize) a string based
upon the character used as the first argument to the explode
function.  For instance, in this usage, the explode command
will take the string returned from the system directory command
and then separate it into individual units according to where
new lines are indicated in the string (recall that the
\n character sequence indicates a new line in PHP).  These
individual units are then stored in an array named "files" which
is iterated through below in order to list each file as a
selection in the drop down menu.
*/
$temp = '$directoryListingCommand';
$files = explode("\n",$temp);
?>
<!DOCTYPE html PUBLIC "-//W3C//DTD XHTML 1.0 Transitional//EN"
"www.w3.org/TR/xhtml1/DTD/xhtml1-transitional.dtd">
<html xmlns="www.w3.org/1999/xhtml">
 <head>
  <title> Simple Content Management System </title>
 </head>
<body>
<h1>Simple Content Management System</h1>
<h2>Select XML File</h2>
<form method="get" action="parse_xml.php">
<?php
if ($files[0] == "")
      {
      echo "There are no XML files to select.  Click <a href=\
      "upload_asset.html\">here</a> to create a library.
      <br />";
      }
else
      {
```

```
      ?>
      <select name="inputFilename">
      <?php
      foreach ($files as $file)
            {
            if ($file <> "") echo "<option value=\"".
            $file."\">".$file."</option>";
            }
      ?>
      </select>
      <input type="submit" value="go">
      <?php
      }
?>
</form>
</body>
</html>
<!– End of "select_xml.php" file –>
```

File 7: parse_xml.php

```
<!–
Project: Content Management System.
Component: Reader/Parser (File 2 of 2).
Filename: "parse_xml.php".
Purpose: Parse the XML document and display the assets
according to type.
–>
<?php
/*
The $rootPath variable is used to indicate the full file path to
the asset files on your hard drive.  This is necessary for the
parser to check that the asset file exists.  This check occurs
later in the script.  The $metaDataDirectory variable holds the
value for the directory containing the XML files.
*/
$rootPath = "c:\\xampp\\htdocs\\xmlparsers\\cms\\";
$metaDataDirectory = "metadata\\";
?>
<!DOCTYPE html PUBLIC "-//W3C//DTD XHTML 1.0 Transitional//EN"
"www.w3.org/TR/xhtml1/DTD/xhtml1-transitional.dtd">
<html xmlns="www.w3.org/1999/xhtml">
 <head>
  <title> Simple Content Management System </title>
 </head>
<body>
```

```
<h1>Simple Content Management System</h1>
<h2>Asset Collection</h2>
<?php
/*
This section of code will check to see if an input file has been
specified through the URI (e.g.
parse_xml.php?inputFilename=test.xml). If not, the parser will
display an error message and will exit. "Exit" is a special
PHP function that will exit the script completely and display a
message if a failure is encountered.
*/
if (isset($_GET["inputFilename"]))
      $inputFilename = $_GET["inputFilename"];
else
      exit("No input file is specified. Please specify an input
      file using by appending a question mark at the
      end of this filename and then including the name of the
      XML file (e.g. parse_xml.php?inputFilename=test.xml). Or,
      you can also use the built-in directory browser
      <a href=\"select_xml.php\">available here</a>.");
?>
<!— This will display the name of the current document being
parsed —>
<hr />
<p>Parsing document named "<?php echo $inputFilename; ?>".
</p>
<?php
/*
The startElement function is the function that will be called
whenever a beginning XML tag is encountered. For instance,
if the parser is currently parsing an element named "<asset_
collection>", the parser will jump to the block of code
specified by the case "asset_collection" condition. In PHP, the
switch statement works like a series of IF statements where the
same value is being compared in each condition. In this case,
we are looking at the $name variable&apos;s value and printing
code to the browser depending on which XML element is currently
being processed.
*/
function startElement($parser, $name, $attributes)
      {
      switch($name)
            {
            case "xml":
                  return;
                  break;
```

```
			case "asset_collection":
				echo "<table border=\"1\"width=\"640\">";
				echo "<tr>";
				echo "<th>Id</th>";
				echo "<th>Filename</th>";
				echo "<th>Asset name</th>";
				echo "<th>Categories</th>";
				echo "<th>Years Used</th>";
				echo "<th>Media</th>";
				echo "<th>Description</th>";
				echo "</tr>";
				break;
			case "asset":
				echo "<tr>";
				break;
			default:
				echo "<td>";
				break;
			}
		}
/*
This function is used to display the character data that is
encapsulated within each element tag. Global is a special
keyword in PHP that enables a function to access a variable
defined outside the scope of the function. In this case, we
need to use the $rootPath variable that is defined at the top of
this script. Next, we use the eregi function to scan the
$value variable and look for any files that end with .jpg, .gif,
.doc, .pdf, or .txt. These will be the asset files themselves
rather than the metadata annotations. We are checking these
files because we need to do some special error checking to make
sure the assets actually exist before we provide links to them
(see next comment below).
*/
function characterData($parser, $value)
		{
		global $rootPath;
		if (eregi("[.]jpg|[.]gif|[.]doc|[.]pdf|[.]txt",$value))
			{
			echo $value;
			$longFileName = $rootPath."assets\\".$value;
/*
This segment of code will check to verify that a file actually
exists on the hard disk drive before providing a link to it.
If a file is missing, the text "missing" will be displayed
instead. This is an advantage to using a custom designed
```

```
parser rather than simple XSL transformations in that we can
use built-in functions like this for additional error checking.
*/
            if (file_exists($longFileName))
                  echo "<br /><a href=\"assets/".$value."\"
                  target=\"_blank\">$value</a>";
            else
                  echo "<br />(missing)";
            }
      else
            {
            echo $value;
            }
      }
/*
The endElement function is the function that will be called
whenever an ending XML tag is encountered.  For instance,
if the parser is currently parsing an element named
"</asset_collection>", the parser will jump to the block of
code specified by the case "asset_collection" condition.
In PHP, the switch statement works like a series of IF
statements where the same value is being compared in each
condition.  In this case, we are looking at the $name
variable&apos;s value and printing code to the browser
depending on which XML element is currently being processed.
If </asset_collection> is parsed, this tells us that the
collection is now entirely scanned and we can go ahead and
close the table using the HTML tag </table>.
*/
function endElement($parser, $name)
      {
      switch($name)
            {
            case "xml":
                  return;
                  break;
            case "asset_collection":
                  echo "</table>";
                  break;
            case "asset":
                  echo "</tr>";
                  break;
            default:
                  echo "</td>";
                  break;
            }
```

```
    }
/*
The function xml_parser_create() is used to create a new parser
in PHP.  It then returns a "resource handle" which is a special
variable that can be referenced from that point on when working
with the parser.  In this case, our resource handle is stored
in a variable named $xml.
*/
$xml = xml_parser_create();
/*
The XML_OPTION_CASE_FOLDING option is used to make sure that
the parser does not transform all elements into uppercase when
parsing an XML document.  Since XML is case-sensitive, leaving
case folding on, which is the default, could cause problems
when parsing the file.  Here, setting this option to "false"
means that the parser will leave lowercase or mixed case
elements as they were originally created.
*/
xml_parser_set_option($xml, XML_OPTION_CASE_FOLDING, false);
/*
This function is used to specify functions that should be
called when starting elements and ending elements are
encountered in the XML hierarchy.  We have defined these
functions, named startElement() and endElement() in the
top portion of this code.  Anytime a new beginning element
(e.g. <asset>) is encountered, the startElement function will
be called and the code can perform conditional actions based
on the name of the element.  Then, when an ending element
(e.g. </asset>) is processed, the endElement function is
called.
*/
xml_set_element_handler($xml, "startElement", "endElement");
/*
This function is very similar to the one above, but instead of
creating a function to handle elements, it creates a function
to handle the character data itself.  For example, when values
that exist between XML element tags are processed, these values
are sent to the function defined here.  In this case, we named
it characterData().
*/
xml_set_character_data_handler($xml, "characterData");
/*
This block of code will read the file into memory and then
attempt to parse it.  If the file is not found, an error message
will be displayed to the user.  Similarly, if the file is unable
to be parsed (which could be due to malformed XML content or
```

```
file corruption) a secondary error message will be displayed.
The ampersand (@) in front of the file_get_
contents() function is used to suppress warning messages.
Since we are already providing error messages in the case of a
failed operation, we don&apos;t need to clutter the user&apos;s
browser with additional warnings from PHP.
*/
if ($xmlString = @file_get_contents($metaDataDirectory.
$inputFilename) or die("Input file not available."))
      {
      xml_parse($xml, $xmlString);
      }
else
      {
      echo "Error. Unable to load input file for parsing.
<br />";
      }
?>
<br />
<a href="select_xml.php">Select Different Asset Database</a>
  
<a href="upload_asset.html">Begin New Asset Collection</a>
  
</body>
</html>
<!— End of "parse_xml.php" file —>
```

File 8: asset_collection_transform.xsl

```
<?xml version="1.0" encoding="utf-8"?>
<xsl:stylesheet version="1.0" xmlns:xsl="www.w3.org/1999/
XSL/Transform">
<!—
Project: Content Management System.
Component: XSL Transform (File 1 of 1).
Filename: "asset_collection_transform.xsl".
Purpose: Transform an XML asset collection into an HTML-based
tabular table format.
—>
<xsl:template match="/">
<html>
 <head>
  <title> Content Management System: Current Asset Collection
</title>
 </head>
      <body>
```

```
        <!– transform assets into a table –>
        <h1>Asset Collection</h1>
        <table border="1">
        <tr>
              <th>Id</th>
              <th>Filename</th>
              <th>Name</th>
              <th>Categories</th>
              <th>Years Used</th>
              <th>Media</th>
              <th>Description</th>
        </tr>
        <xsl:for-each select="asset_collection/asset">
        <tr>
              <td><xsl:value-of select="id"/></td>
              <td><a href="../assets/{filename}"><xsl:value-of
              select="filename"/></a>
              </td>
              <td><xsl:value-of select="name"/></td>
              <td><xsl:value-of select="category"/></td>
              <td><xsl:value-of select="years"/></td>
              <td><xsl:value-of select="media"/></td>
              <td><xsl:value-of select="description"/></td></tr>
        </xsl:for-each>
        </table>
        </body>
</html>
</xsl:template>
</xsl:stylesheet>
<!– End of "asset_collection_transform.xsl" file –>
```

File 9: metadata_buffer.txt (sample)

```
1***asset1.doc***asset1***New Media, Print, Epsilon
Project***2007, 2005***MS-Word Document***This is a sample
brochure idea that was shown to clients before the new Epsilon
2.0 software was released.  It needs a bit more work before it
can be sent to the printer.
2***asset2.pdf***asset2***New Media, Print, Epsilon
Project***2007, 2005***Adobe PDF Document***This is the PDF
version of the Epsilon product brochure.
3***asset3.txt***asset3***New Media, Print, Gamma Project,
Epsilon Project***2007, 2005***Plain text Document***These are
some printing notes from the Gamma Project that might also be
useful for the Epsilon project.
```

```
4***asset4.jpg***asset4***Epsilon Project***2007***Image
File***JPEG logo for the Epsilon project.  Red text with drop
shadow.
```

File 10: asset_collection_Dec192007.xml (sample)

```
<?xml version="1.0" encoding="utf-8"?>
<!— XML Asset Collection Document —>
<!— Automatically generated by Content Management System —>
<!— Created on Dec192007 —>
<?xml-stylesheet type="text/xsl" href="asset_collection_
transform.xsl"?>

<!DOCTYPE asset_collection [
      <!ELEMENT asset_collection (asset)*>
      <!ELEMENT asset (id,filename,name,category,years,media,
      description)>
      <!ELEMENT id (#PCDATA)>
      <!ELEMENT filename (#PCDATA)>
      <!ELEMENT name (#PCDATA)>
      <!ELEMENT category (#PCDATA)>
      <!ELEMENT years (#PCDATA)>
      <!ELEMENT media (#PCDATA)>
      <!ELEMENT description (#PCDATA)>
]>
<asset_collection>
            <asset>
                  <id>1</id>
                  <filename>asset1.doc</filename>
                  <name>asset1</name>
                  <category>New Media, Print, Epsilon Project
                  </category>
                  <years>2007, 2005</years>
                  <media>MS-Word Document</media>
                  <description>This is a sample brochure idea
                  that was shown to clients before the new
                  Epsilon 2.0 software was released.  It needs
                  a bit more work before it can be sent to the
                  printer.</description>
            </asset>
            <asset>
                  <id>2</id>
                  <filename>asset2.pdf</filename>
                  <name>asset2</name>
                  <category>New Media, Print, Epsilon Project
                  </category>
```

```
            <years>2007, 2005</years>
            <media>Adobe PDF Document</media>
            <description>This is the PDF version of the
            Epsilon product brochure.</description>
      </asset>
      <asset>
            <id>3</id>
            <filename>asset3.txt</filename>
            <name>asset3</name>
            <category>New Media, Print, Gamma Project,
            Epsilon Project</category>
            <years>2007, 2005</years>
            <media>Plain text Document</media>
            <description>These are some printing notes
            from the Gamma Project that might also be
            useful for the Epsilon project.</description>
      </asset>
      <asset>
            <id>4</id>
            <filename>asset4.jpg</filename>
            <name>asset4</name>
            <category>Epsilon Project</category>
            <years>2007</years>
            <media>Image File</media>
            <description>JPEG logo for the Epsilon
            project.  Red text with drop shadow.
            </description>
      </asset>
</asset_collection>
```

Appendix D
Source Code for Single Sourcing Demonstration

File 1: add_module.html

```
<!–
Project: Single Sourcing Demo.
Component: Writer (File 1 of 3).
Filename: "add_module.html".
Purpose: Add a documentation module to our master XML file.
–>
<!DOCTYPE html PUBLIC "-//W3C//DTD XHTML 1.0 Transitional//EN"
"www.w3.org/TR/xhtml1/DTD/xhtml1-transitional.dtd">
<html xmlns="www.w3.org/1999/xhtml">
 <head>
  <title> Single Sourcing Demo: Add Documentation Module
</title>
 </head>
<body>
<h1>Single Sourcing Demo</h1>
<h2>Step 1: Add Module</h2>
 <p>Select a feature to document, then enter the skill
 level your documentation is most appropriate for.  Type your
 documentation, then click "Submit Documentation" to add your
 content to the master XML document.</p>
 <form method="post" action="process_module.php">
 <table>
      <tr>
            <td>Feature:</td>
            <td><select name="feature">
                        <option value="introduction">
                        Introduction</option>
                        <option value="tutorial">Accessing the
                        Tutorial</option>
                        <option value="tutorial">Adding Assets
                        </option>
                  </select>
```

```
            </td>
      </tr>
      <tr>
            <td>Skill Level:</td>
            <td><select name="skill_level">
                        <option value="beginner">Beginner
                        </option>
                        <option value="intermediate">
                        Intermediate</option>
                        <option value="advanced">Advanced
                        </option>
                  </select>
            </td>
      </tr>
      <tr>
            <td>Documentation:</td>
            <td><textarea name="documentation" rows="10"
            cols="55"></textarea>
            </td>
      </tr>
</table>
 <br />
 <input type="submit" name="submit" value="Submit
Documentation">
 </form>
 </body>
</html>
<!- End of "add_module.html" file ->
```

File 2: process_module.php

```
<!-
Project: Single Sourcing Demo.
Component: Writer (File 2 of 3).
Filename: "process_module.php".
Purpose: Checks metadata for errors and then writes out
metadata to a temporary file.
->
<!DOCTYPE html PUBLIC "-//W3C//DTD XHTML 1.0 Transitional//EN"
"www.w3.org/TR/xhtml1/DTD/xhtml1-transitional.dtd">
<html xmlns="www.w3.org/1999/xhtml">
<head>
      <title> Single Sourcing Demo: Process Documentation
      Module </title>
</head>
<body>
```

```
<h1>Single Sourcing Demo</h1>
<h2>Step 2: Module Results Page</h2>
<?php
/*
The $bufferFilename variable is used to store the name of the
temporary file we will be writing to. For this exercise, this
file needs to be in the same directory as this script, but it
could be changed by including the directory path as part of
this variable.
*/
$bufferFilename = "ss_buffer.txt";
/*
The first thing we will do is to grab the various
characteristics of our software documentation from the form.
*/
$feature = $_POST["feature"];
$skillLevel = $_POST["skill_level"];
$documentation = $_POST["documentation"];
/*
In this example, rather than using another buffer file to
keep track of the number of the most recently added module,
we will create a module identifier by gluing together the
feature being document with the skillLevel. So, a typical
module ID would look something like "introduction_beginner"
or "introduction_advanced".
*/
$moduleId = $feature."_".$skillLevel;
/* If the feature information is missing, show an error
message. */
if (empty($feature))
      {
      echo "Feature is required!<br />";
      $errors++;
      }
/* If the skill level description is missing, show an error
message. */
if (empty($skillLevel))
      {
      echo "Skill level description is required!<br />";
      $errors++;
      }
/* If the documentation is missing, show an error message. */
if (empty($documentation))
      {
      echo "Documentation is required!<br />";
      $errors++;
```

```
      }
/*
Here is the part where we check the total errors and redirect
the user if need be.
*/
if ($errors > 0)
      {
      echo "There were one or more errors with your data.<br
      />";
      echo "Please <a href=\"javascript:history.back()\">go
      back</a> button and correct your input!<br />";
      }
else
      {
      /*
      What we are going to do now is to "glue" the variables
      together into a single line using the PHP concatenation
      operator, which is the period (.) In order to keep each
      variable distinct, we will use the character sequence
      *** to separate each entry.  Also, a \n (newline escape
      sequence) is used to make each additional entry begin
      on a new line.
      */
      $gluedString =
      $moduleId."***".$feature."***".$skillLevel."
      ***".$documentation."\n";
      /*
      Now, we create a connection to the second buffer text
      file; the first argument is the filename and the second
      argument is the "mode". "a" mode means "append" and new
      text content will be added to the end of the file.
      $fp is a variable that holds the "link" or connection
      to the file that we open and then sends this link to the
      fwrite function for the purpose of writing data to the
      file.
      */
      if ($fp = fopen($bufferFilename,"a"))
            {
            /* This will write the glued string to the file */
            if (fwrite($fp, $gluedString))
                  {
                  echo "Software documentation module added to
                  buffer file!<br />";
                  /* This closes the file after the writing is
                  done */
                  fclose($fp);
```

```
                    }
            else
                    /* This is displayed if the file cannot be
                    written to */
                    {
                    echo "There was an error adding your
                    documentation.<br />";
                    echo "Please check the file permissions for
                    your buffer file.<br />";
                    echo "The buffer filename is: ".
                    $bufferFilename.".<br />";
                    }
            }
        else
            {
            echo "Unable to open buffer file!";
            }
        ?>
        <p>Click here to return to Step 1 and <a href=
        "add_module.html">add another module</a>.</p>
        <p>Click here to proceed to Step 3 and <a href=
        "finalize_modules.php">finalize all documentation
        modules</a> by writing them to an XML file.</p>
        <?php
        }
?>
</body>
</html>
<!— End of "process_module.php" file —>
```

File 3: finalize_modules.php

```
<!—
Project: Single Sourcing Demo.
Component: Writer (File 3 of 3).
Filename: "finalize_modules.php".
Purpose: Loops through metadata file and writes out all
documentation modules to a valid XML file using DOM.
—>
<!DOCTYPE html PUBLIC "-//W3C//DTD XHTML 1.0 Transitional//EN"
"www.w3.org/TR/xhtml1/DTD/xhtml1-transitional.dtd">
<html xmlns="www.w3.org/1999/xhtml">
 <head>
  <title> Single Sourcing Demo: Finalize Modules </title>
 </head>
 <body>
```

```
<h1>Single Sourcing Demo</h1>
<h2>Step 3: Generate XML Document</h2>
<?php
/*
The $bufferFilename variable is used to store the name of the
temporary file we will be reading from.  For this exercise, this
file needs to be in the same directory as this script, but it
could be changed by including the directory path as part of
this variable.  The $xmlOutputFilename will be the name of the
file used to store our XML documentation.
*/
$bufferFilename = "ss_buffer.txt";
$metaDataDirectory = "metadata/";
$xmlOutputFilename = "documentation.xml";
/* This function writes the beginning of the XML file using UTF-
8 (Latin) encoding */
function startXMLdocument($xmlString)
      {
      $xmlString = "<?xml version=\"1.0\" encoding=
      \"utf-8\"?>\n";
      $xmlString .= "<!— XML Single Sourcing Documentation
      File —>\n";
      $xmlString .= "<!— Automatically generated by Single
      Sourcing Demo —>\n";
      $xmlString .= "<!— Created on ".date("MdY")." —>\n";
      $xmlString .= "<?xml-stylesheet type=\"text/xsl\"
      href=\"ss_allmodules_transform.xsl\"?>\n\n";
      return($xmlString);
      }
/*
This function writes the document type definition (DTD) as well
as the top level documentation container to the XML file
*/
function insertDTD($xmlString)
      {
      $xmlString .= "<!DOCTYPE documentation_modules [\n";
      $xmlString .= "\t<!ELEMENT documentation_modules
      (module)*>\n";
      $xmlString .= "\t<!ELEMENT module (id,feature,
      skill_level, documentation)>\n";
      $xmlString .= "\t<!ELEMENT id (#PCDATA)>\n";
      $xmlString .= "\t<!ELEMENT feature (#PCDATA)>\n";
      $xmlString .= "\t<!ELEMENT skill_level (#PCDATA)>\n";
      $xmlString .= "\t<!ELEMENT documentation (#PCDATA)>\n";
      $xmlString .= "]>\n\n";
      $xmlString .= "<documentation_modules>\n";
```

```
      return($xmlString);
      }
/*
This function closes the documentation container
*/
function endXMLdocument($xmlString)
      {
      $xmlString .= "</documentation_modules>\n";
      return($xmlString);
      }
/*
This function will loop through the temporary buffer file
and find each asset and its metadata.  It can tell where one
asset ends and another begins because they are each separated
by newlines (or the \n character sequence).  It can also
tell where one element ends and the next one begins by the
*** character sequence.  Each asset is converted into an XML
element with its values written out into the XML file as
character data.
*/
function addDocumentationCollection($filename,$xmlString)
      {
      if ($filecontents = file($filename))
            {
            $xmlString = startXMLdocument($xmlString);
            $xmlString = insertDTD($xmlString);
            foreach($filecontents as $singleLine)
                  {
                  $line = explode("\n",$singleLine);
                  $units = explode("***",$line[0]);
                  $xmlString .= "\t\t<module>\n";
                  $xmlString .= "\t\t\t<id>".$units[0]."</id>
                  \n";
                  $xmlString .= "\t\t\t<feature>".$units[1]."
                  </feature>\n";
                  $xmlString .= "\t\t\t<skill_level>".
                  $units[2]."</skill_level>\n";
                  $xmlString .= "\t\t\t<documentation>".
                  $units[3]."</documentation>\n";
                  $xmlString .= "\t\t</module>\n";
                  }
            $xmlString = endXMLdocument($xmlString);
            return($xmlString);
            }
      else
            {
```

```
            echo "Unable to open file!";
            }
      }
/*
This function handles the actual writing of the XML document
to a permanent file.  It will use the same metadata directory
and filename as specified by the variables at the top of this
script.
*/
function writeXMLFile($metaDataDirectory, $fileName, $xmlString)
      {
      $fp = fopen($metaDataDirectory.$fileName, "w");
      if ($bytesWrittenToFile = fwrite($fp, $xmlString))
            {
            echo "XML file created successfully!<br />";
            return TRUE;
            }
      else
            {
            echo "There was a problem writing to file.<br />";
            echo "Here is the XML content: <br />
            <pre>$xmlString</pre>";
            return FALSE;
            }
      }

/*
Here is where the addDocumentationCollection is "called" to
create the XML string that will be written to the permanent
file.  For now, the entire XML document will be stored in a
temporary string variable named $xml.
*/
$xml = addDocumentationCollection($bufferFilename, $xmlString);
/*
This segment of code writes the XML string to a permanent file.
If it is unable to create the file, the error message defined in
the writeXMLFile function will be displayed.
*/
if (writeXMLFile($metaDataDirectory, $xmlOutputFilename, $xml))
      {
      echo "<br />";
      echo "Click here to <a href=\"".$metaDataDirectory.
      $xmlOutputFilename."\">view your documentation using an
      XSL transform</a>.<br />";
```

```
      echo "Click here to <a href=\"add_module.html\">begin a
      new documentation collection</a>.<br />";
      echo "Click here to <a href=\"parse_xml.php\">parse
      the documentation file using the single sourcing
      parser</a>.";
      }
?>
</body>
</html>
<!- End of "finalize_modules.php" file ->
```

File 4: parse_xml.php

```
<!-
Project: Single Sourcing Demo.
Component: Reader/Parser (File 1 of 1).
Filename: "parse_xml.php".
Purpose: Creates an interactive browsing system for our
documentation file based on user-selected skill level.
->
<?php
/*
The $rootPath variable is used to indicate the full file path to
the documentation files on your hard drive.  The
$metaDataDirectory variable holds the value for the directory
containing the XML files.
*/
$rootPath = "c:\\xampp\\htdocs\\xmlparsers\\ss\\";
$metaDataDirectory = "metadata\\";
$xmlOutputFilename = "documentation.xml";
/*
These two variables will be used to hold the skill level and
module information once they have been selected by the user
using the unordered (bulleted) list display on the page.
*/
$selectedSkillLevel = $_GET["skill_level"];
$selectedModule = $_GET["module"];
/*
The purpose of this function is to list all skill levels
currently available.  To add new skill levels, simply append
their names to the end of the $skillLevels array.
*/
function listSkillLevels()
      {
      $skillLevels = array("Beginner","Intermediate",
      "Advanced");
```

```
      echo "<p>Please select the skill level that best reflects
      your computing experience.</p>";
      echo "<ul>";
      foreach ($skillLevels as $sl)
            {
            echo "<li>";
            /* This creates a link such that when we click on
            it, the skill level value is passed along with it!
            */
            echo "<a href=\"parse_xml.php?skill_level=
            {$sl}\">";
            echo $sl;
            echo "</a>";
            echo "</li>";
            }
      echo "</ul>";
      }
/*
The purpose of this function is to list all documentation
features (topics) currently available.  Each topic must have
all three skill level modules (beginner, intermediate, and
advanced) available in order for this to work correctly.
To add new topics, simply append their names to the end of
the $completedModules array (e.g. array("Introduction",
"Topic 2","Topic 3")...).
*/
function listTopics($skill)
      {
      $completedModules = array("Introduction");
      echo "<p>Please select a topic to browse.</p>";
      echo "<ul>";
      foreach ($completedModules as $module)
            {
            echo "<li>";
            /* This link will pass along both selected module
            and skill level information to the parser */
            echo "<a href=\"parse_xml.php?module=
            {$module}&skill_level={$skill}\">";
            echo $module;
            echo "</a>";
            echo "</li>";
            }
      echo "</ul>";
      }
if (isset($selectedSkillLevel))
      {
```

```
if (isset($selectedModule))
        {
        /*
        This block of code will execute if BOTH skill
        level and module have been selected by the user.
        For this example, we are creating a DOM parser.
        */
        $xml = new DOMDocument();
        /* Load the file using the path, directory, and XML
        filename specified at the top of this script */
        $xml->load($rootPath.$metaDataDirectory.
        $xmlOutputFilename);
        /*
        With the DOM, we can now use our DTD to validate
        our XML document!
        */
        if ($result = @$xml->validate())
                {
                /* This block of code will "run" only if
                the XML document has successfully passed
                validation. */
                echo "<h1>".$selectedModule."</h1>";
                /* Using the DOM model, this will select all
                DOM nodes with the tag name "module" */
                $modules = $xml->getElementsByTagName
                ("module");
                /* This will loop through each node in an
                iterative fashion. */
                foreach ($modules as $module)
                        {
                        /* First, get any skill level nodes
                        (there is actually only one) associated
                        with each module */
                        $skillLevels =
                        $module->getElementsByTagName
                        ("skill_level");
                        /* Now, get the value associated with
                        that node.  This will be "beginner",
                        "intermediate", or "advanced." */
                        $skillLevel = $skillLevels->item(0)
                        ->nodeValue;
                        /* This will check to see what value
                        was contained in the node. */
                        switch(strtolower($selected
                        SkillLevel))
                                {
```

```
                          /* If beginner skill level is
                          selected, show only that skill
                          level&apos;s content */
                          case "beginner":
                               if ($skillLevel ==
                               "beginner")
                                    {
                                    $node =
$module->getElementsByTagName("documentation");
                                    $doc = $node->
                                    item(0)->nodeValue;
                                         echo
     "<p>".stripslashes($doc)."</p>";
                                    }
                               break;
                          /* If intermediate skill level is
                          selected, show that skill level
                          content plus beginning level
                          content */
                          case "intermediate":
                               if ($skillLevel ==
                               "beginner" || $skillLevel
                               == "intermediate")
                                    {
               $node = $module->getElementsByTagName
               ("documentation");
                                    $doc = $node->
                                    item(0)->nodeValue;
                                         echo "<p>".
                                         stripslashes
                                         ($doc)."</p>";
                                    }
                               break;
                          /* If advanced skill level is
                          selected, show both prior skill
                          levels plus advanced content */
                          case "advanced":
                          if ($skillLevel == "beginner" ||
                          $skillLevel == "intermediate" ||
                          $skillLevel == "advanced")
                                    {
                                    $node = $module->
                                    getElementsByTagName
                                    ("documentation");
                                    $doc = $node->
                                    item(0)->nodeValue;
```

```
                                        echo "<p>".strips
                                        lashes($doc)."</p>";
                                        }
                                   break;
                                          default:
                                   echo "<p>Error: unable to
                                   provide documentation for
                                   skill level $selected
                                   SkillLevel.</p>";
                                   break;
                              }
                         }
                    echo "<a href=\"parse_xml.php\">Back Home
                    </a>";
                    }
               else
                    {
                    echo "<p>Error: the documentation file was
                    unable to validate against its internal DTD.
                    There is likely a problem with the XML
                    documentation file.  Please contact technical
                    support!</p>";
                    }
               }
          else
               {
               /* Show the list of available topics and embed
               the provided skill level into the links */
               listTopics($selectedSkillLevel);
               }
          }
     else
          {
          /* Show the list of currently available skill levels */
          listSkillLevels();
          }
?>
<!— End of "parse_xml.php" file —>
```

File 5: ss_allmodules_transform.xsl

```
<?xml version="1.0" encoding="utf-8"?>
<xsl:stylesheet version="1.0" xmlns:xsl="www.w3.org/1999/
XSL/Transform">
<!—
Project: Single Sourcing Demo.
```

```
Component: XSL Transform (File 1 of 1).
Filename: "ss_allmodules_transform.xsl".
Purpose: List ALL documentation modules in HTML.
->
<xsl:template match="/">
<html>
<head>
     <title> Single Sourcing Demo: All Documentation Modules
     </title>
</head>

     <body>
     <!- list all documentation modules in an unordered list
     ->
     <h1>Documentation Modules</h1>
     <ul>
     <xsl:for-each select="documentation_modules/module">
     <li>Id: <xsl:value-of select="id"/></li>
     <ul>
          <li>Feature: <xsl:value-of select="feature"/></li>
          <li>Skill Level: <xsl:value-of select=
          "skill_level"/></li>
          <li>Documentation: <xsl:value-of select=
          "documentation"/></li>
     </ul>
     </xsl:for-each>
     </ul>
     <br />
     <a href="Javascript: history.back()">Back</a>
     </body>
</html>
</xsl:template>
</xsl:stylesheet>
<!- End of "ss_allmodules_transform.xsl" file ->
```

File 6: ss_buffer.txt (sample)

```
introduction_beginner***introduction***beginner***Asset
Management System 1.0 is a tool for organizing and labeling
collections of organizational documents.  You can use this
system to add asset labels for categories and media and to
assign a list of years during which that asset may have been
used.  In addition, you can provide a detailed description of
the asset and why it may have long term importance for your
organization.  To get started with AMS 1.0, click the
\"Tutorial\" link from this help file.
```

```
introduction_intermediate***introduction***intermediate***This
program supports many features such as user-configurable fields
for categories, year ranges, and media types.  In addition, the
program includes options for clearing all fields or for
restoring defaults using the (Tools, Options, Reset) menus from
the Tools menubar.  Messages concerning program operations are
displayed in the Status: field.  This field is directly above the
Asset Name dropdown menu in the AMS interface.

introduction_advanced***introduction***advanced***Advanced
features include the ability to use keyboard shortcuts and to
import and export settings using XML files.  Keyboard shortcuts
for common commands are found in the Keyboard Shortcuts section
of this help system.  Program options default to verbose mode
for extra warnings and customized messaging from the AMS
system.  To disable verbose mode, see Disabling Verbose Mode in
the list of help topics.
```

File 7: documentation.xml (sample)

```
<?xml version="1.0" encoding="utf-8"?>
<!— XML Single Sourcing Documentation File —>
<!— Automatically generated by Single Sourcing Demo —>
<!— Created on Dec222007 —>
<?xml-stylesheet type="text/xsl"
href="ss_allmodules_transform.xsl"?>
<!DOCTYPE documentation_modules [
      <!ELEMENT documentation_modules (module)*>
      <!ELEMENT module (id,feature,skill_level, documentation)>
      <!ELEMENT id (#PCDATA)>
      <!ELEMENT feature (#PCDATA)>
      <!ELEMENT skill_level (#PCDATA)>
      <!ELEMENT documentation (#PCDATA)>
]>

<documentation_modules>
            <module>
                  <id>introduction_beginner</id>
                  <feature>introduction</feature>
                  <skill_level>beginner</skill_level>
                  <documentation>Asset Management System 1.0
                  is a tool for organizing and labeling
                  collections of organizational documents.
                  You can use this system to add asset labels
                  for categories and media and to assign a list
                  of years during which that asset may have
```

```
        been used.  In addition, you can provide a
        detailed description of the asset and why it
        may have long term importance for your
        organization.  To get started with AMS 1.0,
        click the \"Tutorial\" link from this help
        file.</documentation>
    </module>
    <module>
        <id>introduction_intermediate</id>
        <feature>introduction</feature>
        <skill_level>intermediate
        </skill_level>
        <documentation>This program supports many
        features such as user-configurable fields for
        categories, year ranges, and media types.
        In addition, the program includes options for
        clearing all fields or for restoring defaults
        using the (Tools, Options, Reset) menus from
        the Tools menubar. Messages concerning
        program operations are displayed in the
        Status: field.  This field is directly above
        the Asset Name dropdown menu in the AMS
        interface. </documentation>
    </module>
    <module>
        <id>introduction_advanced</id>
        <feature>introduction</feature>
        <skill_level>advanced</skill_level>
        <documentation>Advanced features include the
        ability to use keyboard shortcuts and to
        import and export settings using XML files.
        Keyboard shortcuts for common commands are
        found in the Keyboard Shortcuts section of
        this help system.  Program options default
        to verbose mode for extra warnings and
        customized messaging from the AMS system.
        To disable verbose mode, see Disabling
        Verbose Mode in the list of help
        topics.</documentation>
    </module>
</documentation_modules>
```

Copyright Credits

Specific Credits

Additional individual credits and permissions are listed below, arranged by chapter.

Cover Image

Cover image by Carole McDaniel.

Chapter 1

Portions reprinted from Applen, J.D. "Technical Communication, Knowledge Management, and Xml." *Technical Communication* 49.3 (2002): 301–13. Used with permission from *Technical Communication*, the journal of the Society for Technical Communication, Arlington, VA, U.S.A.

Chapter 2

Advisory Committee on Human Radiation Experiments, Executive Summary, reprinted with permission from the United States Department of Energy.

Chapter 3

Portions reprinted, with permission, from Applen, J.D. "Extensible Markup Languages and the Remediation of Abstracting and Indexing Strategies." *IEEE Transactions on Professional Communication* 44.3 (2001): 202–6. © 2001 IEEE.

Portions reprinted, with permission, from Applen, J.D. "Disease Classification and the Organization of Large-Scale Web Sites." *IEEE Transactions on Professional Communication* 44.4 (2001): 186–90. © 2001 IEEE.

Portions reprinted, with permission, from Applen, J.D. "Tacit Knowledge, Knowledge Management, and Active User Participation in Web Site Navigation." *IEEE Transactions on Professional Communication* 45.4 (2002): 302–6. © 2002 IEEE.

Portions reprinted from Applen, J.D. "Technical Communication, Knowledge Management, and Xml." *Technical Communication* 49.3 (2002): 301–13. Used with permission from *Technical Communication*, the journal of the Society for Technical Communication, Arlington, VA U.S.A.

Chapter 5

Screenshot of XMLBlueprint used with permission of Mr. Gerbin Abbink.

Images shown in Figures 5.6–5.8 were taken from the DITA Topics Tutorial (www.ditausers.org/training/DITATopics/) by Mr. Bob Doyle. Used with permission.

Chapter 6

Screenshots of XAMPP control panel and XAMPP home page used with permission of Mr. Kai Seidler.

Chapter 7

Interview with Dr. J. Michael Moshell printed with permission.
Interview with Dr. Michael Gourlay printed with permission.
Interview with Dr. Sherry Steward printed with permission.
Interview with Mr. Bill Albing printed with permission.
Interview with Mr. Thomas Gorence printed with permission.

Index

T - #0468 - 101024 - C0 - 229/152/22 - PB - 9780805861808 - Gloss Lamination